FRIEDRICH KOHLER
THE LIQUID STATE

Monographs in Modern Chemistry

Vol. 1: F. Kohler
The Liquid State

Vol. 2: H. Meier
Organic Semiconductors

Vol. 3: H. Kelker
Liquid Crystals

Vol. 4: G. Ertl/J. Küppers
Low Energy Electrons
and Surface Chemistry

Vol. 5: H. Kessler
Dynamic Stereochemistry

The series is to be continued

FRIEDRICH KOHLER
THE LIQUID STATE

In collaboration
with G. H. Findenegg,
J. Fischer, H. Posch and
F. Weissenböck

VERLAG CHEMIE

Prof. Dr. Friedrich Kohler
Institut für Physikalische Chemie
der Universität
A-1090 Wien
Währinger Str. 42

Copy Editing: Dr. Hans F. Ebel

This Book contains 96 figures and 32 tables

ISBN 3-527-25390-4

LIBRARY OF CONGRESS CATALOG CARD No. 75-185277

Copyright © 1972 by Verlag Chemie GmbH, Weinheim/Bergstr.
All rights reserved (including those of translation into foreign languages). No part of this book may be reproduced in any form — by photoprint, microfilm, or any other means — nor transmitted or translated into a machine language without written permission from the publishers.
Alle Rechte, insbesondere die der Übersetzung in fremde Sprachen, vorbehalten. Kein Teil dieses Buches darf ohne schriftliche Genehmigung des Verlages in irgendeiner Form — durch Photokopie, Mikrofilm oder irgendein anderes Verfahren — reproduziert oder in eine von Maschinen, insbesondere von Datenverarbeitungsmaschinen, verwendbare Sprache übertragen oder übersetzt werden.
Registered names, trademarks, etc. used in this book, even without specific indication thereof, are not to be considered unprotected by law.
Printer: Dr. Alexander Krebs, Hemsbach/Bergstr. Bookbinder: Karl Hanke, Großbuchbinderei, Düsseldorf
Printed in Germany

To the Memory of Our American Friends:

Harold E. Affsprung
Theodore S. Gilman

Preface

During the last years there have appeared several books and some collective volumes on special topics of liquid structure. However, we felt that a relatively short monograph is needed which presents the unified views of the physicist and the chemist about the liquid state also to scientists who work in related areas. It is the purpose of this book to fill this gap.

It is clear that our monograph owes much, if not everything, to the scientific literature. The most frequently used sources are given in the list of general references. Special reference should be made to the book by Rowlinson (gen. ref. 11), which comes in several aspects close to the concept of this volume. We have consciously omitted some passages from our monograph which are treated excellently there.

It is my privilege as the senior author to thank my co-authors of the various chapters for the pleasant and rewarding collaboration, in which they have assumed full partnership. We are greatly indebted to Miss C. Kainz for typing the manuscript and its precursors, and to Mrs. J. Schuch for preparing some corrections. Dr. E. Wilhelm was so kind as to read with us all the proofs, which was an invaluable help. Professor T. A. Litovitz contributed several remarks to chapter 11, which is highly appreciated.

Finally we wish to thank all our colleagues who called our attention to pertinent publications, and to thank those colleagues, editors, and publishers who generously permitted the use of their figures.

Vienna, Dec. 14, 1971.   Friedrich Kohler

Contents

1. Introduction (with J. Fischer) 1
1.1. Qualitative Characterization of the Liquid State 1
1.2. Some Thermodynamic Relations 6
1.3. The Canonical Ensemble 10
1.4. The Grandcanonical Ensemble 13
1.5. Qualitative Description of the Various Kinds of Interaction Energy 16
1.5.1. Liquids of Noble Gases 16
1.5.2. Molecular Liquids 17
1.5.3. Ionic Melts 17
1.5.4. Liquid Metals 18
1.5.5. Approximative Pair Potentials 18
1.6. References 19

2. Models for the Liquid State (with F. Weissenböck) 21
2.1. Reason for Models 21
2.2. Lattice Gas Model 21
2.3. Hard Sphere Model 22
2.4. Bernal's Geometric Model 23
2.5. Defective Crystal Model 25
2.6. The Cell Theory and Disorder Models 27
2.6.1. Disorder in Terms of Coordination Defects 28
2.7. References 30

3. Computer Experiments (with J. Fischer) 31
3.1. Introduction 31
3.2. Principles of Monte Carlo Calculations 31
3.3. Molecular Dynamics Calculations 35
3.4. Results for Hard Spheres 37
3.5. References 39

4. Pair Distribution Function (with J. Fischer) 41
4.1. Definition 41
4.2. Pair Distribution Function and Thermodynamic Properties 44
4.2.1. The Situation at the Critical Point. The Direct Correlation Function 46
4.3. Pair Distribution Function and Intensity of Scattered X-rays 47
4.3.1. Problems in the Evaluation of the Measured Intensity 50
4.4. General Remarks about Scattering 52
4.4.1. Static Approximation 52
4.4.2. Inelastic Scattering 55
4.5. Results for the Pair Distribution Function from Scattering Experiments 57
4.6. References 60

5. Pair Potential (with F. Weissenböck) 61
5.1. Introduction 61
5.2. Theory of Pair Interaction 61
5.2.1. Dispersion Energy 61
5.2.2. More About Dispersion Energy 63
5.2.3. Electrostatic and Inductive Energy 64
5.2.4. General Theory of Long Range Interactions 66
5.2.5. Short Range Interactions 68
5.2.6. Empirical Formulae for the Pair Potential 70
5.3. Experimental Determination of Pair Potential 71
5.3.1. Second Virial Coefficient 72
5.3.2. Transport Properties of Gases 73
5.3.3. Scattering of Molecular Beams 74
5.3.4. X-ray Determination of Pair Potentials 75
5.4. Many Body Interactions 75
5.4.1. Effective Pair Potential 76
5.5. Interaction Potentials in Ionic Melts 78
5.6. Interaction Energy in Metals 78
5.7. References 80

6. Thermodynamic Properties of Liquids 83
6.1. The Theorem of Corresponding States 83
6.2. Thermodynamic Properties along the Saturation Curve 85
6.3. Volume Dependence of Thermodynamic Quantities 89
6.4. Melting 91
6.5. The Critical Point (with H. Posch) 96
6.5.1. Classical Theory, Based on Analycity of Free Energy Around the Critical Point . . 99
6.5.2. The Lattice Gas (Ising Model) 101
6.5.3. Correlations in the Critical Region 102
6.5.4. Relations Between the Critical Indices. Scaling Laws 104
6.6. References 107

7. Equilibrium Theories of the Liquid State (with J. Fischer) 111
7.1. Introduction 111
7.2. Cell Model 111
7.2.1. The Lennard-Jones and Devonshire Model 112
7.2.2. The Self-Consistent Model 114
7.3. Pair Distribution Function Theories 116
7.3.1. Yvon-Born-Green Theory 116
7.3.2. Clusters 117
7.3.3. The Cluster Expansion Approach to Pair Distribution Function Theories 119
7.3.4. The Functional Differentiation Approach 122
7.3.5. Numerical Results 124
7.4. Perturbation Theory 127
7.5. References 131

8. Non-Equilibrium Properties: Transport Coefficients (with H. Posch) 133

8.1. Introduction 133
8.2. General Formulation of Transport Coefficients 133
8.2.1. Macroscopic Theory 133
8.2.2. Microscopic Theory 135
8.2.3. The Kinetic Equations 138
8.3. The Kinetic Equations Derived from a Balance of Binary Collisions 138
8.3.1. The Boltzmann-Equation for Dilute Gases 138
8.3.2. Dense Gases: The Enskog-Theory 141
8.4. The Brownian Motion 145
8.4.1. Particle Motion in Configurational Space 145
8.4.2. Kinetic Equations in Phase Space 147
8.4.3. The Langevin Equation 148
8.4.4. Estimation of the Friction Constant 150
8.5. Rice-Allnatt Theory of Transport Properties 151
8.6. Reduced Transport Coefficients 155
8.7. References 155

9. Non-Equilibrium Properties: Liquid Dynamics (with H. Posch) 157

9.1. Introduction 157
9.2. Linear Response Theory 157
9.2.1. Phenomenological Relations 157
9.2.2. Microscopic Relations: Molecular Correlation Functions 160
9.2.3. Electric Conduction 163
9.2.4. Diffusion 164
9.2.5. Experimental Determination of Correlation Functions and Neutron Scattering . . 165
9.3. Single Particle Motion 167
9.3.1. Some Illustrative Examples 168
9.3.2. Computer Calculation for Argon 169
9.3.3. Experimental Determinations of Single Particle Motions 172
9.4. Collective Particle Motions 174
9.4.1. Description of the Behaviour of the Scattering Law $S(Q,\omega)$ 174
9.4.2. Computer Calculations of the Current-Current Fluctuation 177
9.4.3. Models 180
9.5. References 180

10. Polyatomic Molecules (with G. H. Findenegg) 183

10.1. Introduction 183
10.2. Forces between Polyatomic Molecules 183
10.2.1. Angle-Dependent Pair Potentials 185
10.3. Configuration Free Energy 188
10.3.1. Ideal Dipoles 190
10.3.2. Molecules Having a Non-Spherical Shape 191
10.4. Extension of the Theorem of Corresponding States 191
10.5. Melting 194

10.5.1. Rigid Molecules 194
10.5.2. Flexible Molecules 198
10.6. References. 201

11. Molecular Re-Orientation in Liquids (with G. H. Findenegg and H. Posch) 203

11.1. Introduction 203
11.2. Absorption Spectra and Angle Correlation Functions 203
11.3. Dielectric Relaxation and Rayleigh Scattering. 208
11.4. Nuclear Magnetic Resonance Relaxation. 212
11.5. Ultrasonic Relaxation in Fluids 214
11.6. References. 217

12. Associated Liquids (with G. H. Findenegg) 219

12.1. Introduction 219
12.2. Hydrogen Bonds 220
12.3. Thermodynamic Stability of Associates 222
12.4. Study of Association in Liquids 224
12.4.1. Dielectric Methods 224
12.4.2. Ultrasonic Absorption 227
12.5. Some Examples 228
12.5.1. Carboxylic Acids 228
12.5.2. Alcohols 230
12.5.3. Water 234
12.6. References. 234

Appendix 237

A. Fourier-Transformation 237
   References. 240
B. Liouville Operator; Formal Integration of the Liouville Equation 241

List of General References 243
List of Symbols 245
Subject Index 249
Index of Substances 255

1. Introduction

1.1. Qualitative Characterization of the Liquid State

In a popular sense it is not necessary to define a liquid. In our world of room temperature and atmospheric pressure, liquids are different from solids by the absence of rigidity and from gases by having tensile strenght.

Let us investigate the difference between the liquid and solid state in more detail. The absence of rigidity in a liquid means on a microscopic level that the molecules can move much more easily than in a solid. A summary of values of self-diffusion coefficients D at the melting point in the solid and the liquid phase is given in table 1.1. These coefficients are nearly the same for all liquids but differ from those of the solids by several orders of magnitude.

Another consequence of the much higher mobility of the particles in the liquid is the absence of long-range order which is characteristic for the solid. This fact is revealed by X-ray, neutron, or electron scattering. Whereas a crystal shows anisotropy in the sense that scattering occurs only if the Bragg-condition is fulfilled, a liquid behaves isotropically. Moreover, if one compares an X-ray diffraction pattern of a liquid with a Debye-Scherrer pattern of a solid (where the

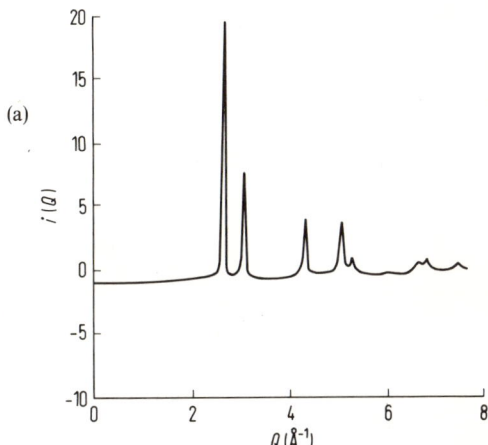

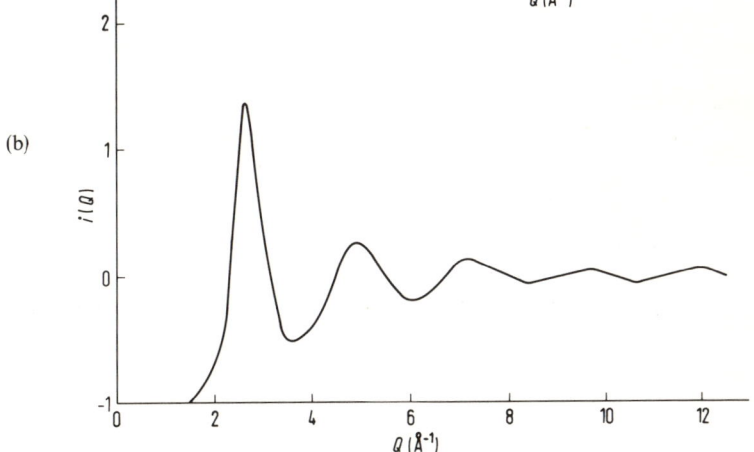

Fig. 1.1. The intensity of X-ray diffraction for a) solid Al at 650°C and b) liquid Al at 670°C (after Ruppersberg and Seemann [16]). The quantity $Q = \frac{4\pi}{\lambda} \sin \theta$ measures the angle of deflection (λ = wave length) and $i(Q)$ is the scattered intensity reduced by that for isolated atoms.

1. Introduction

Table 1.1. Self-diffusion coefficients D at the melting point.

Substance	$T_m(K)$	$D(\text{cm}^2\text{s}^{-1})$ solid	$D(\text{cm}^2\text{s}^{-1})$ liquid
Ar	83.8 [1]	$1 \cdot 10^{-9} - 2 \cdot 10^{-8}$ [6]	$1.75 \cdot 10^{-5}$ [12]
Xe	161.4 [1]	$6.8 \cdot 10^{-10}$ [7]	$1.62 \cdot 10^{-5}$ [12]
CH_4	90.7 [2]	$1.7 \cdot 10^{-9}$ [8]	$1.93 \cdot 10^{-5}$ [12]
C_6H_{12}	279.6 [3]	$7.4 \cdot 10^{-7}$ [3]	$8.0 \cdot 10^{-6}$ [3]
Cl^- (NaCl)	1073 [4]	$6.6 \cdot 10^{-9}$ [9]	$5.9 \cdot 10^{-5}$ [13]
Na	371.0 [5]	$1.64 \cdot 10^{-7}$ [10]	$4.00 \cdot 10^{-5}$ [14]
Sn	505.1 [5]	$1.6 \cdot 10^{-10} - 8 \cdot 10^{-10}$ [11]	$2.1 \cdot 10^{-5}$ [15]

anisotropy is averaged out by the random directions of the many little crystals), the solid gives some very sharp lines of scattered intensity, whereas the liquid produces only broad peaks (fig. 1.1). In chapter 4, this curve of scattered intensity will be related to the probability density of finding another molecule in a certain volume element at a distance R from a given molecule; this probability density, normalized to unity for large distances, is called the pair distribution function $g(R)$. As fig. 1.2 shows, the pair distribution function of solids and liquids are not very

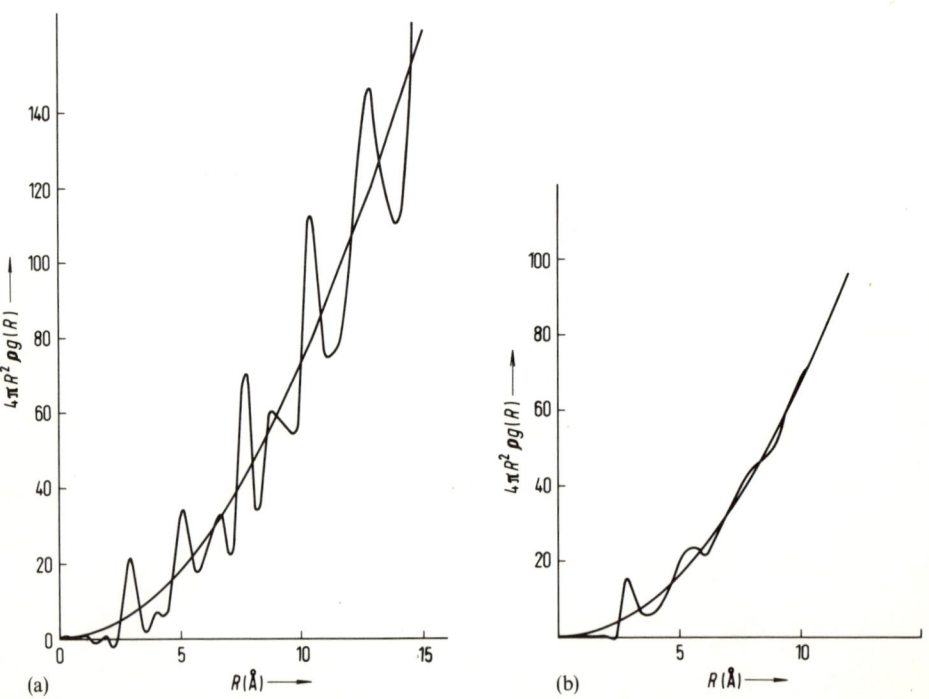

Fig. 1.2. A characteristic plot of the function $4\pi R^2 \rho g(R)$ for a) solid Al at 650°C and b) liquid Al at 670°C (after Ruppersberg and Seemann [16]). The quantity $4\pi R^2 \rho g(R)\,dR$, which is essentially determined by the pair distribution function $g(R)$, gives the mean number of molecules which can be found at a distance between R and $R + dR$ from a given molecule in a substance of density ρ.

different for small distances. That means that a molecule in the liquid is surrounded by nearly the same number of neighbours as in the solid and that the corresponding distances are similarly distributed. But for greater distances $g(R)$ of the solid shows still a great number of distinct peaks, whereas the distribution of distant molecules becomes random in the liquid. This difference is reflected by the sharp or broad diffraction pattern, resp.; thus a liquid still displays short-range order, given by the first peaks of the pair distribution function.

The molar volume does not change much at melting as is expected from the similarity of the pair distribution function at small distances. Examples are given in table 1.2. Apart from few cases like water which are considered to be anomalous, the volume increases for several percents.

Table 1.2. Quantities at the melting point: Heat of melting, relative volume change and energy of vaporization.

Substance	$T_m(K)$	$\dfrac{\Delta H_m}{RT_m}$	$\dfrac{\Delta V}{V_s}$ (%)	$\dfrac{\Delta U_{vap}}{RT_m}$
Ar	83.82 [1]	1.71 [26]	14.3[d] [32]	8.49 [38]
N_2	63.14 [17]	1.37 [17]	8.4 [36]	10.62 [38]
CH_4	90.67 [2]	1.24 [27]	8.2[e] [35]	10.58 [38]
HCl	158.91 [18]	1.51 [18]	–	12.01 [18]
C_2H_2	192.60[a] [19]	3.07 [19]	–	9.14 [19]
CO_2	216.58[b] [20]	4.50 [28]	27.4 [28]	7.52 [38]
SO_2	197.69 [21]	4.50 [21]	–	16.31 [21]
C_6H_6	278.68 [22]	4.25 [23]	13.3[f] [39]	14.08 [23]
n-C_6H_{14}	177.84 [23]	8.81 [23]	11.77[f] [39]	27.82 [23]
n-C_8H_{18}	216.36 [23]	11.53 [23]	14.77[f] [39]	27.94 [23]
KCl	1043 [4]	1.53[c] [29]	17.3[f] [40]	20.09[g] [43]
Hg	234.29 [24]	1.18 [30]	3.6 [41]	30.72 [30]
Na	370.98 [25]	0.84 [31]	2.63 [42]	33.02[g] [44]

a) Triple point value at 1.28 bar.
b) Triple point value at 5.19 bar.
c) This corresponds to half a mole KCl or one mole of ions.
d) The values given in [32] are in good agreement with the liquid density [33], the solid density measured by X-rays [34], and the volume change which can be calculated from the slope of the melting curve and the heat of melting [1, 26, 35].
e) The liquid density is taken from [37], which is between [33] and [38].
f) These values may be somewhat too high, as the volume of the solid was measured some degrees below the melting point.
g) Values calculated for the assumption of monomeric units in the vapour.

The slightly less dense packing of the liquid can be compressed more easily. Table 1.3 compares the isothermal compressibilities β_T for the solid and liquid phase of various substances. When the packing of the liquid has become more dense by the application of external pressure, the compressibilities of solid and liquid become more similar (fig. 1.3).

Table 1.3. Isothermal compressibilities at the melting point.

Substance	T_m(K)	$\beta_T \cdot 10^5$ (bar^{-1}) solid	$\beta_T \cdot 10^5$ (bar^{-1}) liquid
Ar	83.8 [1]	8.91 [34]	19.13[b]
N_2	63.1 [17]	8.05 [45]	19.8 [38]
C_6H_6	278.7 [22]	3.3[a] [46]	8.6 [38]
KCl	1043 [4]	1.61 [50]	3.62 [53]
Na	371.0 [5]	1.65 [51]	1.91 [54]
Al	933 [5]	0.222 [52]	0.238 [55]

a) This value was calculated using also [47, 48, 49].
b) References are given in the text describing table 6.3.

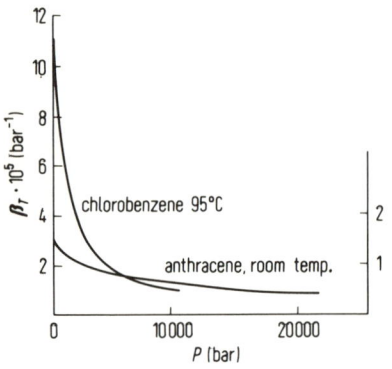

Fig. 1.3. The isothermal compressibility of an organic solid (right scale) and an organic liquid (left scale) as function of pressure [56].

As the mean molecular distance is about the same in liquids and solids, the interaction between the molecules and the resulting cohesion energy is of the same order of magnitude in both phases. This is also shown by table 1.2, where vaporization energies $\Delta U_{vap} (\Delta U_{vap} = \Delta H_{vap} - RT)$ and heats of melting ΔH_m of several substances are compared.

The differences between the liquid and the gaseous phase are at atmospheric pressure more pronounced than those between the solid and the liquid phase. The molar volume, e.g., is greater by a factor 10^2 to 10^3, corresponding to a tenfold increase of the mean molecular distance at boiling. Therefore, the molecular interactions in a gas at this pressure are almost negligible. In view of these discrepancies between liquid and gas at boiling, it is surprising that both phases become identical at somewhat elevated temperatures and pressures, at the so-called critical point (cf. the phase diagram fig. 1.4). As it is possible to pass from any state in the liquid to any state in the gas without a phase transition, it is usual to call both states the fluid state. The region where the adjective "liquid" can be used adequately is given by the dashed area of fig. 1.4.

We will consider the temperature of the triple point relative to the critical temperature as a measure of the liquid range. Usually the difference between triple and melting temperature is within the experimental error. Table 1.4 gives a summary of the temperatures of melting, boiling, and

1.1. Qualitative Characterization of the Liquid State

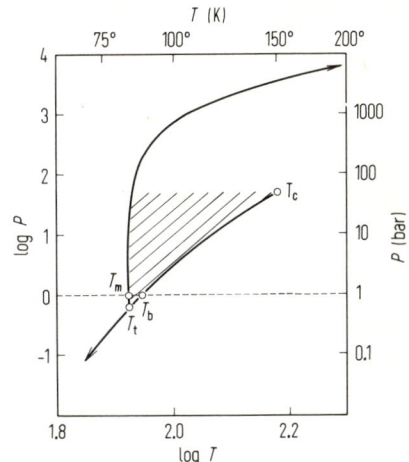

Fig. 1.4. Phase diagram of argon [1, 57, 58].

critical point (T_m, T_b, T_c), and the corresponding ratios, for some typical substances, on which we will comment later in this chapter. Inasmuch as the fraction T_b/T_c is the same for a variety of substances, the interval between melting point and boiling point, which is usually well known for each substance, may be considered as characteristic of the liquid range.

It shall be noted, that an analogue to the critical point has not yet been found on the melting line, despite the fact that already pressures up to several 10^4 bar have been applied. It seems

Table 1.4. Melting, boiling and critical temperatures and their ratios.

Substance	T_m(K)	T_b(K)	T_c(K)	T_m/T_c	T_b/T_c	T_m/T_b
Ar	83.82 [1]	87.30 [26]	150.65 [65]	0.556	0.579	0.960
N_2	63.14 [17]	77.35 [38]	126.1 [38]	0.501	0.613	0.816
CH_4	90.67 [2]	111.66 [2]	190.6 [38]	0.476	0.586	0.812
HCl	158.91 [18]	188.07 [18]	324.6 [61]	0.490	0.579	0.845
C_2H_2	192.60 [19]	189.35[a] [19]	308.33 [19]	0.625	–	–
CO_2	216.58 [20]	194.72[a] [59]	304.19 [62]	0.712	–	–
SO_2	197.69 [21]	263.13 [21]	430.7 [61]	0.459	0.611	0.751
C_6H_6	278.68 [22]	353.25 [23]	562.09 [62]	0.496	0.628	0.789
n-C_6H_{14}	177.84 [23]	341.89 [23]	507.35 [62]	0.351	0.674	0.520
n-C_8H_{18}	216.36 [23]	398.81 [23]	568.76 [62]	0.380	0.701	0.543
KCl	1043 [4]	1686 [24]	3000[b]	0.348	0.562	0.619
Hg	234.29 [24]	629.88 [24]	1763 [64]	0.133	0.357	0.372
Cs	301.79 [24]	958 [24]	2023 [64]	0.149	0.473	0.315
Na	370.98 [25]	1155 [60]	2490[c]	0.149	0.464	0.321

a) sublimation temperature.
b) estimated value [63].
c) calculated by the principle of corresponding states from [64].

that the long range order of the molecules, distinguishing the solid from the fluid state, is a much more inherent property than the mean molecular distance.

The structure and properties of any substance are determined by the interaction energies between its molecules. In principle, the macroscopic properties can then be derived by means of statistical mechanics. On the other hand these properties are interconnected by thermodynamic relations, which may be used to calculate less easily accessible properties from others. In this introduction, we will present a brief review of important thermodynamic relations, then of statistical mechanical methods, and finally characterize qualitatively various kinds of interaction energies.

1.2. Some Thermodynamic Relations [66]

A thermodynamic potential is a quantity, whose knowledge as a function of a certain set of independent variables allows to calculate all other thermodynamic quantities by using the potential itself and its derivatives. Moreover, the system is in equilibrium for a given set of variables, if the corresponding potential has reached an extremum. For a system containing only one kind of molecules a set of three independent variables is necessary.

A frequently used potential is the energy $u^{*)}$ as a function of the entropy s, the volume v and the number of moles n. Its total differential is given by

$$du(s,v,n) = \left(\frac{\partial u}{\partial s}\right)_{v,n} ds + \left(\frac{\partial u}{\partial v}\right)_{s,n} dv + \left(\frac{\partial u}{\partial n}\right)_{s,v} dn = Tds - Pdv + \mu dn, \quad (1.1)$$

where T denotes the temperature, P the pressure and μ the chemical potential. From a given thermodynamic potential others can be obtained by Legendre transformation [66, Callen], replacing one independent variable by another in the total differential. Using e.g. du, we can replace s by T, and get

$$df = d(u - Ts) = -sdT - Pdv + \mu dn, \quad (1.2)$$

which is the total differential of the Helmholtz free energy

$$f = u - Ts = f(T,v,n). \quad (1.3)$$

The total differentials of the Gibbs free energy (or free enthalpy) g, of the enthalpy h and of the potential $(-Pv)$ can be obtained by an analogous procedure

$$dg = d(u - Ts + Pv) = -sdT + vdP + \mu dn, \quad (1.4)$$

$$dh = d(u + Pv) = Tds + vdP + \mu dn, \quad (1.5)$$

$$d(-Pv) = d(u - Ts - n\mu) = -sdT - Pdv - nd\mu. \quad (1.6)$$

In the equations for the total differentials, the coefficients of the righthand side are the partial differentials of the potential with respect to the corresponding independent variables. Thus one has e.g.

$$\left(\frac{\partial f}{\partial T}\right)_{v,n} = -s, \quad \left(\frac{\partial f}{\partial v}\right)_{T,n} = -P. \quad (1.7)$$

---

* Extensive variables are proportional to the amount of substance. We denote the extensive variables by small letters and the intensive variables by capital letters (with the exception of the chemical potential μ). Correspondingly, molar quantities are designed by capital letters.

1.2. Some Thermodynamic Relations

The set of variables connected with the potential g is the most useful for experimental determinations. On the other hand, the set connected with f corresponds to the canonical ensemble in statistical mechanics and the set connected with $(-Pv)$ to the grandcanonical ensemble. (It shall be noted in this connection that in thermodynamics the chemical potential μ is referred usually to one mole, whereas in statistics μ is referred to one molecule. Then the mole number n has to be replaced by the particle number N in all previous equations).

By virtue of eq. (1.2) and (1.4) the Gibbs-Helmholtz equations can be derived, which connect u with f and h with g:

$$u = f + Ts = f - T\left(\frac{\partial f}{\partial T}\right)_{v,n} = -T^2 \frac{\partial}{\partial T}\left(\frac{f}{T}\right), \tag{1.8}$$

$$h = g + Ts = g - T\left(\frac{\partial g}{\partial T}\right)_{P,n} = -T^2 \frac{\partial}{\partial T}\left(\frac{g}{T}\right). \tag{1.9}$$

The Maxwell equations can be deduced from the principle, that mixed second partials do not depend on the order of differentiation. The equality of the mixed second partials of the function f, e.g., means (cf. eq. (1.7))

$$\left(\frac{\partial s}{\partial v}\right)_T = \left(\frac{\partial P}{\partial T}\right)_v \equiv \gamma_v. \tag{1.10}$$

Thus the volume dependence of the entropy is given by the experimentally accessible quantity γ_v, which is called the thermal pressure coefficient. In the same manner one obtains from g

$$-\left(\frac{\partial s}{\partial P}\right)_T = \left(\frac{\partial v}{\partial T}\right)_P. \tag{1.11}$$

This relates the pressure dependence of the entropy to the thermal expansion coefficient

$$\alpha = \frac{1}{v}\left(\frac{\partial v}{\partial T}\right)_P. \tag{1.12}$$

Further quantities, related to the second derivatives of f, g and h are the heat capacity at constant volume

$$c_v = \left(\frac{\partial u}{\partial T}\right)_v = T\left(\frac{\partial s}{\partial T}\right)_v, \tag{1.13}$$

the heat capacity at constant pressure

$$c_P = \left(\frac{\partial h}{\partial T}\right)_P = T\left(\frac{\partial s}{\partial T}\right)_P, \tag{1.14}$$

the isothermal compressibility

$$\beta_T = -\frac{1}{v}\left(\frac{\partial v}{\partial P}\right)_T = -\frac{1}{v}\frac{\partial^2 g}{\partial P^2}, \tag{1.15}$$

and the adiabatic compressibility

$$\beta_s = -\frac{1}{v}\left(\frac{\partial v}{\partial P}\right)_s = -\frac{1}{v}\frac{\partial^2 h}{\partial P^2}. \tag{1.16}$$

The last quantity can be obtained from the easily measurable velocity of sound c by

$$\beta_s = \frac{v}{mc^2}, \tag{1.17}$$

where m denotes the mass of the system.

In principle all the above defined quantities are measurable, but c_v, β_T and γ_v are less readily accessible and one looks for equations relating them to other quantities. A first relation is given by

$$\gamma_v = \left(\frac{\partial P}{\partial T}\right)_v = -\left(\frac{\partial v}{\partial T}\right)_P \bigg/ \left(\frac{\partial v}{\partial P}\right)_T = \frac{\alpha}{\beta_T}. \tag{1.18}$$

A second relation is obtained by differentiating $s(T,P)$ with respect to T at constant v

$$\left(\frac{\partial s}{\partial T}\right)_v = \left(\frac{\partial s}{\partial T}\right)_P + \left(\frac{\partial s}{\partial P}\right)_T \left(\frac{\partial P}{\partial T}\right)_v, \tag{1.19}$$

which gives with eq. (1.11) and (1.18)

$$c_P - c_v = \frac{Tv\alpha^2}{\beta_T}. \tag{1.20}$$

Furthermore the ratio of the compressibilities is the same as that of the heat capacities

$$\frac{\beta_T}{\beta_s} = \frac{c_P}{c_v}. \tag{1.21}$$

The proof can be given by taking s and T as variables of P and v and using eq. (1.13) and (1.14):

$$\left(\frac{\partial v}{\partial P}\right)_s = -\frac{\left(\frac{\partial s}{\partial P}\right)_v}{\left(\frac{\partial s}{\partial v}\right)_P} = -\frac{\left(\frac{\partial s}{\partial T}\right)_v \left(\frac{\partial T}{\partial P}\right)_v}{\left(\frac{\partial s}{\partial T}\right)_P \left(\frac{\partial T}{\partial v}\right)_P} = \frac{c_v}{c_P} \left(\frac{\partial v}{\partial P}\right)_T. \tag{1.22}$$

Finally eq. (1.17), (1.20) and (1.21) can be combined to give

$$\frac{c_P}{c_v} - 1 = \frac{Tv\alpha^2}{c_P \beta_s} = \frac{mTc^2\alpha^2}{c_P}, \tag{1.23}$$

where the ratio of the heat capacity is expressed only by easily measurable properties.

Coexistent phases are characterized by the same chemical potential or, in one-component systems, by the same molar free enthalpy G. If the partials $\partial G/\partial P$ and $\partial G/\partial T$ are different for both phases, the phase transformation is called of the first order. Typical graphs are shown in fig. 1.5, where G is extrapolated for each phase into the unstable region. The angle of intersection of the G-

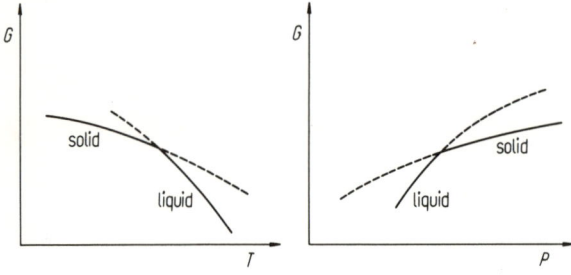

Fig. 1.5. Schematical diagram of the behaviour of $G(P,T)$ for both phases around the phase transformation.

curves is determined by the entropy and volume change at the phase transformation. Sometimes theoretical models make it possible to compare the real phase transformation with a hypothetical continuous sequence of homogeneous states connecting both phases. The first such model was the van der Waals equation, which showed the continuity between liquid and gaseous state. It is illustrative to compare the real transformation with the continuous sequence of homogeneous states between the phases in terms of free energy, pressure, and free enthalpy, shown in fig. 1.6 for constant temperature. The free energy of the heterogeneous portion is given by the double tangent, whereas the pressure of the coexistent phases can be deduced from the P-V-diagram by the criterion that the area under the hypothetical curve has to be the same as that under the real line.

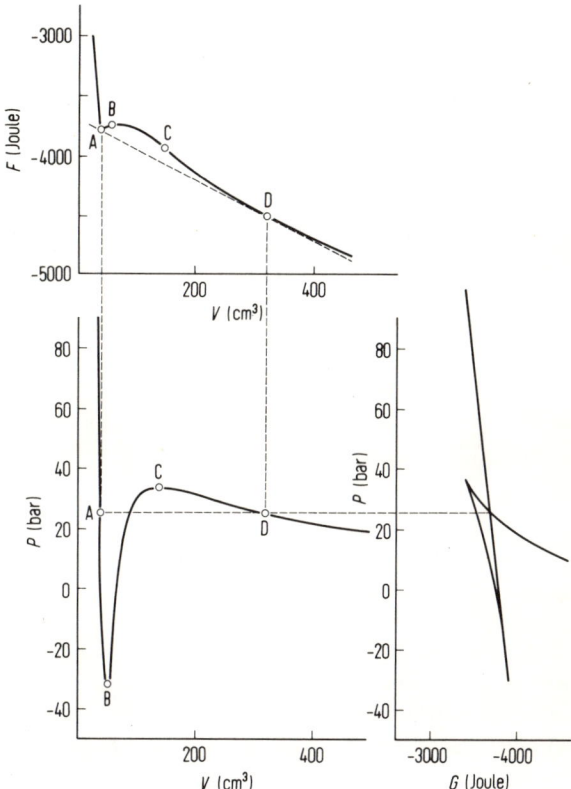

Fig. 1.6. A liquid-gas transition according to a continuous equation of state.

The coexistence curves for two phases in the P-T-plane (cf. fig. 1.4) can be obtained from the Clausius-Clapeyron equation, which follows from the condition of equality of G in both phases (Δ denotes the change accompanied with the transfer from one phase to the other):

$$\frac{dP}{dT} = \frac{\Delta S}{\Delta V} = \frac{\Delta H}{T \Delta V}. \tag{1.24}$$

The last equality follows because $\Delta G = 0$ implies also

$$\Delta H = T\Delta S \tag{1.25}$$

for coexisting phases.

1.3. The Canonical Ensemble [67]

Throughout this book we will deal only with classical statistical mechanics. Thus all liquids, where quantum effects are important (He, H_2, Ne), will not be considered.

The canonical ensemble describes an assembly with given T, v, and N. Each state of the assembly is characterized by a certain set of generalized coordinates of positions $\{q_1, q_2, \ldots, q_f\}$ and momenta $\{p_1, p_2, \ldots, p_f\}$ in phase space and the corresponding Hamiltonian $H = H(q_1, \ldots, q_f, p_1, \ldots, p_f)$. The probability to find the system in a certain state is proportional to its Boltzmann-factor $\exp\left\{-\dfrac{H}{kT}\right\}$ and given by

$$\frac{e^{-\frac{H}{kT}} dq^f dp^f}{\int e^{-\frac{H}{kT}} dq^f dp^f}, \tag{1.26}$$

where the integration in the denominator has to be extended over all possible coordinates and momenta. The ensemble average $\langle \varphi \rangle$ of any function $\varphi = \varphi(q_1, \ldots, p_f)$ can thus be obtained by

$$\langle \varphi \rangle = \frac{\int \varphi e^{-\frac{H}{kT}} dq^f dp^f}{\int e^{-\frac{H}{kT}} dq^f dp^f}. \tag{1.27}$$

The connection with thermodynamics is given via the partition function

$$Z = \frac{1}{h^f N!} \int e^{-\frac{H}{kT}} dq^f dp^f, \tag{1.28}$$

where h is Planck's constant. The partition function is related to the Helmholtz free energy by

$$f = -kT \ln Z. \tag{1.29}$$

As f is the thermodynamic potential with respect to the independent variables T, v, and N, all other thermodynamic quantities can be derived.

Further progress depends on the special form of the Hamiltonian. In the simplest case of monoatomic liquids we have

$$H = \frac{1}{2m} \sum_{i=1}^{N} \vec{p}_i^{\,2} + E(\vec{r}_1, \ldots, \vec{r}_N), \tag{1.30}$$

where m is the mass of an atom, $\vec{p}_i$ the momentum and $\vec{r}_i$ the position of the i-th atom and $E(\vec{r}_1, \ldots, \vec{r}_N)$ the potential energy for the configuration $(\vec{r}_1, \ldots, \vec{r}_N)$. As the Hamiltonian is the sum of the kinetic and potential energy, the partition function can be written as the product

$$Z = (Z_{\text{trans}})^N \frac{Q}{v^N}, \tag{1.31}$$

with
$$Q = \frac{1}{N!} \int_v \ldots \int_v d\vec{r}_1 \ldots d\vec{r}_N \exp\left\{-\frac{1}{kT} E(\vec{r}_1, \ldots, \vec{r}_N)\right\} \tag{1.32}$$

and
$$(Z_{trans})^N = \frac{v^N}{h^{3N}} \int_{-\infty}^{+\infty} \ldots \int_{-\infty}^{+\infty} d\vec{p}_1 \ldots d\vec{p}_N \exp\left\{-\frac{1}{2mkT} \sum_{i=1}^{N} \vec{p}_i^2\right\} \tag{1.33}$$

(Z_{trans} contains the factor v to be dimensionless). As the momenta are independent, the last integral is a product of N identical factors, for which the integration can be performed readily and gives the result

$$Z_{trans} = \left[\frac{2\pi mkT}{h^2}\right]^{3/2} v. \tag{1.34}$$

The quantity Q is called the configurational partition function and its evaluation is the main problem in the equilibrium theory of the liquid state.

At this stage we will evaluate Q for an ideal gas and an Einstein solid. In an ideal gas, potential energy due to particle interaction does not exist, whence

$$Q = \frac{1}{N!} \int_v \ldots \int_v d\vec{r}_1 \ldots d\vec{r}_N = \frac{v^N}{N!}. \tag{1.35}$$

Using the approximation

$$N! = N \ln N - N, \tag{1.36}$$

we get for the Helmholtz free energy

$$f = -NkT \ln\left[\frac{2\pi mkT}{h^2}\right]^{3/2} - NkT \ln \frac{v}{N} - NkT, \tag{1.37}$$

and for the equation of state

$$P = -\left(\frac{\partial F}{\partial v}\right)_T = \frac{NkT}{v}. \tag{1.38}$$

In an Einstein solid a particle can only move around its lattice site; the potential energy of each particle is assumed to be independent of the positions of the other particles and given by a harmonic oscillator potential

$$E(\vec{r}_1, \ldots, \vec{r}_N) = E_s + \sum_{i=1}^{N} \frac{m\omega^2}{2} (\vec{r}_i - \vec{r}_{i0})^2, \tag{1.39}$$

where $\vec{r}_{i0}$ is the position of the lattice site of the i-th particle, $\omega/2\pi$ the Einstein frequency and E_s the energy if all particles are exactly at their lattice sites. Inserting eq. (1.39) into the expression for Q [eq. (1.32)], only a single distribution of the particles on the lattice sites is taken into account. Permuting two particles, the new configuration has also to be counted in Q. Thus considering all possible permutations we get a factor $N!$, so that

$$Q = e^{-\frac{E_s}{kT}} [\int_v e^{-\frac{m\omega^2}{2kT}(\vec{r}_i - \vec{r}_{i0})^2} d\vec{r}_i]^N. \tag{1.40}$$

The integral in eq. (1.40) can be calculated exactly and gives a certain "free volume" $\left(\frac{v_f}{N}\right)$, which is the mean volume accessible for each particle. Naturally $\left(\frac{v_f}{N}\right)$ will be smaller than the molecular volume $\left(\frac{v}{N}\right)$. So we have

$$Q = e^{-\frac{E_s}{kT}} \left(\frac{v_f}{N}\right)^N, \tag{1.41}$$

and
$$f = -NkT\ln\left[\frac{2\pi mkT}{h^2}\right]^{3/2} + E_s - NkT\ln\left(\frac{v_f}{N}\right). \tag{1.42}$$

Comparing this expression with the result for the ideal gas, the difference is in the energy E_s, in the diminished volume v_f, and in the absence of the term $-NkT$. The last difference causes the entropy of an assembly of non-localized molecules to become bigger by a term Nk, which is called the communal entropy. It has been a much discussed question, how and where the communal entropy arises in going from the solid over the liquid to the gaseous state.

It is of interest to consider the general expression for the pressure. Using eq. (1.7), (1.29), and (1.31), we get

$$P = kT\frac{\partial}{\partial v}[\ln(Z_{\text{trans}})^N] + kT\frac{\partial}{\partial v}\left[\ln\left(\frac{Q}{v^N}\right)\right]. \tag{1.43}$$

The differentiation of the translational part gives immediately the value for the ideal gas, whereas the treatment of the configurational part is not so straightforward. Now we make the assumption, that the potential energy is the sum of all pair interactions

$$E(\vec{r}_1, \ldots, \vec{r}_N) = \sum_{i<j} \varepsilon_{ij}, \tag{1.44}$$

and that the pair interactions ε_{ij} depend only on the distance $R_{ij} = |\vec{r}_i - \vec{r}_j|$. With this assumption

$$\frac{Q}{v^N} = \frac{1}{N!v^N}\int_v\ldots\int_v d\vec{r}_1\ldots d\vec{r}_N \exp\left\{-\frac{1}{kT}\sum \varepsilon(R_{ij})\right\}. \tag{1.45}$$

Introducing $\vec{x}_i = \dfrac{\vec{r}_i}{v^{1/3}}$ as new integration variable,

$$\frac{Q}{v^N} = \frac{1}{N!}\int_1\ldots\int_1 d\vec{x}_1\ldots d\vec{x}_N \exp\left\{-\frac{1}{kT}\sum \varepsilon(x_{ij}v^{1/3})\right\}, \tag{1.46}$$

where the integration has to be extended over a unit volume. As this integral contains v only as a parameter, the differentiation can be performed:

$$\frac{\partial}{\partial v}\left(\frac{Q}{v^N}\right) = \frac{1}{N!}\int_1\ldots\int_1 d\vec{x}_1\ldots d\vec{x}_N \exp\left\{-\frac{1}{kT}\sum \varepsilon_{ij}\right\}\left\{-\frac{1}{kT}\sum \frac{\partial}{\partial v}\varepsilon(x_{ij}v^{1/3})\right\}. \tag{1.47}$$

Using the identity

$$\frac{\partial \varepsilon(x_{ij}v^{1/3})}{\partial v} = \frac{\partial \varepsilon(x_{ij}v^{1/3})}{\partial(x_{ij}v^{1/3})}\frac{\partial(x_{ij}v^{1/3})}{\partial v} = \frac{1}{3v}R_{ij}\frac{\partial \varepsilon(R_{ij})}{\partial R_{ij}}, \tag{1.48}$$

and taking again $\vec{r}_i$ as integration variable, the result is

$$\frac{\partial}{\partial v}\left(\frac{Q}{v^N}\right) = -\frac{1}{3vkT}\sum_{i<j}\frac{1}{N!v^N}\int_v\ldots\int_v R_{ij}\frac{\partial \varepsilon(R_{ij})}{\partial R_{ij}} \exp\left\{-\frac{1}{kT}\sum \varepsilon_{ij}\right\}d\vec{r}_1\ldots d\vec{r}_N, \tag{1.49}$$

which yields

$$kT\frac{\partial}{\partial v}\left[\ln\frac{Q}{v^N}\right] = \frac{kT}{\left(\frac{Q}{v^N}\right)}\frac{\partial}{\partial v}\left(\frac{Q}{v^N}\right) = \frac{1}{3v}\left\langle -\sum_{i<j} R_{ij}\frac{\partial \varepsilon(R_{ij})}{\partial R_{ij}}\right\rangle, \quad (1.50)$$

where the definition of the ensemble average according to eq. (1.27) has been used. If one defines the "virial of the intermolecular forces" W by

$$W = -\sum_{i<j} R_{ij}\frac{\partial \varepsilon(R_{ij})}{\partial R_{ij}}, \quad (1.51)$$

the pressure is given by

$$P = \frac{NkT}{v} + \frac{1}{3v}\langle W\rangle. \quad (1.52)$$

This equation results also from the virial theorem [67, Münster].

So far we have dealt with monoatomic liquids [eq. (1.30)]. In the case of polyatomic molecules, the Hamiltonian depends on the momenta of the centres of mass (translation), on the positions of the centres of mass and the orientations of the molecules (configuration), on the angular momenta of the molecules (rotation) and on the intramolecular vibrations (vibration). Thus we write quite formally

$$H = H(\text{trans, conf, rot, vib}). \quad (1.53)$$

In many cases it is possible to write the Hamiltonian as sum of these contributions

$$H = H_{\text{trans}} + H_{\text{conf}} + H_{\text{rot}} + H_{\text{vib}}. \quad (1.54)$$

Then the partition function splits up again into:

$$Z = \frac{1}{N!h^f}\int \exp\left\{-\frac{1}{kT}(H_{\text{trans}} + H_{\text{rot}} + H_{\text{vib}} + H_{\text{conf}})\right\} dq^f dp^f =$$

$$= (Z_{\text{trans}})^N \cdot (Z_{\text{rot}})^N \cdot (Z_{\text{vib}})^N \cdot \frac{Q}{v^N}. \quad (1.55)$$

Whereas Z_{trans} has the same meaning as in eq. (1.34), Q is given now by

$$Q = \left(\frac{1}{4\pi}\right)^N \frac{1}{N!}\int\ldots\int d\vec{r}_1\ldots d\vec{r}_N d\omega_1\ldots d\omega_N \exp\left\{-\frac{1}{kT}E(\vec{r}_1,\ldots,\vec{r}_N,\omega_1,\ldots,\omega_N)\right\}, \quad (1.56)$$

where the $\vec{r}_i$ denote the positions of the centers of mass, the ω_i the orientations of the molecules with respect to some given direction, and E the potential energy for the configuration $(\vec{r}_1,\ldots,\vec{r}_N, \omega_1,\ldots,\omega_N)$. We will postpone a detailed discussion of this partition function to chapter 10. But we can already say that also for polyatomic molecules the essential problem is in most cases the evaluation of the configurational partition function (1.56).

1.4. The Grandcanonical Ensemble [67]

In the canonical ensemble we have considered an assembly with fixed volume, temperature and particle number. But in many real systems, e.g. coexisting phases, the particle number does not remain strictly constant in one phase. So it is more advantageous to take the chemical potential μ as independent variable instead of the particle number N. Assemblies with the independent variables v, T, and μ are represented by the grandcanonical ensemble. Each state of that ensemble is characterized by a certain particle number N, the set of generalized coordinates and momenta of these particles and the corresponding Hamiltonian. For the further purpose we are only interested in the number of particles, and the probability to find a certain N is

$$\frac{e^{\frac{\mu N}{kT}} Z_N}{\sum_{N'=0}^{\infty} e^{\frac{\mu N'}{kT}} Z_{N'}}, \tag{1.57}$$

Here $Z_{N'}$ denotes the canonical partition function for N' particles and its appearance indicates that one has already summed up all different energy states belonging to the same particle number. From the grand partition function

$$\Xi = \sum_{N=0}^{\infty} e^{\frac{\mu N}{kT}} Z_N, \tag{1.58}$$

the thermodynamic potential $-Pv(T,v,\mu)$ can be obtained by

$$-Pv = -kT \ln \Xi. \tag{1.59}$$

We will investigate now the fluctuations of the particle number in a given volume v. By virtue of eq. (1.57) we have for the mean particle number N

$$\langle N \rangle = \frac{\sum_{N=0}^{\infty} N e^{\frac{\mu N}{kT}} Z_N}{\sum_{N=0}^{\infty} e^{\frac{\mu N}{kT}} Z_N}, \tag{1.60}$$

and for the mean square

$$\langle N^2 \rangle = \frac{\sum_{N=0}^{\infty} N^2 e^{\frac{\mu N}{kT}} Z_N}{\sum_{N=0}^{\infty} e^{\frac{\mu N}{kT}} Z_N}. \tag{1.61}$$

The mean square of the particle fluctuation is given by

$$(\Delta N)^2 = \langle (N - \langle N \rangle)^2 \rangle = \langle N^2 \rangle - \langle N \rangle^2, \tag{1.62}$$

and can readily be derived by differentiation of (1.60) with respect to μ

$$\left(\frac{\partial \langle N \rangle}{\partial \mu} \right)_{v,T} = \frac{1}{kT} (\langle N^2 \rangle - \langle N \rangle^2). \tag{1.63}$$

It remains to express this relation by some well-known thermodynamic quantities. As thermodynamic considerations apply to ensemble averages we have

$$\left(\frac{\partial \langle N \rangle}{\partial \mu} \right)_{v,T} = \left(\frac{\partial N}{\partial \mu} \right)_{v,T} = \left(\frac{\partial N}{\partial P} \right)_{v,T} \cdot \left(\frac{\partial P}{\partial \mu} \right)_{v,T}. \tag{1.64}$$

By using eq. (1.6) we get

$$\left(\frac{\partial P}{\partial \mu} \right)_{v,T} = \frac{N}{v}. \tag{1.65}$$

The derivative $\left(\dfrac{\partial N}{\partial P}\right)_{v,T}$ is at constant v, whence

$$\left(\frac{\partial N}{\partial P}\right)_{v,T} = v\left(\frac{\partial \frac{N}{v}}{\partial P}\right)_{v,T}. \tag{1.66}$$

Here $\left(\dfrac{N}{v}\right)$ is an intensive quantity and it means the same if N is kept constant instead of v [67, Davidson, p. 268]:

$$v\left(\frac{\partial \frac{N}{v}}{\partial P}\right)_{v,T} = v\left(\frac{\partial \frac{N}{v}}{\partial P}\right)_{N,T} = Nv\left(\frac{\partial \frac{1}{v}}{\partial P}\right)_{N,T} = N\beta_T. \tag{1.67}$$

From eq. (1.64), (1.65), and (1.67), it follows

$$\left(\frac{\partial N}{\partial \mu}\right)_{v,T} = \frac{N^2}{v}\beta_T, \tag{1.68}$$

and for the relative particle fluctuation

$$\frac{\langle N^2 \rangle - \langle N \rangle^2}{\langle N \rangle^2} = \frac{kT}{v}\beta_T. \tag{1.69}$$

Thus the particle or density fluctuation is related to the isothermal compressibility. As is seen from table 1.3, the particle fluctuation in a liquid is always greater than in a solid.
The particle fluctuations cause two different kinds of light scattering. This fact can be understood by dividing the fluctuations into an adiabatic and an isobaric part:

$$\left(\frac{\partial v}{\partial P}\right)_T = \left(\frac{\partial v}{\partial P}\right)_s + \left(\frac{\partial v}{\partial s}\right)_P \left(\frac{\partial s}{\partial P}\right)_T = \left(\frac{\partial v}{\partial P}\right)_s - \left(\frac{\partial v}{\partial s}\right)_P \left(\frac{\partial v}{\partial T}\right)_P, \tag{1.70}$$

where eq. (1.11) has been used. The adiabatic part is due to propagating fluctuations (hypersonic sound waves, moving for some molecular distances), where there is not time for the particles to transmit their energy out of the particle-rich regions. The isobaric part, on the other hand, consists of non-propagating fluctuations where the denser regions are able to transmit their energy to the less dense regions, so that the kinetic energy per particle is smaller in the dense regions. Particle fluctuations cause optical inhomogenities, which scatter light. The scattered intensity has the same frequency as the incident frequency in the case of nonpropagating (isobaric) fluctuations, and is called Rayleigh scattering. In the case of the propagating (adiabatic) fluctuations the scattered frequency is shifted to both sides by a Doppler effect, and the corresponding lines are called Brillouin scattering. The ratio of Brillouin to Rayleigh scattering is given by [cf. eq. (1.70)]

$$-\frac{\left(\dfrac{\partial v}{\partial P}\right)_s}{\left(\dfrac{\partial v}{\partial s}\right)_P \left(\dfrac{\partial v}{\partial T}\right)_P} = \frac{\beta_s}{\beta_T - \beta_s} \stackrel{?}{=} \frac{c_v}{c_P - c_v}. \tag{1.71}$$

This ratio is known as Landau-Placzek ratio [68]. Whereas in the solid the adiabatic (phonon) part is dominating, the isobaric part becomes of equal magnitude at melting, and increases then steadily up to the critical point, where it is the dominating contribution (for values of c_P and c_v see table 6.3).

1.5. Qualitative Description of the Various Kinds of Interaction Energy

As from the knowledge of the partition function all thermodynamic properties can be derived, the main problem in the equilibrium theory of liquids is the evaluation of the configurational partition function, which depends on the interaction energy $E(\vec{r}_1,\ldots,\vec{r}_N,\omega_1,\ldots,\omega_N)$ of all particles. At this point we give only a qualitative description of the various kinds of interaction energies, whereas a detailed discussion is postponed to chapter 5. It shall be stated, that we will not consider very complicated molecules or complicated aggregates connected by strong interaction energies. This limitation rules out high polymers, glasses and liquid crystals.

It is usual to consider the interaction energy between all particles approximately as a sum of pair interactions [eq. (1.44), cf. chapters 5 and 10], where the pair interactions $\varepsilon_{ij} = \varepsilon(R_{ij},\omega_i,\omega_j)$ depend on the distance, and (for non-spherical particles) on the mutual orientations, but shall not be influenced by the presence of further molecules. The properties of matter show that there acts an attractive force between a pair of molecules at large separations and a repulsive force at small separations. The first fact can be concluded from the existence of condensation of gases, and the second from the finite volume and small compressibility of liquids and solids. The repulsive energy varies sharply with distance and the forces are of short range. The dependence of the attractive intermolecular energy on distance is weaker and the forces are of long range. It has thus become customary to speak of short-range and long-range forces. All forces arise from Coulombic forces between elementary particles. The distinction is merely of practical kind since for the long range forces there occurs no interpenetration of electronic clouds and this simplifies the situation considerably. Some model potentials will be given at the end of the section.

1.5.1. Liquids of Noble Gases

Only the heavier noble gases will be considered here, as for helium and neon quantum mechanical corrections are necessary. The attractive energy is caused by the mutual induction from the electrons fluctuating in the electron clouds. It is called dispersion energy for the analogy to the induction from the oscillating electric field of a light wave. The noble gases are the most simple class of liquids, as the atoms have spherical symmetry, and as the dispersion energy is, apart from relatively small corrections, pairwise additive. Therefore, liquid argon is often considered as "the model liquid". Characteristic properties for these liquids are: small liquid range (table 1.4), volume change at melting of about 15% (table 1.2), entropy change at melting of about $1.7R$ (table 1.2, $\Delta H_m/RT = \Delta S_m/R$), and a large increase of the isothermal compressibility at melting (table 1.3).

1.5.2. Molecular Liquids

As all molecules deviate to some extent from spherical symmetry, their pair potential depends on orientation. Furthermore, the dispersion energy may not be the only kind of attractive energy, as also interactions between electrostatic moments, especially dipole and quadrupole moments, may contribute to the pair potential. The partition function can be written in the most simple case as in eq. (1.55).

In order to survey the great variety of molecules we will distinguish different classes. First there are rigid and flexible molecules. The latter are characterized by the possiblity of intramolecular rotation, which makes the partition function rather complicated. A dipole moment enhances in both cases the orientational dependence of the pair potential, but the properties of polar liquids are not drastically different from nonpolar liquids, if the partial charges are located near the center of the molecule. The situation becomes different, if the partial charges are located on the periphery, thus causing some electron transfer and giving rise to strongly orientated bonds, as e.g. hydrogen bonds. Liquids where such bonds occur are usually termed associated liquids and will be considered as a separate class.

The non-associated rigid molecules are most simple, and might be divided again into two groups. The first consists of molecules with small orientational dependence of their pair potential, as e.g. N_2, HCl, and CH_4, characterized by an orientational order-disorder phase transition in the solid. Because of this transition their volume and entropy change at melting is smaller and their liquid range is usually larger than that of argon. The molecules of the second group have a stronger orientational dependence of their pair potential, such that orientational order is enforced in the solid up to the melting point. Examples are C_2H_2, CO_2, and SO_2, but also discs like benzene. Their liquid range is sometimes larger and sometimes smaller than that of argon. If it is smaller it is usually connected with a large volume change at melting. The entropy change at melting is in any case considerable, as positional as well as orientational disorder occurs at melting.

Flexible molecules comprise most examples of organic chemistry. Their liquid range is medium to large and the entropy change at melting increases with the number of links in the molecular skeleton.

As will be seen in chapter 10, the classification given here is very crude and does not show the whole complexity which is involved in the interaction between polyatomic molecules. This complexity increases in the class of associated liquids, which will be dealt with in chapter 12.

1.5.3. Ionic Melts

The attractive energy in ionic melts and solids is mainly due to the Coulomb interaction of unlike ions. It is much greater than the dispersion energy, which is also present between ions, and thus the cohesion energy is much higher than in the previous cases. Therefore, the melting point is rather high and the critical point has not yet been measured for any substance. Because of these difficulties the experimental material is relatively scarce. So it does not seem justified to discuss the ionic melts in detail, but occasionally their behaviour will be compared to that of other liquids. A further problem is the different kind of interaction in the solid and the gas phase. In the solid, the ions are separate entities, in the gas, highly polar molecules are present in different states of association. The liquid is probably in between. Entropy and volume change on melting is very similar to the noble gases and the liquid range is probably medium to large.

1.5.4. Liquid Metals

Here the difference in the interaction of atoms between the gaseous state and the condensed state is still more pronounced. In the gaseous state, there exist isolated atoms, or associates which might be sometimes termed molecules, depending on the substance. In the condensed state, the interaction is between a multitude of metal ions and the electronic gas. It has been tried to resolve this interaction into a sum of "pseudo"-pair potentials, and many problems have been attacked by means of these in a promising way. However, the pseudo-pair potentials depend on volume and order, so that qualitative statements are very difficult. The most important feature is the long range of the interaction and the softness of the repulsive branch. In view of the difficulty and methodic difference of the theory, and in view of the peculiarity of many experiments, a thorough discussion of liquid metals is outside the scope of this monograph. However, we will refer frequently to them for the sake of comparison. It is interesting to note that metals have the largest liquid range, and that the change of volume, compressibility, and entropy at melting is in general very small.

1.5.5. Approximative Pair Potentials

The first chapters of this book will be devoted to the study of the most simple liquids, the monoatomic liquids. In order to simplify models and theoretical calculations, it is useful to introduce approximative pair potentials for this kind of liquids. The crudest approximation, which neglects the attractive energy at all, is the hard sphere potential (fig. 1.7b)

$$\begin{aligned} \varepsilon(R) &= \infty \quad \text{if} \quad R < \sigma \\ \varepsilon(R) &= 0 \quad \text{if} \quad R \geq \sigma \end{aligned} \quad (1.72)$$

Thus the molecules can be thought of as hard spheres with diameter σ. Many properties of dense liquids can be understood already by considering a system of hard spheres. A more realistic potential, which is used very extensively in the study of noble gas liquids, is the Lennard-Jones or 12/6 potential (fig. 1.7a)

$$\varepsilon(R) = \varepsilon^* \left\{ \left(\frac{R^*}{R}\right)^{12} - 2\left(\frac{R^*}{R}\right)^6 \right\} = 4\varepsilon^* \left\{ \left(\frac{\sigma}{R}\right)^{12} - \left(\frac{\sigma}{R}\right)^6 \right\}, \quad (1.73)$$

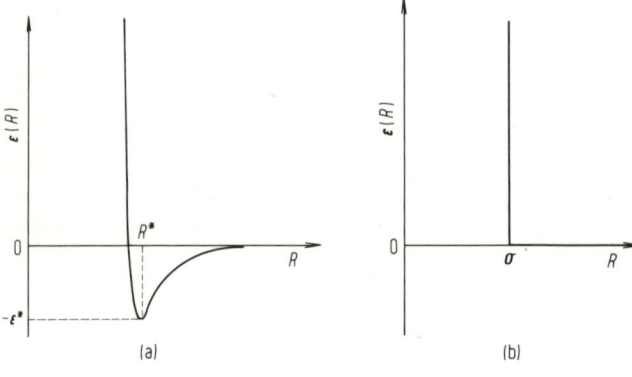

Fig. 1.7. a) The Lennard-Jones potential.
b) The hard-sphere potential.

where R^* and ε^* are the values of the coordinates of the minimum, and $\sigma = 2^{-1/6}\, R^* = 0.891\, R^*$ is the zero of $\varepsilon(R)$. It will be shown in chapter 5 that the dependence of the attractive energy on R^{-6} can be explained theoretically, whereas the dependence of the repulsion on R^{-12} is assumed for mathematical convenience but without physical justification.

1.6. References

1. A. Michels and C. Prins, Physica 28, 101 (1962) (The triple point values have been corrected to give the melting point at 1 atm.).
2. K. Clusius, F. Endtinger, and K. Schleich, Helv. Chim. Acta 43, 1267 (1960).
3. G. M. Hood and J. N. Sherwood, Mol. Cryst. 1, 97 (1966).
4. A. Klemm in: M. Blander, Molten Salt Chemistry, Interscience, New York 1964.
5. National Bureau of Standards, Circ. 500 "Selected Values of Chemical Thermodynamic Properties" (Washington 1952).
6. A. Berné, G. Boato, and M. De Paz, Nuovo Cimento 24, 1179 (1962).
7. W. M. Yen and R. E. Norberg, Phys. Rev. 131, 269 (1963).
8. J. S. Waugh, J. Chem. Phys. 26, 966 (1957).
9. L. W. Barr, I. M. Hoodless, J. A. Morrison, and R. Rudham, Trans. Faraday Soc. 56, 697 (1960).
10. J. N. Mundy, L. W. Barr, and F. A. Smith, Phil. Mag. 14, 785 (1966).
11. G. Pawlicki, Nukleonika 12, 1123 (1967); P. J. Fensham, Austr. J. Sci. R. A 3, 91 (1950), A 4, 225 (1951).
12. J. Naghizadeh and S. A. Rice, J. Chem. Phys. 36, 2710 (1962).
13. J. O'M. Bockris and G. W. Hooper, Disc. Faraday Soc. 32, 218 (1961).
14. R. E. Meyer and N. H. Nachtrieb, J. Chem. Phys. 23, 1851 (1955).
15. K. G. Davis and P. Fryzuk, J. Appl. Phys. 39, 4848 (1968).
16. H. Ruppersberg and H. J. Seemann, Z. Naturforsch. 20a, 104 (1965).
17. K. Clusius, A. Sperandino, and U. Piesbergen, Z. Naturforsch. 14a, 793 (1959).
18. W. F. Giauque and R. Wiebe, J. Am. Chem. Soc. 50, 101 (1928).
19. D. Ambrose, Trans. Faraday Soc. 52, 772 (1956).
20. A. Michels, B. Blaisse, and J. Hoogschagen, Physica 9, 565 (1942).
21. W. F. Giauque and C. C. Stephenson, J. Am. Chem. Soc. 60, 1389 (1938).
22. A. Neckel and H. Volk, Monatsh. Chem. 89, 754 (1958).
23. American Petroleum Institute Research Project 44, "Selected Values of Physical and Thermodynamic Properties of Hydrocarbons and Related Compounds", (Pittsburgh, Pennsylvania, 1953).
24. Landolt-Börnstein, „Zahlenwerte und Funktionen" Vol. II, Part 4, Springer, Berlin 1961.
25. T. B. Douglas, A. A. Ball, D. C. Ginnings, and W. D. Davis, J. Am. Chem. Soc. 74, 2472 (1952).
26. P. Flubacher, A. J. Leadbetter, and J. A. Morrison, Proc. Phys. Soc. (London) 78, 1449 (1961).
27. K. Clusius and L. Popp, Z. Phys. Chem. B 46, 63 (1940).
28. K. Clusius, K. Piesbergen, and E. Varde, Helv. Chim. Acta 43, 1290 (1960).
29. M. Blander in: M. Blander, ref. 4.
30. R. H. Busey and W. F. Giauque, J. Am. Chem. Soc. 75, 806 (1953).
31. D. C. Ginnings, T. B. Douglas, and A. F. Ball, J. Res. Natl. Bur. Std. 45, 23 (1950).
32. W. van Witzenburg, Phys. Letters A 25, 293 (1967).
33. M. J. Terry, J. T. Lynch, M. Bunclark, K. R. Mansell, and L. A. K. Staveley, J. Chem. Thermodyn. 1969, 413.
34. O. G. Peterson, D. N. Batchelder, and R. O. Simmons, Phys. Rev. 150, 703 (1966).
35. K. Clusius and K. Weigand, Z. Phys. Chem. B 46, 1 (1940).
36. K. Clusius, K. Piesbergen, and E. Varde, Helv. Chim. Acta 42, 2356 (1959).
37. A. F. Grigor and W. A. Steele, J. Chem. Phys. 48, 1032 (1968).
38. Gen. ref. 11.
39. H. Sackmann and F. Sauerwald, Z. Phys. Chem. 195, 295 (1950).
40. E. Vogel, H. Schinke, and F. Sauerwald, Z. anorg. und allg. Chemie 284, 131 (1956).
41. P. W. Bridgman, Proc. Am. Acad. Arts Sci. 47, 347, 428 (1911/12).

42. G. W. Thomson and E. Garelis in: M. Sittig, Sodium, its Manufacture, Properties and Uses, New York 1956.
43. S. H. Bauer and R. F. Porter in: M. Blander, ref. 4.
44. W. H. Evans, R. Jacobson, T. R. Munson, and D. D. Wagman, J. Res. Natl. Bur. Std. *55*, 83 (1955).
45. J. W. Stewart, Phys. Chem. Solids *1*, 146 (1956).
46. J. C. W. Heseltine, D. W. Elliot, and O. B. Wilson, Jr., J. Chem. Phys. *40*, 2584 (1964).
47. G. D. Oliver, M. Eaton, and H. M. Huffman, J. Am. Chem. Soc. *70*, 1502 (1948).
48. D. Rozental, Bull. Soc. Chim. Belg. *45*, 585 (1936).
49. A. Eucken and E. Lindenberg, Ber. *75*, 1953 (1942).
50. F. D. Enck, Phys. Rev. *119*, 1873 (1960).
51. R. I. Beecroft and C. A. Swenson, J. Phys. Chem. Solids *18*, 329 (1961).
52. P. M. Sutton, Phys. Rev. *91*, 816 (1953).
53. J. O'M. Bockris and N. E. Richards, Proc. Roy. Soc. (London) *241 A*, 44 (1957).
54. T. E. Pochapsky, Phys. Rev. *84*, 553 (1951).
55. H. J. Seemann and F. Klein, Z. angew. Phys. *19*, 368 (1965).
56. Chlorobenzene: P. W. Bridgman, The Physics of High Pressure (p. 130), Bell and Sons, London 1949, anthracene: P. W. Bridgman, Proc. Am. Acad. Arts Sci. *76*, 9 (1945).
57. A. M. Clark, F. Din, and J. Robb, Physica *17*, 876 (1951).
58. P. W. Bridgman, Phys. Rev. *46*, 930 (1934).
59. W. F. Giauque and C. J. Egan, J. Chem. Phys. *5*, 45 (1937).
60. M. M. Makansi, C. H. Muendel, and W. A. Selke, J. Phys. Chem. *59*, 40 (1955).
61. K. A. Kobe and R. E. Lynn, Jr., Chem. Rev. *52*, 117 (1953).
62. A. P. Kudchadker, G. H. Alani, and B. J. Zwolinski, Chem. Rev. *68*, 659 (1968).
63. F. Kohler, Monatsh. Chem. *103* (1972), in press.
64. F. Hensel and E. U. Franck, Rev. Mod. Phys. *40*, 697 (1968).
65. W. D. McCain, Jr. and W. T. Ziegler, J. Chem. Eng. Data *12*, 199 (1967).
66. Textbooks of thermodynamics: G. Bruhat and A. Kastler, Thermodynamique, sixième edition, Masson, Paris 1968; H. B. Callen, Thermodynamics, Wiley, New York 1960; E. G. Guggenheim, Thermodynamics, second edition, North-Holland, Amsterdam 1959; G. N. Lewis, M. R. Randall, K. S. Pitzer, and L. Brewer, Thermodynamics, second edition, McGraw-Hill, New York 1961; I. Prigogine, R. Defay, and D. H. Everett, Chemical Thermodynamics, Longmans and Green, London 1954.
67. Textbooks of statistical mechanics: R. Becker, Theory of Heat, Springer, Berlin 1967; N. Davidson, Statistical Mechanics, McGraw-Hill, New York 1962; T. L. Hill, gen. ref. 5; T. L. Hill, Statistical Thermodynamics, Addison-Wesley, Reading Mass. 1960; L. D. Landau and E. M. Lifshitz, Statistical Physics, Pergamon Press, London 1958; A. Münster, Statistische Thermodynamik, Springer, Berlin 1956; revised English edition in two volumes: Vol. I appeared in 1969, Vol. II is scheduled for the near future; O. K. Rice, Statistical Mechanics, Thermodynamics and Kinetics, Freeman, San Francisco 1967; G. S. Rushbrooke, Introduction to Statistical Mechanics, Clarendon Press, Oxford 1949.
68. I. L. Fabelinskii, Usp. Fiz. Nauk *63*, 355 (1957); (Engl. Transl.: Adv. Phys. Sci. *63*, 474 (1957)); L. D. Landau and E. M. Lifshitz, Electrodynamics of Continuous Media, Pergamon Press, Oxford 1960; I. L. Fabelinskii, Molecular Scattering of Light, Plenum Press, New York 1968; R. D. Mountain, Rev. Mod. Phys. *38*, 205 (1966).

2. Models for the Liquid State

2.1. Reason for Models

Models for the liquid state aim at a perceptual understanding of the structure of liquids and at evaluations of the configuration integral Q [cf. eq. (1.32)]. Eq. (1.32) cannot be solved exactly and therefore either mathematical approximations or models (possibly in connection with mathematical approximations) have to be used. Even if we would have exact solutions, the use of models would yield valuable physical insight into the nature of the liquid state.

Models have been constructed for which the integral in eq. (1.32) can be evaluated with some accuracy. At present, there exists no model which would reproduce satisfactorily the properties of liquids over the whole density range. But it has been shown that the lattice gas model works well at the critical point and the hard sphere model at the triple point. We will discuss these models in the sections 2.2 and 2.3.

Attempts have been made to understand the facts which distinguish the liquid from the ideal gas and the kind of disorder which distinguishes the liquid from the ideal crystal. The first approach was induced by van der Waals' idea of continuity between liquid and gaseous state. The second approach originated in the much closer similarity of the properties of liquids and solids. Here the formulation of the disorder in the liquid relative to the ordered solid lattice is the main problem. Section 2.4 presents the model experiments of Bernal, which correspond to the packing of a hard sphere assembly at infinite pressure or zero temperature. The following sections comprise approximative mathematical formulations of the disorder problem. Section 2.5 sketches the early attempts of introducing crystal defects and the reason for their failure. In section 2.6 a rigorous formulation of the disorder or "communal entropy" problem is given on the basis of the cell theory. But again approximations based on models are necessary for the numerical evaluation. A disorder model which avoids the disadvantages of the defective crystal model is given in more detail.

Today, with the development of fast computers, it becomes possible to simulate the motion of particles in liquids. "Experiments" of this kind produce valuable information about liquids on a molecular scale. By ascribing certain repulsive and attractive forces to the particles, the effect of molecular interactions can be studied. The most simple interaction is the repulsion of hard spheres, and the possibility to calculate the properties of an assembly of hard spheres easily by computer experiments has greatly increased the importance of this assembly. The hard sphere fluid is also mostly the starting point for the perturbation treatments dealt with in section 7.4. The computer methods will be treated in the next chapter, but we will mention some of the results in this chapter, which are relevant for the models.

2.2. Lattice Gas Model

The lattice gas model is characterized by a division of the liquid volume into cells such that the number of cells is larger than the number of particles. Each cell is centered at one site of a lattice. The number of cells in excess over the number of the particles is related to the volume of the model assembly. For the interaction energy the following assumption is made: the potential energy between two particles is $+\infty$ if the particles are in the same cell, $\varepsilon(<0)$ if they are in a pair of neighbouring cells and zero otherwise. The cells provide a coordinate system and are not thought to constrain the molecular motion.

Lee and Yang [1] showed the mathematical equivalence of the lattice gas model and the Ising model of ferromagnetism. Each cell may be occupied by exactly one particle or be empty. These two possibilities correspond to the two orientations of spin in the Ising model. This relation is important since the properties of the Ising model are accurately known [2]. The lattice gas model separates at lower temperatures into two phases of different density, which become identical at a critical temperature. The details of this critical behaviour, e.g. the critical indices (cf. section 6.5) are compatible with our experimental knowledge of the gas-liquid critical point. It is remarkable that a model as simple as the lattice gas yields such good results.

This agreement suggests that the lattice gas represents adequately many important features of a fluid in the critical region [3]. The most significant characteristic of the critical state is the propagation of correlations through great distances. If now ε were zero there would remain as potential energy only the infinite repulsion between molecules in the same cell. In this case the probability of a cell being occupied is equal for all empty cells. So the long ranged correlation in the lattice gas is solely due to the attractive part of the interaction. This is an indication of the importance of the attractive forces for the properties of a fluid near the critical point. This corresponds to the fact that in the van der Waals-equation the a/V^2-term is necessary for getting critical phenomena and eventually a phase separation between liquid and gas.

Outside the critical region the lattice gas model has not been used. As it can give only one phase transition, its ordered phase is no good description of the liquid state where the differences to the solid are of significance.

2.3. Hard Sphere Model

This model makes use of the fact that the properties of a hard sphere assembly are well known by computer experiments. Furthermore, it is known that the attractive forces between molecules are weak and long ranging compared to the repulsive forces which grow very fast at small distances for approaching molecules.

This suggests that in a fluid as dense as a liquid near its triple point the attractive forces exerted on one molecule by its neighbours largely cancel [3]. A molecule in such a fluid may then be thought of as a hard sphere in a uniform negative background potential. The amount of this potential is proportional to the number of molecules contributing to it and therefore to the density of the fluid. The total potential energy U per mole is then

$$U = -a/V \tag{2.1}$$

where a is a positive constant. The fluid near the triple point is thus pictured as a system of hard spheres embodied in a negative background potential. The equation of state of this model system is

$$P = P_{h.s} - a/V^2 \tag{2.2}$$

where $P_{h.s}$ is the pressure of a hard sphere fluid without a background potential. The effect of the uniform potential is thus merely to lower the pressure by the amount $(\partial U/\partial V) = a/V^2$ following from eq. (2.1).

Since the potential (2.1) is uniform there are no forces exerted on the molecules. There is no change in molecular configurations if one starts from a system of hard spheres and increases gradually the uniform background potential. Therefore, the positional correlation is due entirely to the hard cores of the spheres. If now eq. (2.2) is an adequate equation of state for a real fluid near the triple point, one can conclude that the repulsive forces are determining the structure of a liquid near the triple point.

Computer studies (cf. chapter 3) provide strong evidence that the hard sphere assembly exhibits a first order transition between a more dense phase with crystalline order and a less dense phase without crystalline order. Using these computer calculations it was shown by Longuet-Higgins and Widom (cf. section 6.4) that the hard sphere system with an added background potential can account quite well for the thermodynamic description of the melting point.

2.4. Bernal's Geometric Model

According to Bernal's model the ordering in the liquid state is analogous to that of a randomly packed heap of hard spheres. The strongest support for this model is that the radial distribution of spheres around one central sphere is essentially the corresponding distribution of atoms in a simple monoatomic liquid. Granting the validity of the Bernal model one can explain in a very intuitive way a number of features of the liquid state. All model experiments have been carried out with a "dense" packing of hard spheres [4].

First we will describe how assemblies of randomly dense-packed spheres have been prepared so that geometrical analysis has been possible. The first experiments were done on plasticine spheres by compressing them uniformly so that space was filled completely. Such experiments [5] showed nicely the predominant occurence of pentagonal faces. For a more detailed analysis later experiments were carried out with steel balls. Several thousands of steel balls were vibrated into a dense packing after filling them into containers with irregular walls.

For a geometric description Bernal uses two consepts, the "Voronoi polyhedra" and the "simplical graph". The Voronoi polyhedra are generated by the division of space into regions around the centers of the particles. A Voronoi polyhedron contains all points in space closer to the center of the embodying sphere than to the center of any other sphere. In two dimensions such regions are termed Dirichlet polygons. The planes of the Voronoi polyhedra or the sides of the Dirichlet polygons are perpendicular to certain connective lines between centers of balls. The system of these lines is called the simplical graph and two spheres whose centers are connected are called geometric neighbours. In two dimensions the situation is shown in fig. 2.1.

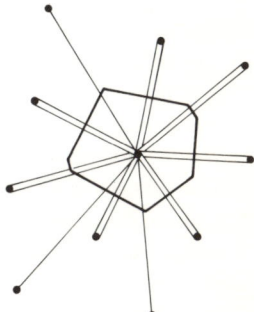

Fig. 2.1. A Dirichlet polygon and the connective lines perpendicular to its sides, which belong to the simplical graph (double lines). Three more distant neighbours, which are not geometric neighbours, are shown also.

2. Models for the Liquid State

The results show that there exists a state of arrangement of hard spheres with definite geometrical properties which is called random close packed structure. There are no states between the random close packed and a regular close packed structure. In the random case the space occupied by the spheres is 64% of the total space whereas it is 74% for the regular arrangement. The ratio of the two densities is nearly the same as the ratio of liquid and solid densities at the triple point of argon (table 1.2). But the Bernal model is the limiting case of a hard sphere fluid for high densities (which is thermodynamically unstable), so that the agreement to the melting process of argon is probably accidental.

Of high interest is the radial distribution of spheres around one chosen sphere. The data are analysed in two ways. Firstly the probability of finding a sphere at a certain distance from any given sphere is determined [6]. This function is closely connected to the pair distribution function $g(R)$ (see chapter 4), which is shown in fig. 2.2 for a random dense packing of hard spheres. Comparison with fig. 4.5 shows that it resembles closely that of atoms in liquid argon. Also given in fig. 2.2 are calculations of $g(R)$ of a hard sphere fluid at lower densities, based on theory and computer experiments, respectively. By this similarity the Bernal model may be regarded as the high density limit of a hard sphere fluid.

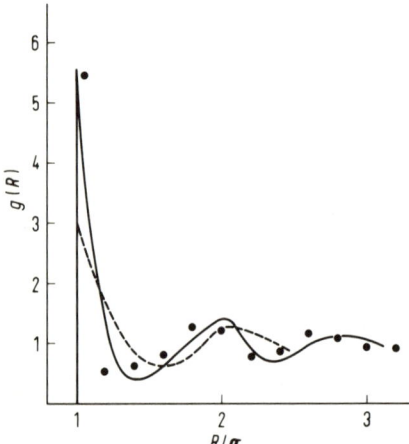

Fig. 2.2. The pair distribution function for hard spheres. The Bernal model (corresponding to a density $\rho/\rho_0 = 0.86$) is indicated by full circles, a theoretical calculation (based on the PY theory, chapter 7) for the density $\rho/\rho_0 = 0.71$ [10] is shown as full line, and a Monte Carlo calculation for the density $\rho/\rho_0 = 0.56$ [15] is shown as dotted line. ρ_0 means the density of closest regular packing. The highest density for the stable hard sphere fluid is $\rho/\rho_0 = 0.67$ [16] (chapter 3).

Secondly Bernal calculates the number of spheres surrounding one chosen sphere within distances of 1.05, $\sqrt{2}$, and 2 times the sphere diameter. As expected one gets for each coordination sphere a variable coordination number. The distribution of the coordination number is shown in fig. 2.3. The mean of the $\sqrt{2}$-coordination numbers is 13.4, very near to the average number of faces of the Voronoi polyhedra, which is 13.6. That means that all particles within a distance of $\sqrt{2}$ times the sphere diameter are geometric neighbours. The relative fluctuation of the coordination number is already small for the 2-diameter sphere, where the mean coordination number approaches the relevant number density. That means that a hard sphere "liquid" exhibits structure only over a relatively narrow range.

An important consequence of the Bernal model concerns the geometry of the short range order. It has long been known from X-ray diffraction studies that liquids exhibit a certain order within small ranges. However, Bernal's model suggests that the arrangement in this region is radically different from that around an atom of a crystal.

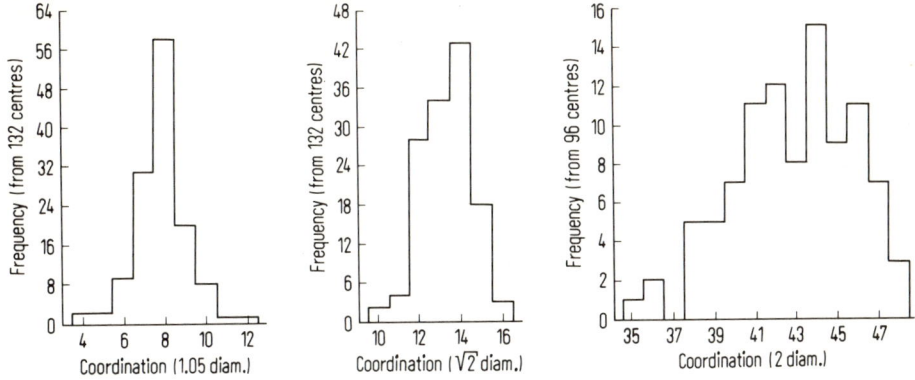

Fig. 2.3. Number of neighbours within 1.05 σ, $\sigma\sqrt{2}$, and 2σ (σ being the hard sphere diameter) in the Bernal model of random close packing. After ref. 5.

The analysis of the Voronoi polyhedra yields as important result the frequent occurence of pentagonal faces. A reduction of the arrangement to two dimensions must give a predominance of pentagons among the Dirichlet polygons. This eliminates any possibility of order over wider ranges and makes necessary a reordering at crystallization. This explains why liquids are undercooled so easily.

2.5. Defective Crystal Model

According to this model the liquid is formed from a crystal by the introduction of a certain amount of crystal defects. Nearly every type of crystal defect has been used. Two special models deserve to be mentioned:

a) The crystal defects are lattice vacancies. The most elaborate quantitative development of this model is due to Ono [7]. At low temperatures he obtains a phase separation into a phase rich in particles (identified with the liquid) and a phase rich in holes (identified with the gas). At increasing temperatures, the "liquid" phase looses particles and gains holes, and just the opposite is true of the "gas" phase. When the phases become identical, the gas-liquid critical point is reached. The law of the rectilinear diameter is readily explained in this way, as the total number of particles and holes remains the same. The entropy gain produced by the introduction of holes is twofold. Firstly there is an entropy of mixture between particles and holes and secondly an increase of the volume available for individual particle motion if a particle is located besides a hole. The "volume available for particle motion" is called free volume.

We will now give an equation which expresses the ideas of Ono and which will be used later. In section 1.3 we obtained a simplified configurational partition function for a solid (neglecting correlations of particle motions):

$$Q = e^{-Nz\varepsilon/2kT}\left(\frac{v_f}{N}\right)^N \tag{2.3}$$

where z is the number of contacts per molecule, and $\bar{\varepsilon}$ the average energy per contact. The configurational entropy is given by $s' = Nk \ln (v_f/N)$. We introduce now holes into the lattice of the solid. Let the number of holes be N_0 and the number of particles N, and x be defined by $N_0/(N + N_0)$. The first assumption we make is that the partition function for the particle motion is given by eq. (2.3) with x-dependent quantities $\bar{\varepsilon}$ and v_f. The increased number of arrangements due to the various distributions of particles and holes is taken into account by a factor $g(x)$*⁾ (which gives rise to an entropy of mixing):

$$Q = e^{-Nz\bar{\varepsilon}(x)/2kT} \left(\frac{v_f(x)}{N}\right)^N g(x). \tag{2.4}$$

The second assumption is that that actual x dependence in $\bar{\varepsilon}(x)$ and $s'(x) = Nk \ln (v_f(x)/N)$ can be approximated by a linear relationship.

$$\bar{\varepsilon}(x) = (1 - x)\bar{\varepsilon} + x\bar{\varepsilon}_0 \tag{2.5}$$

$$s'(x) = Nk \left[(1 - x) \ln \left(\frac{v_f}{N}\right) + x \ln \left(\frac{w_f}{N}\right)\right]. \tag{2.6}$$

$\bar{\varepsilon}$ is here the average potential energy per contact and v_f/N the free volume per particle if all the neighbouring lattice sites are occupied. $\bar{\varepsilon}_0$ and w_f are the corresponding quantities when all nearest lattice sites are empty. Eq. (2.5) and (2.6) lead with eq. (2.4) to:

$$Q = e^{-Nz[(1-x)\bar{\varepsilon}+x\bar{\varepsilon}_0]/2kT} \cdot \left(\frac{v_f^{1-x} w_f^x}{N}\right)^N g(x). \tag{2.7}$$

A detailed examination of this model shows that the entropy increase is not sufficient to overcome the loss in interaction energy [8]. If the free energy is minimized with respect to x, it turns out that the corresponding value of x is very small. Depending on the interaction energy, x is $2 \cdot 10^{-3}$ to $5 \cdot 10^{-3}$, which is thermodynamically insignificant. Apparently, the excess volume generated at melting is much finer divided and distributed than corresponding to holes of molecular size.

Eq. (2.7) can also be used for other models. Then x is the relative number of another kind of defects.

b) The crystal defects are crystallite boundaries**⁾. The thought of having the excess volume distributed in a multitude of grain boundaries looks quite attractive at first sight. In some respect, this would be a finer division than holes of molecular size, as the particle distance over a boundary could be much less than twice the distance to nearest neighbours. The strongest argument against the grain boundary model is the absence of small angle scattering of X-rays, which is typical for a crystallite or gel-like structure up to a crystallite length of 1000 Å.

In one case, it had been tried to amplify such a small angle scattering by using a mixture of hexane + trichlorobenzene, where the less scattering hexane molecules would be located preferably at the inner surface or grain boundaries, for thermodynamic reasons. The result was completely negative.

---

* In the case under consideration $g(x)$ is given by: $\dfrac{(N + N_0)!}{N! N_0!}$.

** This is somewhat equivalent to the model of "paracrystals" put forward by Hosemann [17].

The hexane molecules were obviously distributed at random, and the same has to be assumed for the small holes existing in the somewhat irregular packing of the liquid [9].

2.6. The Cell Theory and Disorder Models

In the cell theory the configuration integral Q is rewritten in a special manner, which is in principle rigorous [11]. For the numerical evaluation, i.e. the calculation of thermodynamic properties, approximations and disorder models are necessary.

The liquid volume is divided into a large number of small cells of equal volume W. We will consider the case where the number of cells is equal to the number of particles. The integration space of each integral occuring in Q is thus divided into subspaces $W_1, \ldots, W_N$. Then Q can be written as a sum of integrals over subspaces:

$$Q = \frac{1}{N!} \sum_{l_1=1}^{N} \cdots \sum_{l_N=1}^{N} \int_{W_{l_1}} \cdots \int_{W_{l_N}} \exp\{-E(\vec{r}_1, \ldots, \vec{r}_N)/kT\} \mathrm{d}\vec{r}_1, \ldots, \mathrm{d}\vec{r}_N. \qquad (2.8)$$

Let us consider the case that there are m_1 particles in cell 1, m_2 in cell 2, etc., and define the quantity:

$$Q^{(m_1, m_2 \cdots m_N)} = \int_{W_1} (m_1) \int_{W_1} \cdots \int_{W_N} (m_N) \int_{W_N} e^{-E/kT} \mathrm{d}\vec{r}_1, \ldots, \mathrm{d}\vec{r}_N. \qquad (2.9)$$

The expression (2.9) occurs as a term in the sum of eq. (2.8). We collect now all the integrals in (2.8) for which $m_1, m_2, \ldots, m_N$ are the same numbers. Each of these integrals has the same value because it does not matter which particle is in which cell. The number of these integrals is equal to

$$\frac{N!}{m_1! m_2! \ldots m_N!}.$$

The configuration integral Q can thus be written

$$Q = \sum_{m_1, \cdots, m_N = 0}^{N} \frac{1}{m_1! \ldots m_N!} Q^{(m_1, m_2, \cdots, m_N)}, \qquad (2.10)$$

where the summation over the m_i is subject to the restrictive condition $\sum_i m_i = N$. Eq. (2.10) can be rewritten in the form

$$Q = Q^{(1, \cdots, 1)} \cdot \sum_{m_1, \cdots, m_N = 0}^{N} \frac{1}{m_1! \ldots m_N!} \frac{Q^{(m_1, m_2 \cdots m_N)}}{Q^{(1, \cdots, 1)}}. \qquad (2.11)$$

By introducing the abbreviations

$$Q^{(1)} = Q^{(1, \cdots, 1)}$$

and

$$\sigma^N = \sum_{m_1, \cdots, m_N = 0}^{N} \frac{1}{m_1! \ldots m_N!} \frac{Q^{(m_1, \cdots, m_N)}}{Q^{(1)}} \qquad (2.12)$$

the final form of eq. (2.10) is obtained:

$$Q = \sigma^N Q^{(1)}. \qquad (2.13)$$

For accurate calculations eq. (2.13) does not have any advantage over eq. (1.32). But if models could be constructed for the approximate evaluation of σ, then the problem of calculating Q

is reduced to the somewhat easier problem of computing $Q^{(1)}$. As $Q^{(1)}$, the configurational partition function for single occupancy, is analogous to that of an ideal crystal where all molecules vibrate around their lattice sites, the factor σ gives the deviation from a completely ordered structure. Comparing to eq. (1.37) and (1.42), the term $NkT\ln\sigma$ may be called "communal free energy", and the models for evaluating σ may be termed disorder models. These disorder models — besides satisfying the intuitive feeling — are nothing else than a mathematical convenience.

As in the ideal crystal $Q^{(m_1,\ldots,m_N)}$ is zero if $(m_1,\ldots,m_N)$ differs from $(1,\ldots,1)$, eq. (2.12) yields $\sigma^N = 1$ for that case. For the ideal gas the ratio $Q^{(m_1,\ldots,m_N)}/Q^{(1)}$ becomes unity, and the sum occuring in (2.12) can be evaluated considering that $N!/m_1!m_2!\ldots,m_N!$ is the general coefficient of the polynomial $(x_1 + x_2 + \ldots x_N)^N$. The result is $\sigma^N = N^N/N!$ or e^N using Stirling's approximation (1.36).

In the case of the liquid state of matter the situation is much less satisfactory. There is no good model which could be used for an approximate evaluation of σ. Frequently defective crystal models have been used in spite of their shortcomings. Two attempts, the introduction of lattice vacancies or of grain boundaries, have already been discussed in section 2.5. In section 2.6.1 a model will be described which allows a finer division of the volume by which a liquid exceeds that of a solid. Two further attempts for evaluating σ are the following:

The cell cluster theory [12] proceeds straightforward from eq. (2.12). Considering only double occupancy of one cell, that means retaining in (2.12) only terms like $Q^{(2,0,1,\ldots,1)}$, and introducing certain approximations for $Q^{(2,0,1,\ldots,1)}$ it arrives at a value for σ. But it turns out that this approach is insufficient.

The tunnel theory [13] introduces a model already at an earlier stage than the factor σ is introduced in the derivation of the cell theory given here. The configuration space is divided into parallel tubes or "tunnels" where the molecule may slide in one dimension. This model takes advantage of the fact that the configurational integral Q can be solved exactly for a one-dimensional system of hard spheres.

2.6.1. Disorder in Terms of Coordination Defects [14]

This disorder model considers only the immediate surroundings of a given molecule. A coordination defect is thought of as the transfer of a particle from the first shell of neighbours into the second shell of neighbours. Expressed in another way, it is the lenghtening of a nearest-neighbour contact from the distance a to the distance $a\sqrt{2}$. Such a lengthened contact will be called a broken contact. As the resulting holes are small and as the local coordination may vary considerably, this model accounts for the fine distribution of order defects and for the irregular arrangement of the particles. The price is the difficulty to view the geometrical situation, which brings about a certain lack of rigour in the mathematical derivation. If the fraction of broken contacts is denoted by x, [cf. eq. (2.7)], the configurational integral is written as

$$Q = Q_{(a)}^{1-x} Q_{(a\sqrt{2})}^{x} g(x). \tag{2.14}$$

Here $Q_{(a)}$ is the configurational integral $Q^{(1)}$ of a cell model with single occupancy and a cell diameter a, and $Q_{(a\sqrt{2})}$ is $Q^{(1)}$ for a cell diameter $a\sqrt{2}$. The factor $g(x)$ accounts for the number of distinguishable arrangements with $Nz(1-x)/2$ normal nearest neighbour contacts (of length a) and $Nzx/2$ broken contacts (z being the coordination number in a regular dense lattice, i.e. 12).

It is interesting to note that this model leads to a distribution of coordination numbers quite similar to that of Bernal's geometrical model, provided that the fraction of particles with an effective coordination number i would be given by a binomial distribution

$$N_i/N = \binom{z}{i}(1-x)^i x^{z-i}. \tag{2.15}$$

Two ways seem possible for the evaluation of $g(x)$. The first is to count the number of interchanges between particles distinguished by their coordination numbers. However, the exchange of coordination defects is somewhat restricted. E.g., considering only molecules with coordination number 12 and 11, these defects can be exchanged only in groups of at least two, as a deficiency in coordination of the central molecule necessitates also a deficiency in coordination of one neighbouring molecule. This correlation of interchange is closely related to the interchange of dimers with solvent molecules. For lower coordination numbers (larger x) the problem becomes very complicated, and other arguments for the evaluation of $g(x)$ are sought. The second possibility is the consideration of the number of ways to interchange broken contacts with normal contacts. The difficulty is that it is necessary to discount all interchanges which will take place by the motion of one particle already taken care of in $Q_{(a\sqrt{2})}$. This correction contains an arbitrary element; on the other hand, a simple expression for $g(x)$ results [14].
For hard spheres of diameter σ_H the configurational integral can then be written

$$Q = \left\{\frac{\pi}{6}(a - \sigma_H)^{3(1-x)}(a\sqrt{2} - \sigma_H)^{3x}\right\}^N g(x), \tag{2.16}$$

where a is connected to the volume per molecule by the relation

$$v/N = a^{3(1-x)}(a\sqrt{2})^{3x}/\sqrt{2} = a^3 2^{(3x-1)/2}. \tag{2.17}$$

The fraction x of broken contacts can be obtained by seeking the maximum of Q (minimum of free energy) for given volume and number of particles. As x enters the relation between a and v/N, it is not possible to give an analytic expression for the disorder parameter σ as function of v/v_0, but this function can be constructed easily point by point.
A comparison of $\ln \sigma$ according to the described disorder model and to the tunnel theory is shown in fig. 2.4 for hard spheres and is contrasted to results of Monte Carlo calculations (see

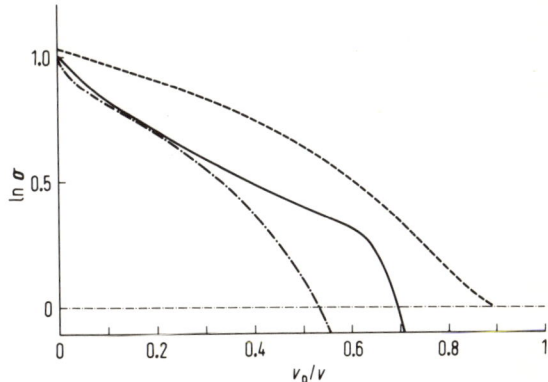

Fig. 2.4. A comparison of $\ln \sigma$ according to Monte Carlo calculations [16] (full line), tunnel theory [13] (dashed-dotted line), and disorder in terms of coordination defects [14] (dotted line).

the following chapter), which are supposed to be exact. The curve for the described disorder model neglects correlations between particle motions in neighbouring cells; thus it is restricted to positive values of ln σ.

2.7. References

1. T. D. Lee and C. N. Yang, Phys. Rev. *87*, 410 (1952).
2. For references see: M. E. Fisher, J. Math. Phys. *5*, 944 (1964).
3. B. Widom in Study Week on Molecular Forces, Pontificiae academiae scientiarium scripta varia. North-Holland, Amsterdam 1967.
4. J. D. Bernal and S. V. King in gen. ref. 12.
5. J. D. Bernal in gen. ref. 7.
6. G. D. Scott, Nature *194*, 956 (1962). Cf. also G. D. Scott and D. L. Mader, Nature *201*, 382 (1964), and G. D. Scott, A. M. Charlesworth, and M. K. Mak, J. Chem. Phys. *40*, 611 (1964).
7. S. Ono, Mem. Fac. Eng. Kyushu Univ. *10*, 195 (1947).
8. cf. J. A. Barker, gen. ref. 1, T. L. Hill, gen. ref. 5 and S. A. Rice, J. Chem. Phys. *24*, 357 (1956).
9. Unpublished work done at the University of Graz, see footnote 28 in F. Kohler, H. Arnold, and R. J. Munn, Monatsh. Chem. *92*, 876 (1961). Also in the case of water, for which cluster models have been proposed, no significant small angle scattering of X-rays could be found: A. M. Levelut and A. Guinier, Bull. Soc. Franc. Mineral. Cristallogr. *90*, 445 (1967). The results of scattering experiments on silicate glasses are also inconsistent with this model: B. E. Warren, J. Appl. Phys. *8*, 645 (1937).
10. G. J. Throop and R. J. Bearman, J. Chem. Phys. *42*, 2408 (1965).
11. J. G. Kirkwood, J. Chem. Phys. *18*, 380 (1950); see also T. L. Hill, gen. ref. 5.
12. J. de Boer, Physica *20*, 655 (1954); J. M. H. Levelt and E. G. D. Cohen in: J. de Boer and G. E. Uhlenbeck, Studies in Statistical Mechanics, Vol. II, North-Holland, Amsterdam 1964.
13. J. A. Barker, gen. ref. 1.
14. F. Kohler, Ber. Bunsenges. *70*, 1068 (1966).
15. M. N. Rosenbluth and A. W. Rosenbluth, J. Chem. Phys. *22*, 881 (1954).
16. W. G. Hoover and F. H. Ree, J. Chem. Phys. *49*, 3609 (1968).
17. Hosemann and K. Lemm in Physics of Non-Crystalline Solids, Proceed. Int. Conf. Delft (1964), North-Holland, Amsterdam 1964; K. Lemm, Mol. Cryst. *10*, 259 (1970).

3. Computer Experiments

3.1. Introduction

The physical properties of classical many-particle systems can either be determined by solving the equations of motion or by applying statistical mechanics. The averaging procedure of statistical mechanics may be done numerically or by analytical methods employing various theories of the liquid state. For the latter methods we refer to chapter 7. The development of high-speed computers with large storage facilities has made it possible to solve the equations of motion and to do numerical statistical mechanics for a limited number of molecules.

In the most direct approach, the so-called method of molecular dynamics, the system is treated by setting up the Newtonian equations of motion, which are coupled by the interaction potential between the particles. This set of equations is then solved numerically to give the position and velocity of all particles as a function of time. The method can be used to examine equilibrium and non-equilibrium properties of a classical system; equilibrium properties are obtained by time averaging.

Another way to get equilibrium properties is the use of ensemble averaging. In most cases a canonical ensemble is used [*]. One might think to create a large number of states by the computer, which are randomly distributed in configuration space, and then to weigh each state with its Boltzmann factor. But from all possible states in configuration space, very few are connected with a Boltzmann factor appreciably different from zero, since for most randomly chosen configurations at least two molecules come so close together, that the potential energy becomes enormously repulsive. This fact suggests to sample the configurations according to their importance, and this is what the Monte Carlo [**] method does. The probability to select a certain state is set proportional to its Boltzmann factor, and in the averaging procedure all states have the same weight.

Both methods have contributed much to understand the physics of many-particle systems. But both methods suffer from two facts. First the number of particles, which can actually be treated by a computer, is very restricted. At the beginning of such computer calculations the use of some ten particles was typical, whereas nowadays a typical number is about thousand. But physical systems normally consist of some 10^{20} molecules, and it is a serious problem to infer from the computer systems to the properties of macroscopic systems. One tries to minimize the difference by introducing periodic boundary conditions. The second problem is that both methods are not able to calculate the entropy directly. Molecular dynamics cannot, because the entropy is not defined in the concept of non-statistical mechanics, and Monte Carlo cannot, because it does importance sampling, instead of evaluating the whole partition function.

3.2. Principles of Monte Carlo Calculations [1–3]

We restrict our discussion to the case of a canonical ensemble of N particles in a cell volume v at temperature T. Call B the number of states in configuration space between which the computer can discriminate, let further φ_i be the value of any function φ which depends on the position

---

[*] For numerical statistical mechanics on other ensembles, see [1].
[**] For certain sampling procedures expressions from the roulette are in use.

coordinates and E_i be the potential energy, if the system is in the state i, then the average of the function φ is given in principle by [cf. eq. (1.26) and (1.27)]

$$\langle\varphi\rangle = \sum_{i=1}^{B} u_i \varphi_i, \qquad (3.1)$$

where u_i means the probability of the i-th state

$$u_i = \frac{\exp(-E_i/kT)}{\sum_{i=1}^{B} \exp(-E_i/kT)}. \qquad (3.2)$$

In spite of B being finite, it is so large that eq. (3.1) cannot be evaluated directly. Therefore, the Monte Carlo method introduces a special random walk, which comprises n out of B possible states, and where the chance for one state i to be chosen is made proportional to its probability u_i. The averaging results then in forming

$$\bar{\varphi} = \frac{1}{n} \sum_{i=1}^{n} \varphi_i. \qquad (3.3)$$

The random walk is characterized by the transition probabilities to get from one state to another. If one has arrived at a state i, let p_{ij} be the probability that the next state will be j, and $p_{ij}^{(m)}$ that one arrives at state j after m steps. Naturally, we have the normalization condition

$$\sum_{j=1}^{B} p_{ij} = 1. \qquad (3.4)$$

The problem is now, whether $\bar{\varphi}$ converges against $\langle\varphi\rangle$ for a sufficiently long random walk. The probability theory [4] shows that for this following conditions are sufficient:
a) Ergodicity condition: it is possible to arrive at any admissible state j (a state with non-vanishing probability u_j) from any admissible state i within a finite number of steps.
b) Steady state condition:

$$\sum_{i=1}^{B} u_i p_{ij} = u_j \quad \text{for all } j. \qquad (3.5)$$

One important step in the theoretical deduction is the result that the m-step probability $p_{ij}^{(m)}$ tends to u_j for large m, independent of the initial state i. That means: starting with any initial state and going through a sufficient large number of steps, every possible state has a chance to be selected as often as its probability indicates.
A slightly more special formulation of the steady state condition is the principle of microscopic reversibility

$$u_i p_{ij} = u_j p_{ji}. \qquad (3.6)$$

Together with the normalization condition eq. (3.4) this gives eq. (3.5). Eq. (3.6) can be considered as the assumption of a dynamic equilibrium between each pair of states. Another expression would be that each random walk could be reversed. Eq. (3.6) gives also a feeling for the fact that the m-step probability $p_{ij}^{(m)}$ tends to u_j. Assuming the validity for the $(m-1)$-step probabilities $p_{ij}^{(m-1)} = u_j$, then

$$p_{ij}^{(m)} = \sum_k p_{ik}^{(m-1)} p_{kj} = \sum_k u_k p_{kj} = \sum_k u_j p_{jk} = u_j \sum_k p_{jk} = u_j. \qquad (3.7)$$

3.2. Principles of Monte Carlo Calculations

The general feature of a Monte Carlo process is thus the set-up of a matrix (p_{ij}) of transition probabilities, whose elements fulfil — besides the ergodicity condition — the eq. (3.4) and (3.6). Then one starts with an admissible configuration, selects the next one in accordance with the transition probabilities, and so on.

The conditions for the choice of the transition probabilities leave a lot of arbitrariness, which is used to make the convergence of the procedure as rapid as possible. The commonly used prescriptions are the following:

Let the v-th sequence of the random walk be the state i with probability u_i. Then one molecule is selected at random and is moved randomly to a new position within a small cube centered at the old position. The random selection is done by applying random numbers*⁾ in the proper interval (number of molecules, or half edge of the cube for each coordinate shift). This gives a possible new state j with energy E_j and probability u_j. If now $u_j > u_i$, then the state j will be accepted as the $(v + 1)$th sequence in the random walk. If $u_j < u_i$, the ratio u_j/u_i is formed and compared to a random number in the interval (0, 1): if the ratio u_j/u_i is greater than this random number, the $(v + 1)$ th sequence in the random walk is the new state j, otherwise the $(v + 1)$ th sequence is the old state i. The small size of the cube giving the limits of the displacement of the one molecule aims at retaining the random walk for some time in the region of important configurations. The best choice of the size is based on empirical experience.

Denoting by $\{Z\}$ the set of states, which are accessible from state i by moving one molecule within the given cube, and calling their number z, the prescription can be summarized in terms of the transition probabilities as follows:

$$p_{ij} = 0 \qquad \text{if } j \text{ outside } \{Z\}$$

$$p_{ij} = \frac{1}{z} \qquad \text{if } j \text{ inside } \{Z\} \text{ and } u_j > u_i$$

$$p_{ij} = \frac{1}{z}\frac{u_j}{u_i} \qquad \text{if } j \text{ inside } \{Z\} \text{ and } u_j < u_i \tag{3.8}$$

$$p_{ii} = 1 - \sum_{j=1}^{z} p_{ij}.$$

Usually the sequence of the random walk is started with a state of regular configuration of the particles. The first part of this sequence is omitted in the averaging procedure in order to get rid of the influence of the special choice at the beginning.

The matrix (3.8) satisfies the conditions (3.4) and (3.6). More difficult to discuss is its consequence for the ergodicity condition. By retaining the system in a region of important configurations (which is achieved, among others, by moving only one particle at a time), the accessibility of other classes of important configurations — if such classes exist — is diminished. The situation is extreme for hard spheres, where some configurations are conceivable (e.g. a primitive cubic packing sphere on sphere), from which there is no access at all to other configurations. A well known example for such difficulties is the existence of two classes of configurations in the solid-liquid transition region, one of high regularity and one without long range order. Even in a random walk of reasonable length there might happen no or very few transitions between the two classes; the first class of configurations will be determined by the starting configuration.

---

* For the generation of random numbers, see Hammersley and Handscomb [2].

In such cases the overall averaging procedure will give poor results. The averaging should be performed rather separately for each class, so that the physical meaning of each class can be investigated.

A macroscopic system can only be simulated by a limited number of molecules with the help of periodic boundary conditions. The cell of volume v is surrounded by identical cells with the same particle configuration (cf. fig. 3.1). Each move of a particle in the central cell is done automatically also by its images in the neighbouring cells. If a particle leaves the central cell, its image enters from the corresponding neighbouring cell. The shape of the unit cell is chosen

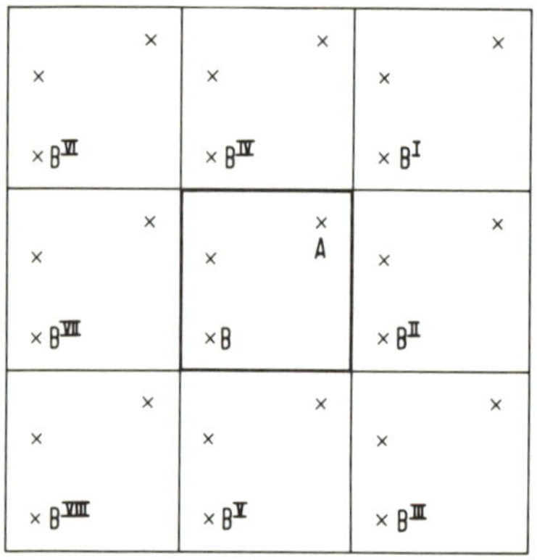

Fig. 3.1. Periodic boundary conditions mean that the central cell is surrounded by cells of the same particle configuration.

in conformity with the symmetry of a regular packing at high densities, and also the number of molecules is chosen from the point of view of such a regular packing (in three dimensions, one refers usually to a cubic dense packing, so a cube is used as unit cell and the number of molecules used is $N = 4g^3$, where g is the ratio of the length of the cell to the length of the crystallographic unit cell).

The periodic boundary conditions are of special importance when for the state i the potential energy $E_i = \sum_{l<k} \varepsilon_{lk}$ is calculated as sum of the pair potentials ε_{lk} between molecules l and k [cf. eq. (1.44)]. Here in principle the energy interactions to all images have to be taken into account. It is then necessary to cut the potential off at a certain distance and to correct the results appropriately. The choice of cell size and cut-off should be done in such a way, that from all possible energy interactions between a particle A (cf. fig. 3.1) and a particle B and its images $B', B'' \ldots$ not more than one interaction has to be taken into account, in order to avoid that the periodicity influences the results. The average value $\langle E \rangle$ for the energy of the ensemble is then taken as the average $\bar{E}$ over all states of the random walk according to eq. (3.3).

Apart from the energy, which must already be calculated for the decision if a configuration can be accepted as a new sequence in the random walk, the pressure and the pair distribution function can be calculated directly with a Monte Carlo calculation. The pressure is given, according to eq. (1.52), by

$$\frac{Pv}{NkT} - 1 = \frac{1}{3NkT} \langle W \rangle. \tag{3.9}$$

The averaging of the virial is again performed according to eq. (3.3). The expression which is equivalent to eq. (1.51) for the discontinuous hard sphere potential will be dealt with in chapter

7. For the calculation of the pair distribution function $g(R)$, which will be discussed in detail in the next chapter, $g_i(R)$ is determined for the state i by counting the number of pairs within certain discrete distances, and averaging is performed again over all sequences of the random walk. The calculation of the pair distribution function is limited to distances smaller than the half edge of the unit cell, because otherwise the periodic boundary conditions would impose an artificial periodicity on the results. Besides the direct evaluation of energy and pressure, these quantities can also be obtained with the aid of the pair distribution function (see chapter 4). Sometimes both methods are used to check the consistence of the calculations [9].

3.3. Molecular Dynamics Calculations

Molecular dynamics calculations have been done either on systems with discontinuous potentials (e.g. hard sphere potential) or on systems with continuous potentials (12/6 potential). The equations of motion, which are solved in the molecular dynamics method, are quite different for both types of potentials, so that we will deal with them separately. For a hard sphere system the particles fly with constant velocity between the collisions, and the change of velocity at a collision can be calculated exactly. On the other hand, for a 12/6 potential the particle trajectory is always influenced by its energy interaction with the other particles and the equations of motion can only be solved approximately.

Again periodic boundary conditions, as described in the previous section, are used. Normally the system is started with the molecules in a regular configuration, and with velocities of equal magnitude but random directions.

a) Method for discontinuous potentials [5, 6]. We will describe the procedure for the first type of potentials by using the hard sphere potential [cf. eq. (1.72)] and particles of identical mass.

If two particles collide, they change their velocities. Therefore, it is of interest to know the time interval from the last collision (or the start) to the next collision. For this all pairs of particles i and k are considered which are able to collide, and a matrix (t_{ik}) is set up which gives their collision times. If t_{lm} is the smallest among all t_{ik}-values, the particles l and m collide at first. Then one calculates the velocities for these particles after their collision and looks for the next collision. For those pairs, where the particles l and m are involved, the collision times have to be recalculated, whilst for all other pairs they are given by $t_{ik} - t_{lm}$. In this manner one observes the evolution of the system as a function of time.

For the calculation of the collision times and the velocity changes one introduces relative coordinates. Let $\vec{r}_{ik}^{\,0}$ be the relative position and $\vec{v}_{ik}^{\,0}$ the relative velocity for the particles i and k at a time $t = 0$, then one has at a time t for the relative position $\vec{r}_{ik} = \vec{r}_{ik}^{\,0} + \vec{v}_{ik}^{\,0} t$. Collision will occur, when $|\vec{r}_{ik}| = \sigma$. Therefore, the collision time t_{ik} can be determined by means of the equation (cf. fig. 3.2)

$$\sigma^2 = (\vec{r}_{ik})^2 = (\vec{r}_{ik}^{\,0})^2 + 2(\vec{r}_{ik}^{\,0} \cdot \vec{v}_{ik}^{\,0}) t_{ik} + (\vec{v}_{ik}^{\,0})^2 t_{ik}^2. \tag{3.10}$$

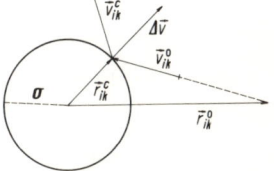

Fig. 3.2. To the mechanics of a collision of hard spheres in relative coordinates.

With the abbreviations $v^2 = (\vec{v}_{ik}^0)^2$, $b = \vec{v}_{ik}^0 \cdot \vec{r}_{ik}^0$, and $c = (\vec{r}_{ik}^0)^2 - \sigma^2$ the result is obtained

$$t_{ik} = [-b - \sqrt{b^2 - cv^2}]/v^2 . \tag{3.11}$$

As the angular momentum and the energy are constants of motion, the relative velocity $\vec{v}_{ik}^c$ after the collision must be symmetrical to the relative velocity $\vec{v}_{ik}^0$ with respect to the collision coordinate $\vec{r}_{ik}^c$ (fig. 3.2). So the change in the relative velocity $\Delta \vec{v} = \vec{v}_{ik}^c - \vec{v}_{ik}^0$ is given by

$$\Delta \vec{v} = -2(\vec{r}_{ik}^c \cdot \vec{v}_{ik}^0)\vec{r}_{ik}^c/\sigma^2 . \tag{3.12}$$

The change of the velocity of each particle can then be derived by using the change in the relative velocity and the conservation of momentum.

If the particles start with velocities of equal magnitude the system is not yet in equilibrium, because then the velocities should have a Maxwell distribution. It turned out that this happens after 2 to 4 collisions per particle [7]. The energy of a hard sphere system has only a kinetic part. This kinetic energy is given at the beginning with the velocities and determines the temperature.

The pressure can again be calculated by forming the average of the pressure virial. Here this is truly time averaging. For the case of hard spheres the forces are expressed as the momentum change per unit time. If one forms the term $m|\Delta \vec{v}|$ after every collision and sums up, one gets the total momentum change parallel to $\vec{r}_{ik}^c$ as a function of time. The virial $\langle W \rangle$, which measures the momentum change per unit time, is then given by

$$\langle W \rangle = m\sigma \frac{\partial}{\partial t} \sum_{\text{all coll.}} |\Delta \vec{v}| . \tag{3.13}$$

Access to the time-dependent properties is usually obtained by the mean square displacement and by the velocity autocorrelation function (cf. chapter 9).

Frequently the detailed information on molecular positions and velocities is recorded on magnetic tape and analysed later for the various properties of the system. The particle trajectories have been also displayed on a cathode-ray tube in order to get a feeling for the situation in such a system (cf. next section).

b) Method for continuous potentials [8, 9]. Investigations with interest for the liquid state have only been done on a system with a 12/6 interaction [cf. eq. (1.73)]. Knowing the positions and velocities of all particles at a time t, one looks for the positions at time $t + \Delta t$. Let the minimum of the 12/6 potential and its position be a measure for a typical kinetic energy and a typical distance (cf. table 6.1), then one has a typical time of 10^{-12} s. Furthermore the discussion in chapters 8 and 9 shows that the short-time behaviour of many systems is governed by characteristic times of 10^{-13} s. In order to observe the essential features of a particle trajectory, one chooses for $\Delta t = 10^{-14}$ s. The Newtonian equation of motion at a time t is then

$$\ddot{\vec{r}}_i(t) = -12\frac{\varepsilon^*}{m} \sum_k{}' \frac{1}{R_{ik}^2} \left\{ \left(\frac{R^*}{R_{ik}}\right)^{12} - \left(\frac{R^*}{R_{ik}}\right)^6 \right\} \vec{r}_{ik}(t) . \tag{3.14}$$

If it would be a sufficient approximation to keep the acceleration over the whole interval Δt equal to $\ddot{\vec{r}}_i(t)$, then

$$\vec{r}_i(t + \Delta t) = \vec{r}_i(t) + \dot{\vec{r}}_i(t)\Delta t + \ddot{\vec{r}}_i(t)(\Delta t)^2/2 . \tag{3.15}$$

A reasonable choice for the velocity $\dot{\vec{r}}_i(t)$ is the value $[\vec{r}_i(t + \Delta t) - \vec{r}_i(t - \Delta t)]/2\Delta t$. This gives the result [9]

$$\vec{r}_i(t + \Delta t) = -\vec{r}_i(t - \Delta t) + 2\vec{r}_i(t) + \ddot{\vec{r}}_i(t)(\Delta t)^2 . \tag{3.16}$$

As in Monte Carlo work, the potential is also cut off at a certain distance, and the same problems arise.

At the start only the sum of kinetic and potential energy is known, and it is necessary to wait for attainment of equilibrium in order to determine them separately. Thus one does not know a priori the temperature for which the calculation is performed. Therefore, the kinetic energy is sometimes kept constant artificially by multiplying all velocities with a constant factor at the beginning. The pressure can be calculated by forming the virial W after every time interval Δt and all other properties can be determined as in the hard sphere case.

3.4. Results for Hard Spheres

For a long time the abstraction of an ideal gas or an ideal solid has been well established. But it was only with the help of machine calculations that a system of hard spheres has been named an ideal liquid. One peculiarity of this system is, that its configurational partition function is independent of the temperature. Hence its equation of state can be described by a single curve, as the ratio $\dfrac{P}{T}$ is always the same for a given density [eq. (1.43), cf. also eq. (7.64)].

Nowadays there are many attempts to derive the properties of systems with more realistic potentials by perturbation theories from the hard sphere system. In view of this importance of the hard sphere system and the many careful studies on it with both the Monte Carlo and the molecular dynamics method, we present some results here. Results on other systems are given in comparison to experimental or theoretical studies at other places of this book. For a general review see [10].

In the hard sphere system N particles of an interaction according to eq. (1.72) are treated in a cubic box of volume v. The volume is given in units of v_0, where $v_0 = N\sigma^3/\sqrt{2}$ denotes the volume of closest packing. Nearly all studies start with the spheres in a regular configuration according to a f.c.c. packing and use periodic boundary conditions. The first study was made by Rosenbluth and Rosenbluth (1954) with Monte Carlo [11]. They made calculations at 20 different densities, used 256 particles and moved each particle about 100 times. Their equation of state showed no first-order phase transition in the whole density range.

The next was a molecular dynamics study by Alder and Wainwright (1957) with only 32 particles but an observation of a great number of collisions [12, 6]. In the density region of $1.5 < v/v_0 < 1.7$ they found the following striking results: if the system was started with a regular configuration the pressure had a certain value for some hundred or thousand collisions per particle and then jumped suddenly to a much higher value. There it stayed for a while and then jumped back to the lower value and so on. In order to investigate the phenomenon more closely, the authors made cathode-ray pictures of the particle trajectories whilst the system was in the one class of states or in the other. Though these pictures have already been shown very often, they are so instructive that we are reproducing them again (see fig. 3.3 − 3.4). The low pressure class is characterized by a regular configuration as in the solid, whilst in the high pressure class the long range order is broken down as is imagined for the liquid. In order to show that the phase transition is not produced by the small number of particles, Alder and Wainwright later made calculations with more particles also, e.g. with 108 or 256, and found the same phase transition. As the jumps between the "solid" and "liquid" class do not occur frequently within a reasonable time of calculation, it is not possible to estimate the importance of the two classes by the time of their stability. Therefore, an over-all averaging does not make much sense and one is not able to predict the exact location of the phase transition. The averaging is performed on either class separately. The equation of state is shown in fig. 3.5 for 108 particles. There is some change in the equation of state when passing from 32 to 108 particles, but very little when using 256 instead of 108 particles.

Fig. 3.3. The trajectories of 32 hard spheres in a cubic box with volume $v/v_0 = 1.525$, obtained by a molecular dynamics calculation and projected into a plane. The trajectories are shown for 3000 collisions in the ordered arrangement corresponding to a f.c.c. packing. After ref. 7.

Fig. 3.4. The system of fig. 3.3 after a jump into the disordered state. Again the trajectories are shown for 3000 collisions. After ref. 7.

Naturally, the discovery of the "solid-liquid" phase transition was a challenge to reinvestigate the hard sphere system by the Monte Carlo method. Wood and Jacobson (1957) did so and used also a system of 32 spheres, but moved each particle about 300000 times in the density region of the transition [13, 1]. Their results were indeed in good agreement with that of molecular dynamics. Whilst in the case of 32 particles the system jumped from one phase to the other and back again, a later investigation of Rotenberg (1964) on a 864 sphere system showed only jumps from the ordered to the disordered class of states [14].

In order to classify the exact location of the phase transition, more sophisticated work has been done recently [15], which uses the thermodynamic equilibrium condition of equal pressure and chemical potential at the transition point. For the determination of the free energy along the fluid-isotherm the equation of state was integrated from the ideal gas state to high densities. Now the problem had to be solved to arrive at a free energy value along the solid isotherm. For this, Hoover and Ree made a Monte Carlo calculation for the single-occupancy cell model (cf. section 2.6) over the whole density range. The isotherm of this model (Pv_0/NkT vs. v/v_0) proved to be a smooth extension of the solid state isotherm to low densities. For very low densities, it is clear that the free energy of the single-occupancy cell model is just for the term NkT more positive than the free energy of the ideal gas (cf. sections 1.3 and 2.6). Starting from this reference state, the free energy could be calculated by integration to high densities, and taken as the free energy of the solid state. The results for the transition are the following: $Pv_0/NkT = 8.27$, solid density $v_0/v = 0.736$, fluid density $v_0/v = 0.667$. This is a volume change at "melting" of 10.3%. The corresponding entropy change is $\Delta S = 1.16R$.

Finally a brief comparison between the Monte Carlo and the molecular dynamics method should be carried out. The equations of state obtained from the best results of both methods [6, 14] are shown in fig. 3.5. There is no appreciable difference between both results. The slight discrepancies in the phase transition region are not of great importance. Thus both methods

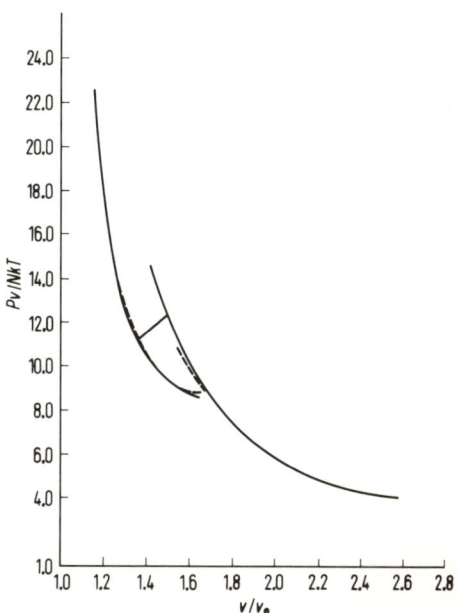

Fig. 3.5. Equation of state for hard spheres. The results are obtained by a Monte Carlo calculation with 864 particles ———— [14], and by a molecular dynamics calculation with 108 particles ------ [6]. The only difference appears near the phase transition. The line of the phase transition is drawn according to [15]. Mainly after Rotenberg [14] (due acknowledgement is expressed to him and to the A.E.C.).

seem to be equally efficient in simple cases, where they can be applied without any further approximation. Both methods are difficult to handle in the neighbourhood of phase transitions. Perhaps molecular dynamics is better suited to pass from one ergodic class to another, as all particles move at the same time. For systems with continuous potentials the Monte Carlo procedure is in principle exact – besides the cut off of the potential, used in either method –, whilst molecular dynamics has to use approximations. Also the computational effort seems to be always smaller in the Monte Carlo method. On the other hand, molecular dynamics has the great advantage to give also time-dependent and time-averaged properties.

3.5. References

1. W. W. Wood, chapter 5 in gen. ref. 12 (This article gives also many further references).
2. J. M. Hammersley and D. C. Handscomb, Monte Carlo Methods, Methuen, London 1964.
3. N. Metropolis, A. W. Rosenbluth, M. N. Rosenbluth, A. H. Teller, and E. Teller, J. Chem. Phys. *21*, 1087 (1953).
4. W. Feller, An Introduction to Probability Theory and its Applications, Wiley, New York 1950.

5. J. R. Beeler, Jr., The Technique of High Speed Computer Experiments. In: E. Meeron, Physics of Many-Particle Systems, Gordon and Breach, New York 1964. (This article gives also many further references).
6. B. J. Alder and T. E. Wainwright, J. Chem. Phys. *31*, 459 (1959) and *33*, 1439 (1960).
7. T. E. Wainwright and B. J. Alder, Nuovo Cimento *9*, Supplement, 116 (1958).
8. A. Rahman, Phys. Rev. *136 A*, 405 (1964).
9. L. Verlet, Phys. Rev. *159*, 98 (1967) and *165*, 201 (1968).
10. I. R. McDonald and K. Singer, Quart. Rev. *24*, 238 (1970).
11. M. N. Rosenbluth and A. W. Rosenbluth, J. Chem. Phys. *22*, 881 (1954).
12. B. J. Alder and T. E. Wainwright, J. Chem. Phys. *27*, 1208 (1957).
13. W. W. Wood and J. D. Jacobson, J. Chem. Phys. *27*, 1207 (1957).
14. A. Rotenberg, 1964, Monte Carlo Studies of Systems of Hard Spheres, Report NYO-1480-3, AEC Comp. and Appl. Math. Center (Courant Inst. Math. Sci., New York University, New York).
15. W. G. Hoover and F. H. Ree, J. Chem. Phys. *49*, 3609 (1968).

4. Pair Distribution Function

4.1. Definition

The pair distribution function $g(R)$ has already been defined in chapter 1. Starting from a given molecule at position 0, the quantity $\rho g(R) d\vec{r}$ gives the probability to find another molecule in the volume element $d\vec{r}$ a distance R apart from 0. At very short distances, $g(R)$ is zero as the given molecule cannot be penetrated by others. At about the distance, where the pair potential has its minimum, $g(R)$ has a distinct peak from the first shell of neighbours. At very large distances, $g(R)$ tends to unity as any heterogenity of the space filling due to structure will average out. In the study of liquids $g(R)$ plays an important role for three reasons. a) The thermodynamics of the system can be derived from the knowledge of $g(R)$ (section 4.2). b) From scattering experiments with X-rays or neutrons $g(R)$ can be determined directly with a certain accuracy (sections 4.3 – 4.5). c) The function $g(R)$ gives some insight into the structure of liquids, as will be discussed below.

We will start with the definition of distribution functions on the basis of a canonical ensemble. The probability to find a certain configuration of the N particles, which means to find simultaneously particle 1 in a volume element $d\vec{r}_1$ at $\vec{r}_1$, particle 2 in $d\vec{r}_2$ at $\vec{r}_2$, and so on, is given by

$$\frac{e^{-\frac{E}{kT}} d\vec{r}_1 \ldots d\vec{r}_N}{\int \ldots \int e^{-\frac{E}{kT}} d\vec{r}_1 \ldots d\vec{r}_N}. \tag{4.1}$$

The probability for a certain configuration of the first s particles is obtained by integration over all positions of the other particles

$$\frac{d\vec{r}_1 \ldots d\vec{r}_s \int \ldots \int e^{-\frac{E}{kT}} d\vec{r}_{s+1} \ldots d\vec{r}_N}{\int \ldots \int e^{-\frac{E}{kT}} d\vec{r}_1 \ldots d\vec{r}_N}. \tag{4.2}$$

As real molecules are not labelled, the above expression has no physical meaning. But we can define a s-particle distribution function $n(\vec{r}_1, \ldots, \vec{r}_s)$ as the probability density to find simultaneously any particle at $\vec{r}_1$ and any particle at $\vec{r}_2$ and so on. Then we have [*]

$$n(\vec{r}_1, \ldots, \vec{r}_s) = \binom{N}{s} s! \frac{\int \ldots \int e^{-\frac{E}{kT}} d\vec{r}_{s+1} \ldots d\vec{r}_N}{\int \ldots \int e^{-\frac{E}{kT}} d\vec{r}_1 \ldots d\vec{r}_N}. \tag{4.3}$$

The first combinational factor stems from the fact, that such a configuration can be produced by any set of s particles out of the N particles, and the second factor is necessary, as every permutation of the s particles with respect to the different positions has to be counted. The normalizing condition for the distribution function is given by

$$\int \ldots \int n(\vec{r}_1, \ldots, \vec{r}_s) d\vec{r}_1 \ldots d\vec{r}_s = \binom{N}{s} s! = \frac{N!}{(N-s)!}. \tag{4.4}$$

Let us deal at first with the one particle distribution function $n(\vec{r}_1)$. In a solid the molecules are always located near their lattice sites and thus $n(\vec{r}_1)$ will be a periodic funtion in space with maxima at the lattice sites. In the fluid such a structure does not exist and in the ensemble

---

[*] Sometimes instead of $n(\vec{r}_1, \ldots, \vec{r}_s)$ also the notation $n(1, \ldots, s)$ will be used.

average the probability to find any molecule at a certain position will be the same for all positions. Thus $n(\vec{r}_1)$ will be a constant, and from eq. (4.4) follows

$$n(\vec{r}_1) = \frac{N}{v} = \rho. \tag{4.5}$$

Secondly, we will consider the s-particle distribution function for the case of no intermolecular forces acting. Then it becomes

$$n(\vec{r}_1, \ldots, \vec{r}_s) = \frac{N!}{(N-s)!} \frac{v^{N-s}}{v^N} = \left(\frac{N}{v}\right)^s \left(1 + 0\left\{\frac{1}{N}\right\}\right), \tag{4.6}$$

or in the limiting case $N \to \infty$, $v \to \infty$, $N/v = \rho = \text{const.}$,

$$n(\vec{r}_1, \ldots, \vec{r}_s) = \rho^s. \tag{4.7}$$

This result is expected, as in a system without forces the particles are randomly distributed, and the probability to find s particles somewhere is just the product of the one-particle distribution functions. The effect of forces between particles brings the system away from the random distribution and thus it seems useful to define the s-tuplet distribution function by

$$g(\vec{r}_1, \ldots, \vec{r}_s) = \frac{1}{\rho^s} n(\vec{r}_1, \ldots, \vec{r}_s), \tag{4.8}$$

which becomes unity for a random distribution. By combining eq. (4.8), (4.3), and (1.32) this definition can be written in the alternative form

$$g(\vec{r}_1, \ldots, \vec{r}_s) = \frac{1}{(N-s)!} \frac{1}{\rho^s} \frac{1}{Q_N} \int \ldots \int e^{-\frac{E}{kT}} d\vec{r}_{s+1} \ldots d\vec{r}_N. \tag{4.9}$$

It is also necessary to define the distribution functions in the grandcanonical formalism, as frequently, e.g. in scattering experiments, the number of particles fluctuates within the volume considered (e.g., irradiated). If we denote the s-particle distribution function of a canonical ensemble with N particles by n_N, the grandcanonical distribution function is the weighted average

$$n(\vec{r}_1, \ldots, \vec{r}_s) = \sum_{N=s}^{\infty} p_N n_N(\vec{r}_1, \ldots, \vec{r}_s), \tag{4.10}$$

where p_N, the probability to find N particles in a given volume, is given by eq. (1.57). The normalizing condition then becomes

$$\int n(\vec{r}_1, \ldots, \vec{r}_s) d\vec{r}_1 \ldots d\vec{r}_s = \sum_{N=s}^{\infty} p_N \frac{N!}{(N-s)!} = \left\langle \frac{N!}{(N-s)!} \right\rangle, \tag{4.11}$$

where eq. (4.4) has been used and the angular brackets mean grandcanonical averaging. The difference between canonical and grandcanonical expressions can certainly be neglected if fluctuations about the mean particle number $\langle N \rangle$ are small. As eq. (1.69) shows, these fluctuations can always be made sufficiently small in the case of a finite compressibility, if the volume is made sufficiently large at constant density. The situation becomes more complicated in the neighbourhood of the critical point, which will be discussed in section 2.

We turn now to the pair distribution function $g(\vec{r}_1, \vec{r}_2)$. In a fluid this quantity does not depend on a special choice of $\vec{r}_1$ and $\vec{r}_2$ but only on their relative distance, so that we can write

$$g(\vec{r}_1, \vec{r}_2) = g(R) \quad \text{with} \quad R = |\vec{r}_2 - \vec{r}_1|. \tag{4.12}$$

According to the statistical definitions (4.3) and (4.8) the quantity $\rho^2 g(R)$ is the probability density to find simultaneously two molecules at two given positions, which are a distance R apart. This probability can be considered as the product of the probability to find one molecule at a given position and of the probability to find another molecule a distance R apart from the given molecule. The first probability is given by ρ [eq. (4.5)], so that the statistical definition gives for the latter $\rho g(R)$ in accord with our statement at the beginning of this section.*⁾
For a very dilute gas, where the influence of other particles on a certain pair can be neglected, we have

$$g(R) = \exp\left\{-\frac{\varepsilon(R)}{kT}\right\} \quad \text{(low density limit)} \tag{4.13}$$

(fig. 4.1 a). For somewhat higher densities, even the influence of further particles can be handled explicitely. As will be shown in chapter 7 [eq. (7.32)], a density expansion of $g(R)$ can be made, where the first order term considers the effect of one further particle, the second order term that of two further particles and so on. But for densities in the real liquid range this expansion cannot be used, as the effect of too many particles is present.

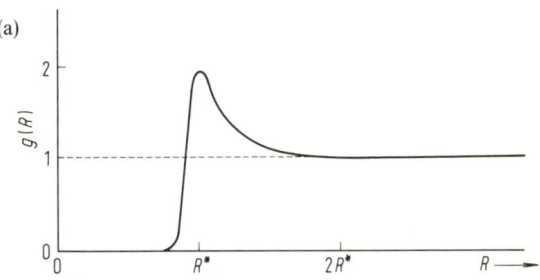

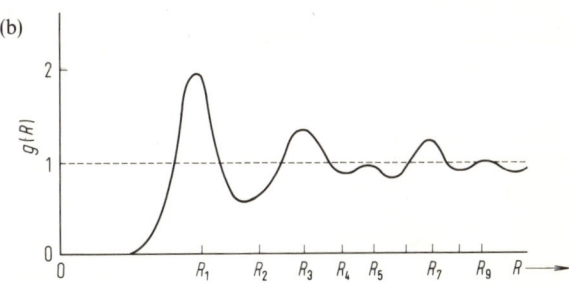

Fig. 4.1. a) The pair distribution function for a low density gas according to eq. (4.13). For $\varepsilon(R)$ a 12/6 potential has been taken and the temperature $kT/\varepsilon^* = 1.5$ is somewhat above the critical.
b) The pair distribution function for a model-solid. The δ-function of eq. (4.14) was replaced by a Gaussian with $b = 0.2$.

The other extreme is that of an idealized solid, where all particles are thought to be exactly at their lattice sites (this case is never realized in nature because of the quantum-mechanical zero-point motion). Then one particle has z_1 nearest neighbours at a distance R_1, z_2 next-nearest neighbours at a distance R_2 and so on, so that (by using Dirac's δ-function)

$$g(R) = \sum_i z_i \frac{\delta(R - R_i)}{4\pi R^2 \rho} \quad \text{(idealized solid)}. \tag{4.14}$$

---

* Sometimes it will be convenient to use the notation $g(\vec{r})$. In accordance with the above statements this means $g(\vec{r}) = g(R)$ with $R = |\vec{r}|$.

Thus, whereas $g(R)$ has only one broad peak for the dilute gas, we find here an infinite number of very sharp peaks at the distances R_i. But it must be noted, that on geometrical reasons the distances between the different peaks become smaller with increasing R. For a f.c.c. packing e.g., with a unit cell of length $a\sqrt{2}$ we have $R_i = a\sqrt{i}$, and thus $\lim_{i \to \infty}(R_{i+1} - R_i) = \lim_{i \to \infty} a(\sqrt{i+1} - \sqrt{i}) = 0$.
It is now easy to see what happens if the particles can move around the lattice sites. Then the sharp peaks will broaden and may be described, e.g., by a Gaussian $\frac{1}{\sqrt{\pi b}} \exp\{-(R - R_i)^2/b^2\}$.
The first peaks, which are somewhat remote from each other, will still be distinguishable. But for larger values of R the peaks will overlap and only oscillations in $g(R)$ will be found (fig. 4.1 b). This picture is confirmed by experimentally obtained $g(R)$-curves; such a curve is shown in fig. 1.2a [it shall be noted that this curve does not give $g(R)$ directly, but $4\pi R^2 \rho g(R)$].
For a liquid a simple picture of $g(R)$ is harder to obtain. A given molecule is closely surrounded by some others, which approach as far as the electron clouds allow. This causes a first peak in $g(R)$. Furthermore on account of the dense packing a certain short range order is exhibited, which was seen from the models of chapter 2, and thus eventually a second and third peak in $g(R)$ will arise. But for somewhat larger distances $g(R)$ will approach the value unity characteristic for random distribution. Indeed, an experimentally obtained curve, as that in fig. 1.2b, shows the expected behaviour. An exact theoretical expression could not be derived for $g(R)$ of a three-dimensional system at liquid densities; attempts are described in chapter 7. However, the one-dimensional case could be solved for some simple potentials [1].
We may ask now for the structural insight given by the pair distribution function. For a liquid the number of nearest neighbours of a molecule, the coordination number, can be determined from the first peak in $g(R)$. Several different methods have been suggested [2]. But a similar determination for the next-nearest neighbours involves usually too much arbitrariness. Moreover every molecular structure is a structure in three dimensions, whereas $g(R)$ gives only a probability averaged over a spherical shell. Thus for a given $g(R)$ the structure in space is not determined uniquely. For a solid the description of structure in terms of a Bravais lattice is much more instructive than that given by a $g(R)$-curve. The question remains still open, whether a similarly instructive description can also be found for the liquid. The geometric and disorder models discussed in chapter 2 are some attempts in this direction. More insight than from the pair distribution function could be obtained from the higher distribution functions, but these are hardly accessible.

4.2. Pair Distribution Function and Thermodynamic Properties

There are three routes from the pair distribution function to the thermodynamic properties of a system, which give the potential energy, the pressure and the compressibility, respectively. The mean potential energy $\langle E \rangle$ is given according to eq. (1.27) by

$$\langle E \rangle = \frac{\int \ldots \int (\sum \varepsilon_{ik}) \exp\left\{-\frac{1}{kT} \sum \varepsilon_{ik}\right\} d\vec{r}_1 \ldots d\vec{r}_N}{\int \ldots \int \exp\left\{-\frac{1}{kT} \sum \varepsilon_{ik}\right\} d\vec{r}_1 \ldots d\vec{r}_N}. \tag{4.15}$$

As the sum contains $N(N-1)/2$ terms it may be written as

$$\langle E \rangle = \frac{N(N-1)}{2} \frac{1}{(N!Q_N)} \iint \varepsilon_{12} \left[\int \ldots \int \exp\left\{ -\frac{1}{kT} \sum \varepsilon_{ik} \right\} d\vec{r}_3 \ldots d\vec{r}_N \right] d\vec{r}_1 d\vec{r}_2 . \quad (4.16)$$

The bracketed term can be expressed by the pair distribution function, whence

$$\langle E \rangle = \frac{\rho^2}{2} \iint \varepsilon_{12} g(\vec{r}_1, \vec{r}_2) d\vec{r}_1 d\vec{r}_2 = \frac{N}{2} \rho \int \varepsilon(R) g(R) d\vec{r} . \quad (4.17)$$

The total energy is obtained by adding the mean kinetic energy, so that we have, e.g., for a monoatomic fluid

$$u = \frac{3}{2} NkT + \frac{N}{2} \rho \int \varepsilon(R) g(R) d\vec{r} . \quad (4.18)$$

The result of eq. (4.17) is immediately evident, as $\rho g(R) d\vec{r}$ denotes the probability to find another molecule at the volume element $d\vec{r}$ a distance R apart from a given molecule, and thus $\rho g(R)\varepsilon(R) d\vec{r}$ gives the mean interaction energy with that volume element. The factor $N/2$ arises, as we can start from any molecule, but then count every interaction twice.

The pressure is determined according to eq. (1.52) by the mean value of the virial $\langle W \rangle$. For this we get by a similar calculation as above

$$\langle W \rangle = -\frac{N}{2} \rho \int R \frac{\partial \varepsilon(R)}{\partial R} g(R) d\vec{r} , \quad (4.19)$$

and hence as pressure equation

$$\frac{Pv}{NkT} = 1 - \frac{\rho}{6kT} \int R \frac{\partial \varepsilon}{\partial R} g(R) d\vec{r} . \quad (4.20)$$

This equation is rather sensitive to small errors in the integral, as, e.g., $Pv/NkT = 2.4 \cdot 10^{-3}$ at the triple point of argon and this result should be obtained as the difference between 1 and the integral. The most important contribution to this integral and to the potential energy (eq. 4.17) comes from distances, where $\varepsilon(R)$, $\partial \varepsilon/\partial R$, and $g(R)$ are important, i.e. around the first sharp increase of $g(R)$. Both results, eq. (4.17) and (4.19), could have also been obtained with a grandcanonical formalism.

The third route can only be derived in the grandcanonical formalism. As $\rho = \langle N \rangle / v$, eq. (4.11) yields for $s = 2$

$$\langle N \rangle \rho \int g(R) d\vec{r} = \langle N^2 \rangle - \langle N \rangle . \quad (4.21)$$

Insertion of this into eq. (1.69) then gives

$$1 + \rho \int [g(R) - 1] d\vec{r} = \langle N \rangle \frac{kT}{v} \beta_T . \quad (4.22)$$

With the help of the total correlation function, which is simply defined by

$$h(R) = g(R) - 1 , \quad (4.23)$$

and the definition of the compressibility [eq. (1.15)], the compressibility equation may be written as

$$1 + \rho \int h(R) d\vec{r} = kT \left(\frac{\partial \rho}{\partial P} \right) . \quad (4.24)$$

Contrary to eq. (4.17) and (4.19), the above equation is strongly influenced by the behaviour of $h(R)$ for great R. Moreover, whereas the other equations could only be derived with the assumption that the energy interactions between the molecules are pairwise additive [eq. (1.44)], this assumption is not necessary for the compressibility equation.

As the thermodynamic quantities, which are directly available from $g(R)$, are only derivatives of the free energy, the calculation of the latter and of the entropy involves integration over density or over $1/T$. So we obtain, e.g. from the pressure equation, for the free energy per particle

$$F(\rho) = F(\rho_0) + kT \int_{\rho_0}^{\rho} \frac{Pv}{NkT} \frac{d\rho}{\rho}, \tag{4.25}$$

where the low density limit is usually taken as well defined reference state.

4.2.1. The Situation at the Critical Point. The Direct Correlation Function

Usually $h(R) = g(R) - 1$ will tend to zero within some molecular distances. But at the critical point $\partial \rho / \partial P$ becomes infinite, and thus according to eq. (4.24) the integral $\int \rho h(R) d\vec{r}$ diverges [*], so that $h(R)$ has still finite values for large distances. This is an indication for the long-ranging correlated fluctuations at the critical point. But then $g(R)$, or $h(R)$, does not say much about the situation around a certain molecule, as it averages about its location in a dense region as well as about that in a less dense region. This has led to the introduction of the direct correlation function, which gives the connection between density fluctuations in neighbouring volume elements [3].

We assume for all volume elements around the volume element at the point 0 an instantaneous density $\rho(\vec{r})$. This density distribution will produce at 0 a mean density $\overline{\rho(0)}$, which is a functional [**] of the density in the surroundings; $\overline{\rho(0)} = F(\rho(\vec{r}))$. By means of a Taylor expansion about ρ, the mean density in the total volume, and retaining only the first order term, we get

$$\overline{\rho(0)} = \rho + \int \rho c(\vec{r}) [\rho(\vec{r}) - \rho] d\vec{r} \tag{4.26}$$

with $$\rho c(\vec{r}) = \frac{\delta F}{\delta \rho(\vec{r})}. \tag{4.27}$$

This relation means that the density at 0 is directly correlated to the density at $\vec{r}$ by the direct correlation function $c(\vec{r})$. Correlations between third volume elements are neglected. From this definition it can be assumed that $c(\vec{r})$ will be short ranging. Moreover, it follows from the isotropy in a fluid that $c(\vec{r}) = c(R)$ with $R = |\vec{r}|$.

To find a connection with the pair distribution function we assume a fixed particle at point 1 (vector $\vec{s}$, fig. 4.2). The density distribution will then be determined by the pair distribution around 1

$$\rho(\vec{r}) = \delta(\vec{r} - \vec{s}) + \rho g(\vec{r} - \vec{s})$$
$$\overline{\rho(0)} = \rho g(\vec{s}), \tag{4.28}$$

---

[*] The critical point should only be discussed in a grandcanonical formalism. In a canonical ensemble the relation $1 + \rho \int h(R) d\vec{r} = 0$ is always valid [eq. (4.23) and (4.4)], as the canonical formalism cannot account for particle fluctuations.

[**] The function $F(i)$ is called a functional of $\varphi(j)$, if knowledge of φ for all j defines F for all i [4].

where the δ-function considers the fixed particle at 1. Insertion of this into eq. (4.26) and replacing $g(\vec{r})$ by $h(\vec{r}) + 1$ yields

$$\rho h(\vec{s}) = \rho c(\vec{s}) + \rho^2 \int c(\vec{r})h(\vec{r}-\vec{s})d\vec{r} \tag{4.29}$$

or

$$h(0,1) = c(0,1) + \rho \int c(0,2)h(1,2)d\vec{r}_2 . \tag{4.30}$$

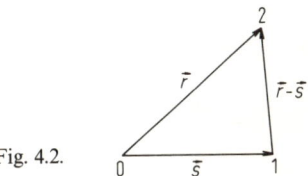

Fig. 4.2.

This relation is known as the Ornstein-Zernike equation, which expresses the total correlation function in terms of a direct effect plus the feedback of third molecules which influence directly the position 0. Eq. (4.30) determines $c(\vec{r})$ uniquely, as can be seen from the Fourier-transform of the above equation. With the definition

$$\tilde{h}(\vec{Q}) = \int h(\vec{r})e^{-i\vec{Q}\cdot\vec{r}}d\vec{r} \tag{4.31}$$

we obtain by using the folding theorem (cf. appendix A)

$$\tilde{h}(\vec{Q}) = \tilde{c}(\vec{Q}) + \rho\tilde{c}(\vec{Q})\tilde{h}(\vec{Q}), \tag{4.32}$$

and then

$$\tilde{c}(\vec{Q}) = \frac{\tilde{h}(\vec{Q})}{1 + \rho\tilde{h}(\vec{Q})} . \tag{4.33}$$

We can rewrite now the compressibility equation in terms of $c(R)$. With the definition (4.31) we have $\int h(R)d\vec{r} = \tilde{h}(0)$, and then obtain by taking the inverse of eq. (4.24)

$$\frac{1}{kT}\left(\frac{\partial P}{\partial \rho}\right) = \frac{1}{1 + \rho\tilde{h}(0)} - 1 - \rho\tilde{c}(0) \tag{4.34}$$

or

$$\frac{1}{kT}\left(\frac{\partial P}{\partial \rho}\right) = 1 - \rho \int c(R)d\vec{r} . \tag{4.35}$$

Thus we have at the critical point $\rho \int c(R)d\vec{r} = 1$, which shows that $c(R)$ has a finite range there. Later on it will turn out that the direct correlation function is not only useful in the critical region, but is in general a useful concept in the theory of fluids (chapter 7).

4.3. Pair Distribution Function and Intensity of Scattered X-Rays

An X-ray interacts with an electron either by incoherent Compton scattering or by causing the electron to be the source of a coherent spherical wave. Normally many electrons are present in a system and then the coherently scattered waves interfere and produce a certain intensity pattern.

The incident wave is described by its wave vector $\vec{k}_0$, with $|\vec{k}_0| = k = \dfrac{2\pi}{\lambda}$ (λ wavelength), and hence its momentum and energy is given by

4. Pair Distribution Function

$$\vec{p}_0 = \hbar \vec{k}_0 \tag{4.36}$$

$$E_0 = \hbar \omega_0 = \hbar c k. \tag{4.37}$$

As will be shown later, the energy exchange with the irradiated system in coherent X-ray scattering is very small. Thus the wave vector $\vec{k}_1$ of the scattered wave has the same magnitude $|\vec{k}_1| = k$. Using the definition

$$\vec{Q} = \vec{k}_1 - \vec{k}_0 \tag{4.38}$$

the momentum transfer at scattering is $\hbar \vec{Q}$ and only determined by the change in direction

$$|\vec{Q}| = Q = \frac{4\pi \sin \theta}{\lambda}, \tag{4.39}$$

where 2θ denotes the scattering angle (cf. fig. 4.3).

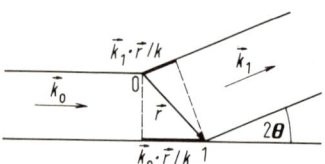

Fig. 4.3.

If we consider now two scattering electrons, one at a point 0 and the other at point $\vec{r}$ (fig. 4.3), then the path difference of the scattered waves is $-\vec{r} \cdot (\vec{k}_1 - \vec{k}_0)/k = -\vec{r} \cdot \vec{Q}/k$. Thus if the first wave has an amplitude $a e^{-i\omega_0 t}$, at a certain point of observation, the amplitude of the second wave is

$$a e^{-i\omega_0 t} e^{-i\vec{Q} \cdot \vec{r}}. \tag{4.40}$$

In a real system we have many scattering electrons, whose configuration in space can be described by an electron density function $\rho_e(\vec{r})$. Then the total amplitude of all scattered waves is given by

$$A = a e^{-i\omega_0 t} \int \rho_e(\vec{r}) e^{-i\vec{Q} \cdot \vec{r}} d\vec{r}. \tag{4.41}$$

Usually the total electron density can be written as the sum of $\rho_i(\vec{r} - \vec{r}_i)$, the electronic density around the center of atom i at $\vec{r}_i$,

$$\rho_e(\vec{r}) = \sum_i \rho_i(\vec{r} - \vec{r}_i). \tag{4.42}$$

Inserting this into eq. (4.41) yields

$$A = a e^{-i\omega_0 t} \sum_i e^{-i\vec{Q} \cdot \vec{r}_i} \int \rho_i(\vec{r} - \vec{r}_i) e^{-i\vec{Q} \cdot (\vec{r} - \vec{r}_i)} d\vec{r}. \tag{4.43}$$

The quantities

$$f_i(\vec{Q}) = \int \rho_i(\vec{r}) e^{-i\vec{Q} \cdot \vec{r}} d\vec{r} \tag{4.44}$$

depend only on the electronic distributions in the various atoms and hence are called atomic scattering factors[*]. By using them we have

$$A = a e^{-i\omega_0 t} \sum_i e^{-i\vec{Q} \cdot \vec{r}_i} f_i(\vec{Q}). \tag{4.45}$$

---

[*] It shall be mentioned, that $f(\vec{Q})$ can be calculated from the electron wave-function of an atom [5]. If Z is the period number of the atom, $f(0) = Z$.

Now we will consider the most simple case of a monoatomic system. Then all scattering factors are equal and the above equation reduces to

$$A = a e^{-i\omega t} f(Q) \sum_i e^{-i\vec{Q}\cdot\vec{r}_i}. \tag{4.46}$$

A scattering experiment does not measure the amplitude of a wave but only the time-average of the corresponding intensity. The intensity is given as the square of the amplitude which depends on time firstly by the factor $e^{-i\omega t}$ of eq. (4.46) and secondly by the positions $\vec{r}_i$ which change in the course of time. But for X-rays the atoms remain practically on their positions for one period of the scattered wave, so that time averaging over the intensity can be divided into averaging over the rapidly varying part and that over the slowly varying part. Thus taking the square of the total amplitude and averaging over one wave period, we get

$$I(\vec{Q},\{\vec{r}_i\}) = j|f(Q)|^2 \sum_i \sum_j e^{-i\vec{Q}\cdot(\vec{r}_i-\vec{r}_j)}. \tag{4.47}$$

Here j is the coherently scattered intensity due to an isolated electron[*], $j|f(Q)|^2$ the intensity due to an isolated atom and $I(\vec{Q},\{\vec{r}_i\})$ the total intensity that would be obtained if the atoms were fixed at the positions $\vec{r}_i$. But as the atoms move around a second time averaging is necessary, which can be replaced by ensemble averaging:

$$I(\vec{Q}) = \langle I(\vec{Q},\{\vec{r}_i\})\rangle = j|f(Q)|^2 \left\langle \sum_i \sum_j e^{-i\vec{Q}\cdot(\vec{r}_i-\vec{r}_j)} \right\rangle. \tag{4.48}$$

In treating the double sum we distinguish two cases. If $i = j$, we have $\exp\{i\vec{Q}\cdot(\vec{r}_i-\vec{r}_j)\} = 1$, and this occurs N times. If $i \neq j$ the average can be evaluated with help of the pair distribution function in analogy to the derivation of eq. (4.18). So we have

$$\left\langle \sum_i \sum_j e^{-i\vec{Q}\cdot(\vec{r}_i-\vec{r}_j)} \right\rangle = N + N\rho \int g(\vec{r}) e^{-i\vec{Q}\cdot\vec{r}} d\vec{r}, \tag{4.49}$$

an expression which depends for an isotropic fluid only on the magnitude of Q and not on its direction, as $g(\vec{r}) = g(R)$. The intensity produced individually by the N atoms of the assembly may be abbreviated by

$$I_0(Q) = Nj|f(Q)|^2. \tag{4.50}$$

By inserting eq. (4.49) and (4.50) in (4.48) we obtain

$$I(Q) = I_0(Q)(1 + \rho \int g(R) e^{-i\vec{Q}\cdot\vec{r}} d\vec{r}). \tag{4.51}$$

This important relation connects the scattered intensity with the structure of the scattering object and was obtained under the assumption of no multiple scattering. It suggests to calculate $g(R)$ by a Fourier transformation. But first one has to remove the singularity of $I(Q)$ for $Q = 0$, which corresponds to forward scattering. If the electron density due to the NZ electrons of the system would be homogeneous, the total scattered amplitude would be [cf. eq. (4.41)]

$$A_h = a e^{-i\omega t} Z\rho \int e^{-i\vec{Q}\cdot\vec{r}} d\vec{r} \tag{4.52}$$

---

[*] For polarized incident beams, j depends on $\vec{Q}$. We will consider in the following only non-polarized beams and include averaging over all polarizations in j, which depends then only on the scattering angle given by the magnitude of Q.

and the corresponding intensity*⁾

$$I_h = NjZ^2 \rho \int e^{-i\vec{Q}\cdot\vec{r}} dr = I_0(Q)\rho \int e^{-i\vec{Q}\cdot\vec{r}} d\vec{r} \ . \tag{4.53}$$

As the integral is given by $(2\pi)^3 \delta(\vec{Q})$ (see appendix A), I_h is singular for $Q = 0$ (for an infinite volume) and zero otherwise, i.e. restricted to forward scattering. Defining now the scattering function

$$i(Q) = \frac{I(Q) - I_h - I_0(Q)}{I_0(Q)} \tag{4.54}$$

we get the relation

$$i(Q) = \rho \int h(R) e^{-i\vec{Q}\cdot\vec{r}} d\vec{r} \ . \tag{4.55}$$

This equation yields after Fourier transformation

$$\rho h(R) = \frac{1}{(2\pi)^3} \int i(Q) e^{i\vec{Q}\cdot\vec{r}} d\vec{Q} \ . \tag{4.56}$$

As $i(\vec{Q}) = \rho \tilde{h}(\vec{Q})$, one realizes from eq. (4.33) that also the direct correlation function can be obtained immediately from the scattered intensity

$$\rho c(R) = \frac{1}{(2\pi)^3} \int \frac{i(Q)}{(1 + i(Q))} e^{i\vec{Q}\cdot\vec{r}} d\vec{Q} \ . \tag{4.57}$$

Finally, the 3-dimensional Fourier transforms can be reduced easily to 1-dimensional integrals, e.g.,

$$\rho h(R) = \frac{1}{2\pi^2} \int_0^\infty i(Q) \frac{Q \sin QR}{R} dQ \ . \tag{4.58}$$

It is important that the limiting behaviour of $i(Q)$ for $Q \to 0$ and $Q \to \infty$ can be derived from the above relations. By setting $\vec{Q} = 0$ in eq. (4.55) and using eq. (4.24) we get

$$i(0) = -1 + kT\left(\frac{\partial \rho}{\partial P}\right) = -1 + \frac{NkT\beta_T}{v} \ . \tag{4.59}$$

Also, it follows immediately from eq. (4.55) that

$$\lim_{Q \to \infty} i(Q) = 0 \ . \tag{4.60}$$

4.3.1. Problems in the Evaluation of the Measured Intensity

Though the total correlation function $h(R)$ is simply the Fourier transform of the scattering function $i(Q)$, some procedures are involved in the experimental evaluation which limit the precision of the result.

The incident X-ray beam must be monochromatized. The scattered intensity is measured at different angles of deflection by quantum counters and then corrected for the absorption in the sample and the sample holder and for the scattering of the sample holder. The total intensity $I_{exp}(Q)$ finally obtained is the sum of the coherent and of the incoherent intensity and still given

---

* The relation $f(0) = Z$ is used here.

in arbitrary units. But $I_0(Q)$ as well as the incoherent intensity $I_{inc}(Q)$ can be calculated from first principles in absolute units [5]. Thus the problem remains to find the normalizing factor c which converts $I_{exp}(Q)$ into absolute units:

$$cI_{exp}(Q) = I(Q) - I_h + I_{inc}(Q), \tag{4.61}$$

where I_h has been substracted, as the zero angle scattering is not included in $I_{exp}(Q)$. By inserting the above relation into the definition of $i(Q)$ [eq. (4.54)] we get

$$i(Q) = c\frac{I_{exp}(Q)}{I_0(Q)} - \frac{I_0(Q) + I_{inc}(Q)}{I_0(Q)}. \tag{4.62}$$

The limiting equations (4.59) and (4.60) could be used for the determination of c, if the intensity could be measured for very small and very large values of Q. In absence of special small angle equipment, measurements can only be made for deflection angles $>2°$, as the undeflected beam is difficult to eliminate. The wavelengths of K_α-lines are of about 1 Å (0.5–2 Å), so that the smallest values for $Q = \frac{4\pi \sin \theta}{\lambda}$ are about 0.2 Å^{-1}. Extrapolation to $Q = 0$ then becomes rather uncertain. On the other hand the maximum deflection angle is that for backward scattering, so that $Q_{max} < \frac{4\pi}{\lambda} \approx 12.6$ Å^{-1} for a wavelength of 1 Å. Despite the fact that $i(Q)$ has still small oscillations around zero at these Q-values, c is sometimes determined by the approximation $i(Q_{max}) = 0$, which yields

$$c = \frac{I_0(Q_{max}) + I_{inc}(Q_{max})}{I_{exp}(Q_{max})}. \tag{4.63}$$

Another method uses the limit of eq. (4.58) for $R \to 0$ [6, 7]. This is given by

$$\int_0^\infty i(Q)Q^2 dQ = -2\pi^2 \rho, \tag{4.64}$$

and by inserting eq. (4.62) we obtain

$$c = \left\{-2\pi^2\rho + \int_0^\infty \frac{I_0(Q) + I_{inc}(Q)}{I_0(Q)} Q^2 dQ\right\} \Big/ \int_0^\infty \frac{I_{exp}(Q)}{I_0(Q)} Q^2 dQ. \tag{4.65}$$

As the integration should be extended from zero to infinity, the limited interval of Q-values causes an error with this procedure, too.

After the conversion of the measured intensity to absolute units and the determination of the scattering function $i(Q)$, Fourier transformation needs an integration from zero to infinity, whereas our knowledge of $i(Q)$ is restricted to, e.g., 0.2 Å^{-1} < Q < 12.6 Å^{-1}. The limiting value $i(0)$ is given by eq. (4.59), which helps in the extrapolation from the smallest measured Q-values. On the other hand, the Fourier integrals are truncated at some value Q_{max}. Both approximations are critical in the experimental determination of $h(R)$. Whereas an error in $i(Q)$ for small Q-values causes erroneous long ranging oscillations in $h(R)$, the truncation of $i(Q)$ at a value Q_{max} leads to short ripples in $h(R)$ [8]. In this connection a study is of interest where the Fourier integral was truncated for a measured function $i(Q)$ at different values of Q_{max} [9]. The resulting curves are shown in fig. 4.4. A truncation at small Q-values produces a total correlation function of no physical meaning. Truncation at sufficiently large Q-values does not effect the essential features of $h(R)$, but details are still varying.

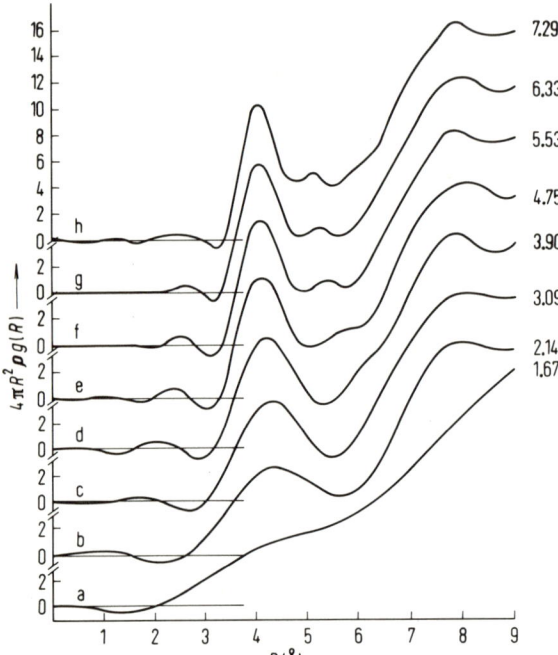

Fig. 4.4. The different results for the function $4\pi R^2 \rho g(R)$, obtained from the same function $i(Q)$, if the Fourier integral is truncated at different values $Q_{\max}$ (these are listed on the right scale in units of Å^{-1}). After ref. 9.

4.4. General Remarks about Scattering

Scattering experiments can be carried out with all kinds of radiation. Here we will outline some general features to indicate the applicability of scattering experiments to the study of fluids. For further details we refer to [10].

4.4.1. Static Approximation

Let us first consider the case of an incident plane wave $e^{i[\vec{k}_0 \cdot \vec{r} - \omega_0 t]}$ meeting a scattering potential $V(\vec{r})$, which does not change in the course of time ("static approximation"). In general, the interaction between the radiation and the scattering potential is a complicated quantum-mechanical problem. But if the interaction is small compared to the energy of the radiation, the Schrödinger equation can be solved by a perturbation method, known as Born approximation. In this case the eigenfunction is given by [11]

$$\psi(\vec{r},t) = \left[e^{i\vec{k}_0 \cdot \vec{r}} + f(\Omega) \frac{e^{ikR}}{R} \right] e^{-i\omega_0 t} \tag{4.66}$$

with
$$f(\Omega) = \text{const.} \int e^{-i\vec{Q}\cdot\vec{r}} V(\vec{r}) d\vec{r}. \tag{4.67}$$

Thus the scattered amplitude at a given point of observation is

$$A = \text{const.} \, e^{-i\omega_0 t} \int V(\vec{r}) e^{-i\vec{Q}\cdot\vec{r}} d\vec{r}, \tag{4.68}$$

which reduces to eq. (4.41) in the case of X-rays, where the scattering potential is the electron density function $\rho_e(\vec{r})$. In general, if the scattering potential can be written as sum of individual potentials due to each atom

$$V(\vec{r}) = \sum V_i(\vec{r} - \vec{r}_i) \tag{4.69}$$

a quantity d_i can be introduced by

$$d_i = \text{const.} \int V_i(\vec{r} - \vec{r}_i) e^{-i\vec{Q}\cdot(\vec{r}-\vec{r}_i)} d\vec{r}, \tag{4.70}$$

so that
$$A = \text{const.} \, e^{-i\omega_0 t} \sum_i d_i e^{-i\vec{Q}\cdot\vec{r}_i}. \tag{4.71}$$

If the d_i are identical for all atoms, $d_i = d$, then the calculation of the coherent intensity is straightforward as in the X-ray case. As the scattered intensity depends on the scattering angle, it can be expressed by $\left(\dfrac{d\sigma}{d\Omega}\right)_{\text{coh}}$, the differential cross section per unit solid angle for coherent scattering. This is given by (omitting the singularity of the scattered intensity for $Q = 0$)

$$\left(\frac{d\sigma}{d\Omega}\right)_{\text{coh}} = N|d|^2 S(Q) \tag{4.72}$$

with
$$S(Q) = 1 + i(Q) = 1 + \rho \int e^{-i\vec{Q}\cdot\vec{r}} h(R) d\vec{r}. \tag{4.73}$$

The quantity $|d|^2$ is the differential cross section per atom and — besides some constants — determined by eq. (4.70). For X-rays a detailed treatment would give

$$|d|^2 = \left(\frac{e^2}{m_e c_L^2}\right)^2 |f(Q)|^2 \frac{1 + \cos^2 2\theta}{2} \tag{4.74}$$

(e charge, m_e mass of an electron, c_L velocity of light).

For neutrons $|d|^2$ is not angle-dependent, as will be shown below. Thus, within the validity of the above assumptions, the scattered intensity is always determined by the so-called structure factor $S(Q)$, which is essentially the Fourier transform of the total correlation function.

Considering now the usefulness of the various radiations for structure determinations, we are restricted to wavelengths of the order of interatomic distances, i.e., a few Å. Thus X-rays, thermal neutrons and electrons could be used. But in order to use the Born approximation for electrons, their wavelength should be about 0.1 Å and then the important scattering occurs for rather small angles, so that it is hard to resolve. Considering now our second assumption, the static approximation, both X-rays and neutrons of 1 Å wavelength have the same momentum, but different energies ($E_0^X = 12\,\text{keV}$, $E_0^n = 82\,\text{meV}$), and hence different times for one wave period ($\tau^X = 3\cdot 10^{-19}\,\text{s}$, $\tau^n = 5\cdot 10^{-14}\,\text{s}$). As a typical time period for atomic motions is $10^{-13}\,\text{s}$, an X-ray does not see a change in atomic positions, but for neutrons the situation is critical. It is clear that for neutrons of still smaller energy, i.e. wavelengths of 4–6 Å, the static approximation is not justified. On the other hand, this fact opens the way to study also atomic motions. For neutrons of 1 Å wavelength the static approximation can still be used, possibly with some corrections [12].

As neutrons are, besides X-rays, the only means for structure study, it is of interest to discuss their behaviour in more detail [12]. Neutrons are only scattered by the nuclei and as the extension of a nucleus is negligible, the scattering potential may be written as

$$V_i(\vec{r} - \vec{r}_i) = \text{const.}\, \delta(\vec{r} - \vec{r}_i). \tag{4.75}$$

Then it follows from eq. (4.70), that d_i is an angle-independent constant. This quantity which cannot be calculated from first principles in this case is called scattering length b_i. It varies between isotopes and eventually also between different spin states, and can be positive or negative [13]. If the same scattering length b can be used for all atoms, we have according to eq. (4.72),

$$\left(\frac{d\sigma}{d\Omega}\right) = N b^2 S(Q). \tag{4.76}$$

But usually isotopic mixtures with different b_i values occur. Then the averaging procedure leading to the intensity must be rechecked. Starting from eq. (4.71), one gets

$$\left(\frac{d\sigma}{d\Omega}\right) = \left\langle \sum_{i,j} b_i b_j e^{-i\vec{Q}\cdot(\vec{r}_i - \vec{r}_j)} \right\rangle. \tag{4.77}$$

Those terms in the sum, for which $i = j$ (self-terms), yield always b_i^2, so that their total contribution is given by the mean value $N\langle b^2 \rangle$. For the evaluation of the other terms (different terms), it is necessary to assume random distribution of the b_i. Then

$$\left\langle \sum_{i \neq j} b_i b_j e^{-i\vec{Q}\cdot(\vec{r}_i - \vec{r}_j)} \right\rangle = \langle b \rangle^2 \left\langle \sum_{i \neq j} e^{-i\vec{Q}\cdot(\vec{r}_i - \vec{r}_j)} \right\rangle, \tag{4.78}$$

so that we obtain finally

$$\left(\frac{d\sigma}{d\Omega}\right) = N\langle b^2 \rangle + N\langle b \rangle^2 \rho \int h(R) e^{-i\vec{Q}\cdot\vec{r}} d\vec{r}. \tag{4.79}$$

Comparing with the general formulation for coherent scattering [eq. (4.72) and (4.73)], we see that the scattered intensity has also an incoherent contribution:

$$\left(\frac{d\sigma}{d\Omega}\right) = \left(\frac{d\sigma}{d\Omega}\right)_{\text{inc}} + \left(\frac{d\sigma}{d\Omega}\right)_{\text{coh}} \tag{4.80}$$

with

$$\left(\frac{d\sigma}{d\Omega}\right)_{\text{coh}} = N\langle b \rangle^2 S(Q) \tag{4.81}$$

and

$$\left(\frac{d\sigma}{d\Omega}\right)_{\text{inc}} = N(\langle b^2 \rangle - \langle b \rangle^2). \tag{4.82}$$

Hence the incoherently scattered intensity, which is angle-independent, has to be subtracted before the structure factor can be evaluated. As the b_i can be positive or negative, $\langle b \rangle$ can be small, so that incoherent scattering can be the dominant contribution in neutron scattering. It shall be mentioned that for neutrons, contrary to X-rays, multiple scattering can be important, which is also in the main angle-independent.

The relative merits of X-ray or neutron scattering for structure studies are mainly given by the difference between the scattering length b and the atomic scattering factor $f(Q)$. The angle-independence of b leads to reasonable scattered intensities also at large angles. The strongly varying values b_i for different isotopes complicate the work for monoatomic liquids, and in some cases the incoherent scattering becomes so large for the normal isotopic mixture, that pure isotopes must be chosen. On the other hand, this fact may be used to investigate alloys and chemical compounds and this is an important possibility, which shall be described shortly.

We consider a mixture of two different atomic species with N_1 atoms of species 1 and N_2 of species 2 ($N = (N_1 + N_2)$, $x_i = N_i/N$, $\rho = N/v$). Then three different pair distribution functions occur, $h_{11}(R)$, $h_{12}(R)$ and $h_{22}(R)$, which should be determined. Defining the structure factors by

$$S_{ij} = 1 + \rho \int h_{ij}(R) e^{-i\vec{Q}\cdot\vec{r}} d\vec{r}, \tag{4.83}$$

a straightforward calculation from eq. (4.71) leads to

$$\left(\frac{d\sigma}{d\Omega}\right) = N[x_1 b_1^2 + x_2 b_2^2 + x_1^2 b_1^2 (S_{11} - 1) + x_2^2 b_2^2 (S_{22} - 1) + 2x_1 x_2 b_1 b_2 (S_{12} - 1)], \tag{4.84}$$

where b_i is the scattering length of species i. The three unknown structure factors can be obtained from the above equation, if the scattering experiment is made with three different combinations of isotopes characterized by different b_i's. Such experiments have already been done for alloys [14]. If this method is not used, an interpretation of the scattering results for a molecular liquid becomes rather arbitrary [15].

4.4.2. Inelastic Scattering [16, 17]

Hitherto we have assumed that the scattering atoms do not move around or that their motion is comparatively slow and cannot be seen by the radiation. For thermal neutrons this condition holds only for those with shorter wavelength. Thus we will now consider briefly the more general case of a time dependent scattering system, where the static approximation is not allowed.

The scattering potential is now $V(\vec{r}, t)$; with the help of a Fourier analysis we can imagine this potential to be built up of partial waves $e^{i(\vec{Q}\cdot\vec{r} - \omega t)}$. For a liquid those partial waves will be the most important which have wave vectors $\vec{Q}$ of a few Å^{-1} and energies $E = \hbar\omega$ of about 10 meV. The incoming radiation is described by $e^{i(\vec{k}_0\cdot\vec{r} - \omega_0 t)}$, where the energy $E_0 = \hbar\omega_0$ and the momentum $\vec{p}_0 = \hbar\vec{k}_0$ are connected by the well known energy-momentum relation. If the radiation is scattered by the system, the time-dependence of the scattering potential leads to a Doppler-shift and the scattered wave is given by

$$e^{i[(\vec{k}_0 \pm \vec{Q})\cdot\vec{r} - (\omega_0 \pm \omega)t]}, \tag{4.85}$$

where the energy-momentum relation holds now between $\vec{p}_1 = \hbar\vec{k}_1 = \hbar(\vec{k}_0 \pm \vec{Q})$ and $E_1 = \hbar\omega_1 = \hbar(\omega_0 \pm \omega)$. In addition to the momentum change, which occured also in the static approximation, we have now an energy change due to the scattering process (inelastic scattering). The momentum change $\hbar\vec{Q}$ is not any longer determined solely by the scattering angle 2θ, but its amount depends also on the energy change $\hbar\omega$ or on the energy E_1:

$$Q^2 = k_1^2 + k_0^2 - 2k_0 k_1 \cos 2\theta \tag{4.86}$$

$\left(\text{for particles } k_1^2 = \frac{2mE_1}{\hbar^2}, \text{ for photons } k_1 = \frac{E_1}{\hbar c}\right).$

Therefore, it is necessary to measure the scattered intensity as a function of both the energy change and the scattering angle. Only if the incident energy $E_0 = \hbar\omega_0$ is large compared to the transferred energy $\hbar\omega$, E_1 is nearly the same as E_0, or $k_1 \approx k_0$, and then [cf. eq. (4.86)] $Q = 2k_0 \sin\theta$

is practically independent of the energy change and only determined by the scattering angle. This is always the case for X-rays, where $\hbar\omega_0 \approx 10\,\text{keV}$, compared to $\hbar\omega \approx 10\,\text{meV}$. On the other hand, neutrons of 4 Å have $\hbar\omega_0 \approx 5\,\text{meV}$ and then the energy transfer makes k_1 appreciably different from k_0 so that for a given scattering angle different momenta occur.

The total intensity of the scattered system can again be calculated by solving the Schrödinger equation, but now in its time dependent formulation. The solution can be best expressed by using the time dependent generalization of the pair distribution function $g(\vec{r})$, the van Hove correlation function $G(\vec{r},t)$. This function gives the probability to find at time t any particle at position $\vec{r}$, if at $t = 0$ a particle has been at 0. $G(\vec{r},t)$ is the key in the study of fluid dynamics and will be discussed extensively in chapter 9. As a special case we have $G(\vec{r},0) = \delta(\vec{r}) + \rho g(\vec{r})$.

If the scattered intensity is expressed by $\dfrac{d^2\sigma}{d\Omega d\omega}$, the differential cross section per unit solid angle per unit energy transfer, for neutrons the result is obtained

$$\left(\frac{d^2\sigma}{d\Omega d\omega}\right)_{\text{coh}} = N\langle b\rangle^2 \frac{k}{k_0} S(\vec{Q},\omega) \qquad (4.87)$$

with $\qquad S(\vec{Q},\omega) = \dfrac{1}{2\pi}\int e^{-i(\vec{Q}\cdot\vec{r} - \omega t)}[G(\vec{r},t) - \rho]d\vec{r}\,dt. \qquad (4.88)$

The function $S(\vec{Q},\omega)$, called scattering law, is the space-time Fourier transform of the van Hove correlation function $G(\vec{r},t)$. Thus the most general procedure in a scattering experiment is to measure the scattered intensity as a function of energy and momentum transfer, which gives the scattering law $S(\vec{Q},\omega)$ and then to make a double Fourier transformation in order to obtain $G(\vec{r},t)$. Neutrons of 4–6 Å wavelength are thus suited to study the structure as well as the dynamics of a fluid. The latter aspect will be dealt with in chapter 9.

For theoretical reasons some properties of $S(\vec{Q},\omega)$ have been found [18], and one important relation is

$$S(\vec{Q}) = \int S(\vec{Q},\omega)d\omega, \qquad (4.89)$$

which means that the structure factor is obtained by integration over all transferred energies at constant momentum change. It should be noted that for very slow neutrons intensity measurement at a given scattering angle without energy determination does not give $S(Q)$, as the integration has to be performed at constant Q and not at constant angle [cf. eq. (4.86)].

So far we have dealt only with the coherent scattered intensity. But we have had an incoherent contribution already in the static approximation if $\langle b^2\rangle$ is not equal to $\langle b\rangle^2$ [eq. (4.82)]. The term with $\langle b^2\rangle$ has come from the self-terms in eq. (4.77) ($i = j$). In the more general case of inelastic scattering these self-terms, which can be evaluated by studying the incoherent contribution, yield information on the motions of single atoms. The result can be best expressed by using the van Hove self-correlation function $G_s(\vec{r},t)$, which expresses the probability to find the same particle, which was at 0 at $t = 0$, at position $\vec{r}$ at time t. Then a self-scattering law can be defined by

$$S_s(\vec{Q},\omega) = \frac{1}{2\pi}\int e^{-i(\vec{Q}\cdot\vec{r} - \omega t)} G_s(\vec{r},t)d\vec{r}\,dt \qquad (4.90)$$

and the differential cross section for incoherent scattering is given by

$$\left(\frac{d^2\sigma}{d\Omega d\omega}\right)_{\text{inc}} = N(\langle b^2\rangle - \langle b\rangle^2)\frac{k}{k_0} S_s(\vec{Q},\omega). \qquad (4.91)$$

This relation reduces to eq. (4.82) for the static approximation, as it can be shown that the following relation holds:

$$1 = \int S_s(\vec{Q}, \omega) d\omega. \tag{4.92}$$

4.5. Results for the Pair Distribution Function from Scattering Experiments

Extensive work on the structure determination of monoatomic as well as of polyatomic liquids has been done by X-ray and neutron scattering, which is reviewed elsewhere [19–21]. We will shortly discuss some recent results, which are most instructive for our purpose.

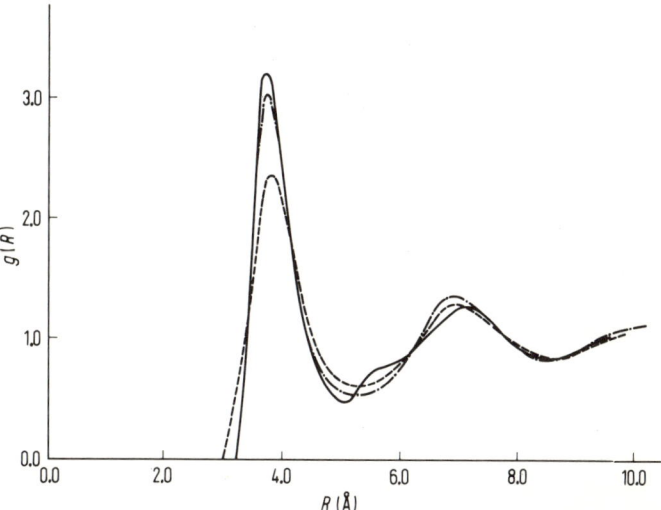

Fig. 4.5. The pair distribution function of liquid argon near the triple point obtained with X-rays (———) and neutrons (- - - -) [22, 23]. For comparison the molecular dynamics result for a 12/6-liquid at a corresponding temperature and density is shown (- · - · - · - · -) [28].

It is interesting to compare the neutron [22] and X-ray results [23] at the triple point of argon, as this gives some feeling about the accuracy of the pair distribution function obtained from scattering experiments. Fig. 4.5 shows, that the first peak in $g(R)$ obtained from neutron scattering is lower and broader than that obtained by X-rays. The scattering function $i(Q)$ obtained by neutrons showed only three peaks, whereas that from X-ray work showed two additional small peaks; if the X-ray data would have been Fourier-transformed without these two peaks, the result would have been similar to the neutron result. But both curves show the same nearest-neighbour coordination number of 8.5, if the function $4\pi R^2 \rho g(R)$ is made symmetrical with respect to its first maximum and the area under this symmetric peak is used to compute the coordination number. For comparison fig. 4.5 shows also the pair distribution function of a molecular dynamics calculation for a 12/6 liquid at a temperature and density corresponding to the triple point [28].

A neutron study [9] on the pair distribution functions of krypton for five different states along the vapour pressure curve is evaluated in fig. 4.6 for the nearest neighbour distance and the coordination number. It is seen that the expansion of the liquid is caused primarily by the decrease of the coordination number and not by changes in the nearest neighbour distance.

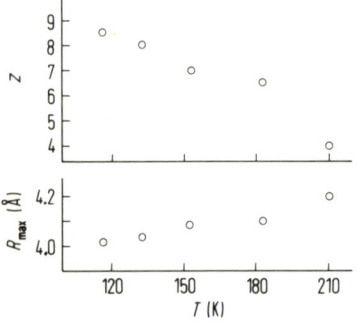

Fig. 4.6. The nearest neighbour distance R_{max} (taken from the first maximum of the $4\pi R^2 \rho g(R)$ curves), and the number of nearest neighbours z, for liquid krypton along the vapour pressure curve [9].

The most extensive study on argon was made by X-rays at 19 different states lying on a temperature and density grid (temperatures between 108 and 163 K and densities between 0.280 g/cm³ and

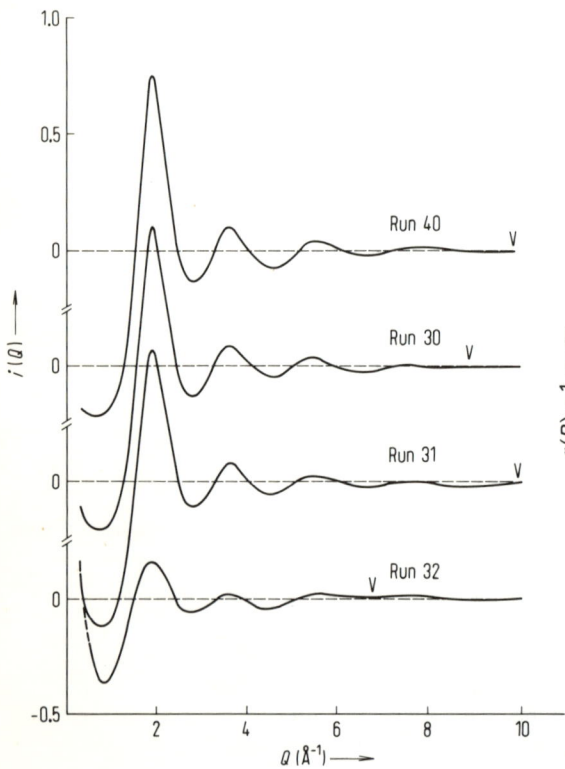

Fig. 4.7. The scattering functions $i(Q)$ for argon at 148 K obtained with X-rays. The different runs correspond to different densities: 0.982 g/cm³ (run 40), 0.910 g/cm³ (run 30), 0.780 g/cm³ (run 31) and 0.280 g/cm³ (run 32). The arrows designate the truncation points in the Fourier transformation. After ref. 24.

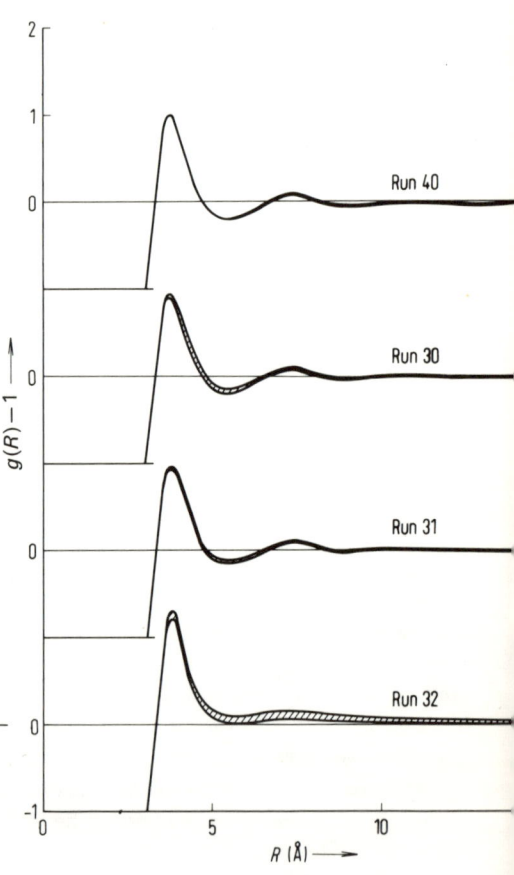

Fig. 4.8. The pair distribution functions for argon at 1 obtained from the $i(Q)$-curves of fig. 4.7. After ref. 24.

4.5. Results for the Pair Distribution Function from Scattering Experiments

1.261 g/cm^3) [24, 25]. For a temperature very close to the critical $i(Q)$, $g(R)$ and $c(R)$ are reproduced in fig. 4.7 to 4.9. The shaded areas account for the uncertainties due to different truncation points, normalization errors and experimental errors. The direct correlation function $c(R)$ has, as expected, only one peak and is short-ranged. Moreover, an analysis of this work showed that the variation of the coordination number is mainly an effect of density and not so much of temperature.

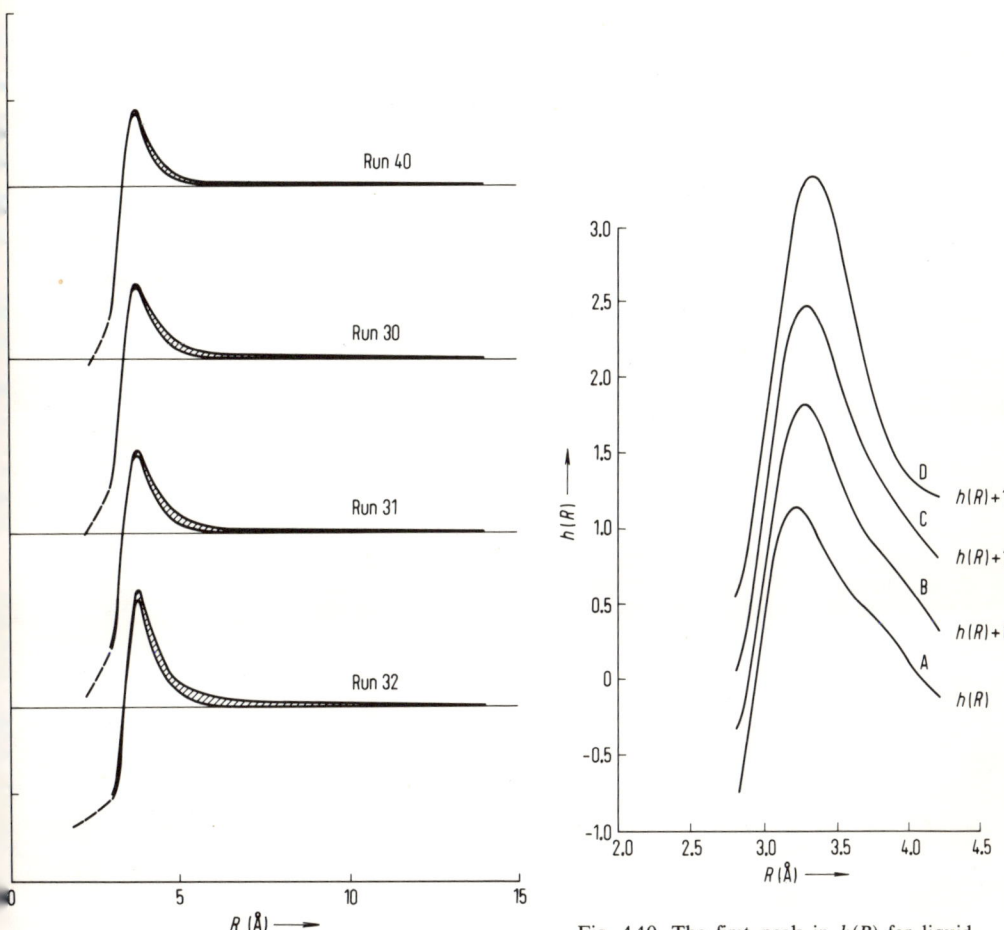

Fig. 4.9. The direct correlation functions for argon at 148 K obtained from the $i(Q)$-curves of fig. 4.7. After ref. 24.

Fig. 4.10. The first peak in $h(R)$ for liquid Pb at different temperatures: 1373 K (A), 1053 K (B), 873 K (C) and 613 K (D). After ref. 26.

Another interesting feature is that the first peak in $g(R)$ is frequently even shifted to slightly smaller R-values with increasing temperature (fig. 4.10). Moreover, the position of the first peak has been used to calculate the packing density [27], which is higher for liquid metals than for noble gases.

4.6. References

1. Z. W. Salsburg, R. W. Zwanzig, and J. G. Kirkwood, J. Chem. Phys. *21*, 1098 (1953).
2. P. G. Mikolaj and C. J. Pings, Phys. Chem. Liquids, *1*, 93 (1968).
3. L. S. Ornstein and F. Zernike, Proc. Acad. Sci. Amsterdam *17*, 793 (1914).
4. V. Volterra, Theory of Functionals, Dover, New York 1959.
5. C. H. Mac Gillavry and G. D. Rieck, International Tables of X-Ray Cristallography, Vol. 3, Kynoch Press, Birmingham 1962.
6. J. Krogh-Moe, Acta Cryst. *9*, 951 (1956).
7. N. Norman, Acta Cryst. *10*, 370 (1957).
8. J. Wasser and V. Schomaker, Rev. Mod. Phys. *25*, 671 (1953).
9. G. T. Clayton and L. Heaton, Phys. Rev. *121*, 649 (1961).
10. Chapters 6 and 8 in gen. ref. 3.
11. L. I. Schiff, Quantum Mechanics, p. 324, third edition, McGraw-Hill, New York 1968.
12. J. E. Enderby, Chapter 14 in gen. ref. 12.
13. G. E. Bacon, Neutron Diffraction, second edition, Clarendon Press, Oxford 1962.
14. J. E. Enderby, D. M. North, and P. A. Egelstaff, Phil. Mag. *14*, 961 (1966).
15. J. G. Powles, Discussion Meeting of the Bunsengesellschaft für Physikalische Chemie on "Molecular Motions in Liquids", Herrenalb, Germany, 1970, and private communication.
16. W. M. Lomer and G. G. Low, chapter 1 in: P. A. Egelstaff (Ed.), Thermal Neutron Scattering, Academic Press, London 1965.
17. L. van Hove, Phys. Rev. *95*, 249 (1954).
18. P. G. de Gennes, Physica *25*, 825 (1959).
19. S. Steeb, Fortschr. chem. Forsch. *10*, 473 (1968).
20. R. F. Kruh, Chem. Rev. *62*, 319 (1962).
21. K. Furukawa, Rept. Progr. Phys. *25*, 395 (1962).
22. D. G. Henshaw, Phys. Rev. *105*, 976 (1957).
23. N. S. Gingrich and C. W. Tompson, J. Chem. Phys. *36*, 2398 (1962).
24. P. G. Mikolaj and C. J. Pings, J. Chem. Phys. *46*, 1401, 1412 (1967).
25. S. C. Smelser and C. J. Pings, Acta Cryst. *A 25*, S 3, 19 (1969).
26. D. M. North, J. E. Enderby, and P. A. Egelstaff, J. Phys. C *1*, 1075 (1968).
27. H. Ruppersberg, Z. f. Naturforsch. *24 a*, 1034 (1969).
28. L. Verlet, Phys. Rev. *165*, 201 (1968).

5. Pair Potential

5.1. Introduction

The part of the internal energy of an assembly, which is due to intermolecular forces, is usually comprehended as sum over all pair potentials [cf. eq. (1.44), and section 1.5]:

$$E(\vec{r}_1,\ldots,\vec{r}_N,\omega_1,\ldots,\omega_N) = \sum_{i<j} \varepsilon(R_{ij},\omega_i,\omega_j). \tag{5.1}$$

An outline of the theory of pair interactions is given in section 5.2, mainly for neutral molecules, including semi-empirical evaluations. The treatment includes in principle also the case of orientation-dependent potentials. In the following sections we restrict ourselves to simple particles with potentials which depend only on distance. Section 5.3 is dedicated to the experimental determination of potential energy curves. In section 5.4 we will discuss questions of additivity of pair potentials. A short account of interaction energies in ionic melts and metals is given in sections 5.5 and 5.6, resp.

5.2. Theory of Pair Interaction [1]

5.2.1. Dispersion Energy

This kind of interaction is the only long range interaction between neutral nonpolar molecules and occurs also between any other kind of molecules. The origin of these forces is a certain correlation of the electronic motions of the interacting molecules. In order to show the mechanism we will give a simple picture [2]. Fluctuations in the electron cloud produce instantaneous dipoles which in turn polarize the electron cloud of neighbouring molecules. These dipoles are in phase and give rise to attractive forces. It will be shown below that calculations based on this picture lead to the correct result. The close similarity of this effect to the polarisation of the electron clouds by the periodically changing electric field of a light wave — which causes refraction and dispersion — has given the name dispersion energy.

Quantum mechanics yields [3] the following eq. (5.2) for the polarizability α_a of molecule a as a function of the frequency v of a changing electric field:

$$\alpha(v) = \frac{2}{h}\sum_j \frac{f_{aj}}{v_{aj}^2 - v^2} \tag{5.2}$$

where h is Planck's constant, v_{aj} the resonance frequency for a transition from the ground state of molecule a to the excited state j and f_{aj} the corresponding oscillator strength [*]:

$$f_{aj} = v_{aj}\mu_{aj}^2/3. \tag{5.3}$$

Here $\vec{\mu}_{aj}$ is the transition dipole moment for the transition $a \to j$. In quantum mechanical language the transition moment is defined by

$$\vec{\mu}_{aj} = e\int \psi_a^*(\sum \vec{r}_i)\psi_j \mathrm{d}\tau = e\langle a|\sum_i \vec{r}_i|j\rangle. \tag{5.4}$$

---

[*] It is customary to define f_{aj} with an additional factor $2\pi^2 m_e/e^2 h$, where e is the charge and m_e the mass of electron. For reasons of simplicity this factor is omitted here.

In eq. (5.4) e is the electronic charge and $\vec{r}_i$ the vector leading to the i-th electron. $|a\rangle$ denotes the ground state of molecule with eigenfunction ψ_a and $|j\rangle$ its j-th excited state with eigenfunction ψ_j [4].

The interaction energy of an electric field with the field strength $\vec{F}$ and a molecule with polarizability $\alpha(v)$ is:

$$\varepsilon = -\alpha(v) \cdot F^2/2. \tag{5.5}$$

$\vec{F}$ is now assumed to be produced by the fluctuating electron cloud of a neighbouring molecule b being a distance R apart. These fluctuations can be described by "virtual transitions" $b \to k$ with frequencies v_{bk} and transition moments $\vec{\mu}_{bk}$. If the angle between dipole $\vec{\mu}_{bk}$ and the line joining the two molecules is ϑ_{bk}, the field produced by $\vec{\mu}_{bk}$ at the point of molecule a is:

$$\vec{F}_{bk} = \vec{\mu}_{bk}(1 + 3\cos^2 \vartheta_{bk})^{1/2}/R^3. \tag{5.6}$$

The interaction energy according to eq. (5.5) is then:

$$\varepsilon_{bk,a} = \frac{\mu_{bk}^2}{3hR^6}(1 + 3\cos^2 \vartheta_{bk}) \sum_j \frac{\mu_{aj}^2 v_{aj}}{v_{aj}^2 - v_{bk}^2}. \tag{5.7}$$

Averaging over all angles ($\langle \cos^2 \vartheta_{bk} \rangle = 1/3$) and summation over all virtual transitions $b \to k$ gives the interaction energy of molecule a with the field produced by molecule b. Adding the interaction energy of molecule b with the field produced by molecule a gives the total interaction energy ε_{ab}:

$$\varepsilon_{ab} = -\frac{2}{3hR^6}\sum_{j,k} \mu_{aj}^2 \mu_{bk}^2 \left(\frac{v_{aj}}{v_{aj}^2 - v_{bk}^2} + \frac{v_{bk}}{v_{bk}^2 - v_{aj}^2}\right) = -\frac{2}{3hR^6}\sum_{j,k} \frac{\mu_{aj}^2 \mu_{bk}^2}{v_{aj} + v_{bk}}. \tag{5.8}$$

Eq. (5.8) can be derived rigorously from quantum mechanics [5] along the lines shown in section 5.2.4. The smaller the energy difference $E_j - E_a = hv_{aj}$ between the ground state and the most important excited states, the larger is α and ε and vice versa. Coloured substances and fluorocarbons are on the two ends of the scale.

Eq. (5.8) has been used to determine the coefficient of R^{-6} for a number of pairs. Exact quantum mechanical calculations have been confined to H and He. For semi-empirical evaluations the frequencies can be obtained from spectra (for the rare gases with high accuracy). Experimental determinations of oscillator strengths are not accurate, but improved sets of oscillator strengths can be obtained by using certain sum rules:

$$S(k) = \sum_j f_{aj}/(E_j - E_a)^k. \tag{5.9}$$

$S(0)$ is proportional to the number of electrons in the molecule, $S(2)$, $S(4)$, and $S(6)$ are connected to refractive index and Verdet constant.

Quite accurate values of coefficients of R^{-6} for pairs of rare gases have been obtained with this method [6]. Recently Bell has introduced approximations which make it possible to write eq. (5.7) in terms of the $S(k)$'s [7].

In cases where not sufficient data for the oscillator strengths are available other methods are still useful. One may assume that all important electronic transitions have practically the same energy differences $E_j - E_a (\approx \Delta E_a)$ and $E_k - E_b (\approx \Delta E_b)$, so that hv_{aj} and hv_{bk} can be taken outside the sum of eq. (5.2) and (5.8). The static polarizability is then given by [cf. eq. (5.4)]:

$$\alpha_a(0) = \frac{2e^2}{3\Delta E_a}\sum_j \langle a|\sum_i \vec{r}_{ai}|j\rangle^2. \tag{5.10}$$

Using the quantum mechanical sum rule*) this can be transformed into:

$$\alpha_a(0) = \frac{2e^2 A_a}{3\Delta E_a} \qquad (5.11)$$

where $A = \langle 0|\left(\sum_i \vec{r}_{ai}\right)^2|0\rangle$.

In the same manner eq. (5.8) can be transformed into:

$$\varepsilon_{ab} = -\frac{2}{3}\frac{e^4}{R^6}\frac{A_a A_b}{\Delta E_a + \Delta E_b}. \qquad (5.12)$$

Elimination of the A's in eq. (5.12) by eq. (5.11) leads to the approximate formula of London:

$$\varepsilon_{ab} = -\frac{3}{2}\frac{1}{R^6}\alpha_a(0)\alpha_b(0)\frac{\Delta E_a \Delta E_b}{\Delta E_a + \Delta E_b}. \qquad (5.13)$$

Retaining instead the quantum mechanical averages yields the formula of Salem [8].

$$\varepsilon_{ab} = -\frac{e^2}{R^6}\frac{\alpha_a(0)\alpha_b(0)}{\alpha_a(0)/A_a + \alpha_b(0)/A_b}. \qquad (5.14)$$

The quantity $\langle 0|\left(\sum_i \vec{r}_i\right)^2|0\rangle$ is made up of two terms, the sum of the squares and the sum of the cross terms:

$$A = \sum_i \langle 0|(\vec{r}_i)^2|0\rangle + \sum_{i \neq k} \langle 0|(\vec{r}_i \cdot \vec{r}_k)|0\rangle. \qquad (5.15)$$

The first term can be obtained from experimental values of the diamagnetic susceptibility. The second term expresses the correlation between electronic positions and may give an appreciable contribution. Calculations depend on the kind of wave functions used; insertion of Hartree-Fock wave functions gives somewhat too high values of the R^{-6} coefficient of ε.

Table 5.1 gives results for the coefficient of R^{-6} obtained by some of the methods mentioned above. For comparison also the results of the variational calculation of Slater and Kirkwood is given [9].

Table 5.1. Coefficient of the R^{-6}-term in units of 10^{60} erg cm^6 for argon.

Oscillator strength		Var. calc. S. K. [9]	Salem [8] with Hartree-Fock-A's
Kingst. [6]	Bell [7]		
−65.6	−62.2	−64.1	−86.5

5.2.2. More About Dispersion Energy

In the preceding paragraph the discussion has been confined to the dipole-dipole dispersion energy. But this is only the leading term in an expansion of the dispersion energy between two neutral nonpolar molecules with respect to inverse powers of the interatomic distance R:

$$\varepsilon = C_6/R^6 + C_8/R^8 + C_{10}/R^{10} + \cdots \qquad (5.16)$$

---

* $\sum_{j \neq 0} \langle 0|\sum_i r_i|j\rangle \cdot \langle j|\sum_i r_i|0\rangle = \langle 0|\left(\sum_i r_i\right)^2|0\rangle$

This expansion only holds for large R where no overlap of the electron cloud occurs. In the region around the minimum of the pair potential other contributions (second order exchange forces) become important (see section 5.2.5).

The first term in eq. (5.16) is the dipole-dipole term which has been discussed in the preceding section. The R^{-8} term arises from the dipole-quadrupole interaction. It is important in the case of hydrogen and might not be negligible in the case of argon. On the basis of preliminary estimations of C_8 [10], the R^{-8} term becomes important for argon in a region of 1.3 R^*, where overlap of the electron clouds may occur such that the series expansion (5.16) is invalid.

In the case of interaction of two different molecules where at least one is not spherical, coupling occurs of the dipole-dipole and dipole-quadrupole term. This gives rise to a R^{-7} term, which is already present between methane and argon, and which is orientation dependent. This term might influence the energy of a crystal but will probably average to zero in calculations of gaseous equilibrium properties.

Casimir and Polder [11] discovered that eq. (5.16) is not valid for very large distances R. In the derivation of section 5.2.1 the effect of the finite velocity of propagation of electromagnetic radiation (retardation effect) has been neglected. With other words, it has been assumed that the photons generated by the fluctuating dipoles have the same phase at all points of the two molecules. If, however, R is larger than a small fraction of the photon wave length, retardation effects become important and the series of eq. (5.16) starts with an R^{-7}-term. The important transition frequencies of eq. (5.2) and (5.8) are located in the ultraviolet corresponding to wavelengths of about 2000 Å, so that retardation is not negligible for distances greater than 200 or 300 Å. Retardation effects are also important for experiments with bodies of larger than molecular size [12], which have proved the correctness of the general theory.

A generalization of the theory of dispersion energy to the interaction between two systems A and B, which might be molecules, bodies of larger sizes, or a continuous dielectric, is the recently developed susceptibility theory [13]; it relates in a general way the interaction energy between A and B to the response of each separate system to external perturbations. There seems to be some hope that this susceptibility theory will not only give a unifying view of interaction between different systems but also open new paths for understanding the cohesive energy of chemically more complicated substances.

In the case of more complicated and larger molecules the series eq. (5.16) converges at small distances slowly or not at all. Then it may be a good approximation to write the dispersion energy as a sum of dispersion energies between pairs of atoms, one in each molecule. But this assumption of "local additivity" of the dispersion energies implies that electron correlation dies rapidly away with distance between different parts of the molecule [14]. Local additivity may not hold for large conjugated systems.

One important potential model – the Kihara potential – has emerged from similar considerations. Here the interacting regions are thought to be placed outside of "hard cores" of the interacting atoms.

5.2.3. Electrostatic and Inductive Energy

The electrostatic energy (also called orientation energy) arises from interaction of permanent multipoles. Each multipole also induces multipoles in polarizable neighbouring molecules. The interaction energy of a permanent and an induced multipole is called inductive energy.

We will now consider in more detail the electrostatic energy between two dipoles. For a certain orientation it is given by:

$$\varepsilon_{ab,or}(\vartheta,\varphi) = -\mu_b F_a \cos\varphi = -\frac{\mu_a \mu_b}{R^3}(1 + 3\cos^2\vartheta)^{1/2}\cos\varphi. \tag{5.17}$$

In eq. (5.17) the electric field strength caused by the dipole of the polar molecule a on a position b is denoted by $\vec{F}_a$ [cf. eq. (5.6)], ϑ being the angle between $\vec{\mu}_a$ and the line joining the centres of the two molecules, φ is the angle between $\vec{\mu}_b$ and the direction of $\vec{F}_a$. The question of averaging over orientation-dependent pair potentials is a problem of statistical mechanics, which will be dealt with in chapter 10. We will indicate here only the averaging by weighing every orientation by its Boltzmann factor, which has led to the oldest concept of van der Waals interactions [15]:

$$\bar{\varepsilon}_{ab,or} = \int \varepsilon_{ab,or}(\vartheta,\varphi) e^{-\frac{\varepsilon_{ab,or}(\vartheta,\varphi)}{kT}} d\tau \Big/ \int e^{-\frac{\varepsilon_{ab,or}(\vartheta,\varphi)}{kT}} d\tau, \tag{5.18}$$

where $\int d\tau = \int_0^\pi \int_0^\pi \frac{1}{2}\sin\vartheta\, d\vartheta \frac{1}{2}\sin\varphi\, d\varphi$ goes over all orientations. The integral can be evaluated by a series expansion, provided that

$$c = \frac{\mu_a \mu_b}{R^3 kT} \tag{5.19}$$

is small compared to unity. The result is

$$\bar{\varepsilon}_{ab,or} = -\frac{2}{3}\frac{\mu_a^2 \mu_b^2}{R^6 kT}. \tag{5.20}$$

For $\mu_a = \mu_b = 1\,\mathrm{D}$, $R = 5\,\text{Å}$, and $T = 300\,\mathrm{K}$, one obtains $c = 0.2$. Retaining the next term in the series expansion gives a factor $(1 - 7c^2/75)$ to the r.h.s. of eq. (5.20). Thus for sufficiently low moments eq. (5.20) can be used down to nearest neighbour distances in condensed systems. Eq. (5.17) and the following has been based on the assumption that the dipole moment is located in the centre of the corresponding molecule and is of negligible length compared to molecular distances. If this assumption is justified, we will speak of ideal dipoles. It may be worthwhile to state that $\bar{\varepsilon}_{ab,or}$ increases only by a factor 1.05 if the dipole length is one fifth of R, but by a factor of 2 if the dipole length becomes two third of R, compared to $\bar{\varepsilon}_{ab,or}$ of ideal dipoles. Large non-ideal dipoles lead, therefore, to strong specific interactions, which will be considered in more detail in chapter 12.

The inductive interaction between a permanent ideal dipole and a polarizable molecule can be described by

$$\varepsilon_{a\to b,ind}(\vartheta) = -\alpha_b F_a^2/2 = -\frac{\alpha_b \mu_a^2}{2R^6}(1 + 3\cos^2\vartheta). \tag{5.21}$$

If a preference for certain orientations can be neglected, averaging yields $\langle 3\cos^2\vartheta \rangle = 1$; if also the interaction $b \to a$ is taken into account, the result for the induction energy is

$$\varepsilon_{ab,ind} = -(\mu_a^2 \alpha_b + \mu_b^2 \alpha_a)/R^6. \tag{5.22}$$

Usually this contribution is smaller than the dipole-dipole orientation energy. An important exception is the interaction between a non-ideal dipole and a molecule where some spots can be especially easily polarized. An example will be discussed in section 12.5.1.

5.2.4. General Theory of Long Range Interactions

All long range forces, including also forces between charged particles, can be derived from a uniform theory. We will indicate only the principles of this theory, but refer for details to the literature [16].

The problem is the calculation of the quantum mechanical mean over all instantaneous interactions between electrons and nuclei of two interacting molecules. As long as the energy difference with respect to two isolated molecules is small, perturbation theory can be used. The interaction Hamiltonian H is given by

$$H = \sum_{i,l} \frac{e_{ai} e_{bl}}{|\vec{R} + \vec{r}_{bl} - \vec{r}_{ai}|}, \tag{5.23}$$

where the sum goes over all Coulomb interactions between charges i of molecule a and charges l of molecule b; $\vec{R}$ is the intermolecular distance, $\vec{r}_{ai}$ denotes the locus of charge i with respect to the center of mass of molecule a, $\vec{r}_{bl}$ the locus of charge l with respect to the center of mass of molecule b. The quantum mechanical averaging goes over the internal quantum states, thus giving the interaction energy for fixed molecular position and orientation. The problem of averaging over molecular orientations is left to chapter 10. Perturbation theory yields

$$\varepsilon = \langle ab|H|ab \rangle - \sum_{j,k} \frac{\langle ab|H|jk \rangle^2}{E_j - E_a + E_k - E_b}, \tag{5.24}$$

where E_a is the energy eigen value of molecule a in the ground state and E_j the eigen value in some excited state of molecule a. Similarly $E_k - E_b$ is the energy difference between excited state and the ground state of molecule b. The sign $|jk\rangle$ denotes the product $|j\rangle|k\rangle$ of wave functions of the two isolated (unperturbed) molecules.

The first term in eq. (5.24) arises from first order perturbation theory, and gives the mean over the Coulomb interaction for the electronic configurations of the ground states. It is usual to expand the interaction Hamiltonian (5.23) with respect to the $\vec{r}_{ai}$ and $\vec{r}_{bl}$. Expansion by means of a Taylor series leads to the so-called multipole expansion in inverse powers of R, where the coefficients are formed by combinations of products of components of $\vec{r}_{ai}$ with charges e_{ai}. These coefficients are used to define the corresponding multipole moments.*[)]

Another expansion uses the spherical harmonics Y_l^m [cf. eq. (10.8)], and leads for molecules with rotational symmetry to

---

* We will briefly review the situation up to the quadrupole moment [17]. As we refer to one molecule the index a is omitted. Let the vector $\vec{r}_i$ have the components r_{ki}, $k = 1, 2, 3$.
The total charge is given by $q = \sum_i e_i$, and is zero for neutral molecules. The sum $\mu_k = \sum_i e_i r_{ki}$ defines the k-th component of the dipole moment. It is independent of the choice of the origin of the coordinate system only when $q = 0$.

The sum $\theta_{jk} = \frac{1}{2} \sum_i e_i (3 r_{ji} r_{ki} - r_i^2 \delta_{jk})$ defines the matrix element of a tensor called quadrupole moment.

Here $r_i^2 = \sum_{k=1}^{3} r_{ki}^2$, and δ_{jk} is the Kronecker δ. $\vec{\theta}$ is a traceless symmetric tensor. It is always possible to find a coordinate system such that $\theta_{jk} = 0$ for $j \neq k$. There are three principal quadrupole moments $\theta_{11}, \theta_{22},$ and θ_{33}, but only two are independent. If the molecule has a symmetry axis of order greater than two, which is identified with the z-axis, then $\theta_{11} = \theta_{22}$ and the quantity $\theta = \theta_{33} = -2\theta_{11} = -2\theta_{22}$ is called quadrupole moment. It is independent of the choice of the origin of the coordinate system only when the molecule has no charge and no dipole moment.

$$H = \sum_{l=0}^{\infty} \sum_{l'=0}^{\infty} \sum_{m=-l_s}^{l_s} X^{ll'm}(R) Y_l^m(\vartheta_a \varphi_a) Y_{l'}^m(\vartheta_b \varphi_b). \tag{5.25}$$

Here l_s is the smaller of l or l'. The coordinate systems of the two molecules are chosen such that each symmetry axis is z-axis of the respective coordinate system. The angles $(\vartheta_a \varphi_a)$ and $(\vartheta_b \varphi_b)$ give the orientations of the two coordinate systems as shown in fig. 5.1.*) The quantities

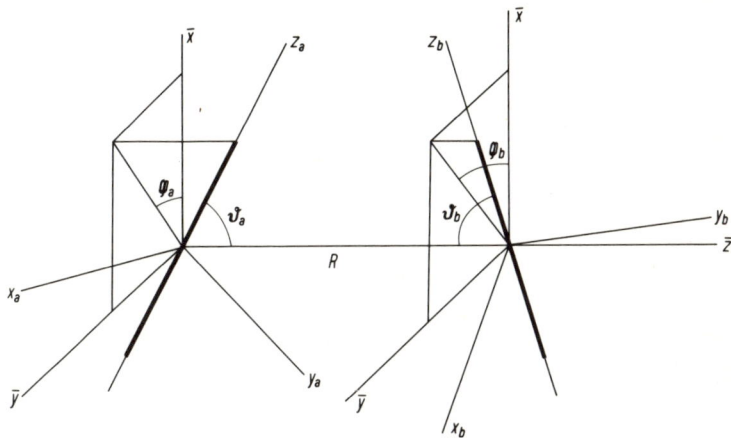

Fig. 5.1. The coordinate systems of the molecules relative to a fixed coordinate system for an expansion in spherical harmonics.

$X^{ll'm}$ contain products of multipoles and inverse powers of R (all multipoles of charge distributions with rotational symmetry are characterized by one component of their moments). For the first triplets $ll'm$ the $X^{ll'm}$ are given in table 5.2.

Table 5.2. Expansion coefficients in eq. (5.25).

l	l'	m	$X^{ll'm}$
0	0	0	$q_a q_b / R$
1	0	0	$\mu_a q_b / \sqrt{3} R^2$
0	1	0	$q_a \mu_b / \sqrt{3} R^2$
1	1	0	$2\mu_a \mu_b / 3 R^3$
1	1	1	$\mu_a \mu_b / 3 R^3$
2	0	0	$\theta_a q_b / \sqrt{5} R^3$
0	2	0	$q_a \theta_b / \sqrt{5} R^3$
2	1	0	$\sqrt{3} \theta_a \mu_b / \sqrt{5} R^4$
2	1	1	$\theta_a \mu_b / \sqrt{5} R^4$
1	2	0	$\sqrt{3} \mu_a \theta_b / \sqrt{5} R^4$
1	2	1	$\mu_a \theta_b / \sqrt{5} R^4$
2	2	0	$6 \theta_a \theta_b / 5 R^5$
2	2	1	$4 \theta_a \theta_b / 5 R^5$
2	2	2	$\theta_a \theta_b / 5 R^5$

---

* Averaging over the orientations by means of the spherical harmonics will be dealt with in section 10.3.

The first term of eq. (5.24) is (after the expansion of H) built up by terms $\langle a|\mu|a\rangle \langle b|\mu|b\rangle$, i.e. by permanent moments. It is different from zero only if both molecules have non-zero permanent multipole moments.

The second term of eq. (5.24) is obtained by second order perturbation theory. Its numerator contains products of matrix elements (after the expansion of H) like $\langle a|\mu|j\rangle \langle b|\mu|k\rangle$ where the μ's are the instantaneous moments of the corresponding molecules. The case of either $j = a$ or $k = b$ occurs only if the corresponding molecule has permanent multipoles. The sum of such terms constitutes the induction energy:

$$\varepsilon_{ab,\text{ind}} = -\sum_k \frac{\langle ab|H|ak\rangle^2}{E_k - E_b} - \sum_j \frac{\langle ab|H|jb\rangle^2}{E_j - E_a}.$$

If both molecules have no permanent moments then the only terms left in eq. (5.24) are the terms with $j \neq a$ and $k \neq b$ which constitute the dispersion energy:

$$\varepsilon_{ab,\text{disp}} = -\sum_{j \neq a, k \neq b} \frac{\langle ab|H|jk\rangle^2}{E_j - E_a + E_k - E_b}. \tag{5.26}$$

From eq. (5.26) the series eq. (5.16) can be obtained [5], and it can be shown that C_6/R^6 arises from (instantaneous) dipole-dipole, C_8/R^8 from dipole-quadrupole and C_{10}/R^{10} from the quadrupole-quadrupole and dipole-octupole interaction. Also the R^{-7}-term discussed in section 5.2.2 can be derived [16] from eq. (5.26). Eq. (5.26) can be interpreted (as done in section 5.2.1) as if the perturbation caused by the presence of another molecule consists in "virtual transitions" to the excited states of the unperturbed molecule.

5.2.5. Short Range Interactions

Relatively little general theoretical information is available in the range of small and intermediate separations. This range of separation is characterized by the overlap of the electronic clouds of the interacting particles. This causes the occurence of the so-called exchange and overlap integrals [18], which complicates the calculations considerably. Whereas perturbation methods are satisfactory for large distances (the isolated atoms taken as reference system) and for very small distances (the united atom taken as reference system), variation methods applied on approximate wave functions with many adjustable parameters are required for small and intermediate distances. Variation methods yield the total energy and the interaction energy has to be determined as difference between the total energy and the energy of the two isolated atoms. Thus only very accurate calculations can give reasonable interaction energies.

The repulsive forces for small interatomic distances are a consequence of the Pauli exclusion principle, which requires considerable distortions of the electron clouds when the region of overlap is large. These distortions result in a decrease of the electron density in the region between the nuclei. The effect is a reduction of the screening of the nuclear charges by the electrons and therefore a repulsion of the nuclei. In terms of the theory of the chemical bond [19], the situation at small separations is somewhere between the isolated atoms and the united atom. If the isolated atoms are saturated (all electronic levels doubly occupied), then the orbitals of the combined atoms will be half bonding and half antibonding, so that the net effect is antibonding.

The only case of interacting rare gas atoms thoroughly investigated is the helium-helium interaction. Let us consider here the total potential energy at small distances. More and more penetration of the electron clouds will finally result in the electron configuration of the united atom, i.e. of beryllium. Apart from the Coulomb repulsion between the nuclei, the interpenetration becomes more attractive with decreasing distance (see fig. 5.2), especially at very small distances where

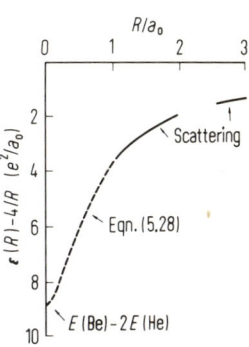

Fig. 5.2. The interaction energy of helium atoms without the contribution of nuclear repulsion (partly after Buckingham [20]). Distance is in terms of the Bohr radius a_0 and energy in units of e^2/a_0. The minimum of the helium pair potential is given by $R^* = 5.5\, a_0$ and $\varepsilon^* = 3.22 \times 10^{-5}\, e^2/a_0$.

the situation corresponds almost to the united atom. At such small distances the potential energy is given by [20]:

$$\varepsilon = \frac{Z^2}{R} + E_0 + E_2 R^2 + E_3 R^3 + \cdots , \tag{5.27}$$

where Z is the charge of the nucleus, E_0 the electronic energy of the united atom (i.e. $E_{Be} - 2E_{He}$), E_2 and E_3 can be calculated in principle by a perturbation method. Buckingham proposed a composite formula to cover a wider range of R:

$$\varepsilon = \frac{Z^2}{R}(1 + p_1 R + p_2 R^2 + p_3 R^3 + p_4 R^4)e^{-\alpha R} , \tag{5.28}$$

where the coefficients p_1 to p_4 and the constant α are interrelated with the calculable coefficients of the small R expansion (5.27). An equation of the form (5.28) (p_2 and p_4 negative) has given a satisfactory fit for He-He from $R = 0$ to the potential minimum at R^* and has failed only for large R giving a faster decrease of ε than the obligatory R^{-6} law. At distances beyond the validity of the expansion (5.27), Phillipson [21] has made a very thorough variational calculation which has been confirmed recently for small R and which covers the region up to $R^*/2$.

Results of this theoretical calculation are shown in fig. 5.3, whereas results of scattering experiments (see below) on He and Ar are given in fig. 5.3 and 5.4. The curves for some empirical potentials, which will be discussed in the next paragraph, are also included in the figures. The remarkable softening of the repulsion for small distances, e.g. compared to a 12/6-potential, occurs at distances below $R = 0.65\, R^*$, which are usually not of interest in connection with properties at readily accessible temperatures and pressures.

In the intermediate range of separations, around the minimum of the pair potential, the short-range and long-range forces merge into each other. Here second-order exchange energies are important. They occur at separations where the electron clouds overlap and stem from the exchange of electrons between the interacting systems. They show up in second-order theories [22].

70   5. Pair Potential

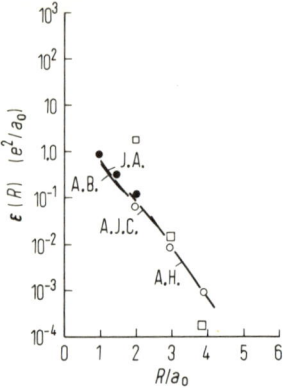

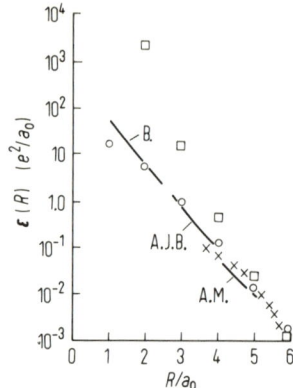

Fig. 5.3. The short range interaction of He atoms: Comparison of scattering experiments (initials refer to the authors in ref. 32) with the variational calculation of Phillipson [21] (dots), and model potentials (□ 12/6, ○ exp/6) (Partly after Abrahamson [21]). For the scale cf. fig. 5.2.

Fig. 5.4. The short range interaction of Ar atoms: Comparison of scattering experiments (initials refer to the authors in ref. 32) with model potentials (□ 12/6, ○ exp/6, × numerical [27]) (Partly after Abrahamson [21]). The minimum of the argon pair potential is given by $R^* = 7.23 \, a_0$ and $\varepsilon^* = 4.5 \times 10^{-4} \, e^2/a_0$.

5.2.6. Empirical Formulae for the Pair Potential

The most commonly used approximation for the pair potential is the so-called Lennard-Jones or 12/6 potential [eq. (1.73)]. Sometimes a R^{-8} term is included:

$$\varepsilon = AR^{-12} - BR^{-6} - CR^{-8}. \tag{5.29}$$

A relation of the constants A, B, C to ε^* and R^* is only possible by postulating a certain ratio of the coefficients C/B.

The Kihara potential (core model) is usually written

$$\varepsilon = 4\varepsilon^* \left\{ \left(\frac{1-\gamma}{R/\sigma - \gamma} \right)^{12} - \left(\frac{1-\gamma}{R/\sigma - \gamma} \right)^{6} \right\}, \tag{5.30}$$

where $\gamma\sigma$ is the diameter of the core. Again, by postulating a certain value for γ, the Kihara potential can be made to be a two-parameter potential. The value for γ depends on the nature of the molecule (e.g., it can be related to the third parameter in the corresponding states treatment of Holleran, section 10.4), but for rare gases $\gamma = 0.08$ holds approximately.

Finally, the potential with the exponential repulsive branch introduced by Buckingham should be mentioned

$$\varepsilon = Ce^{-aR} - BR^{-6}. \tag{5.31}$$

Though this potential is more satisfactory than the 12/6 potential with respect to the analytic form of the repulsive branch, it is more complicated mathematically and has one parameter more. So it has not been used as widely as the other potentials.

Parameters for the various empirical model potentials have been reported by a number of authors [23], [24], [25], [31]. The best overall fit of the properties of Ar using a 12/6 potential may be obtained with the parameters $\varepsilon^*/k = 117.2$ K and $\sigma = 3.405$ Å [25].

5.3. Experimental Determination of Pair Potential

It is not possible to determine pair potentials directly. One is forced to measure properties which can theoretically be related to integrals over functions of the pair potential. Accurate pair potentials can only be obtained by a consideration of several properties, for which good data and a good theory exists.

The best source for our knowledge of pair potentials are still macroscopic properties of dilute gases. Other methods make use of molecular beam scattering and X-ray diffraction.

It is a commonly used procedure to assume an analytic form for ε, which depends on a number of parameters (e.g. ε^* and σ for the 12/6 potential) and calculate dilute gas properties with it. Comparison with experiment yields numerical values for the parameters. But the obtained parameters depend in general on the property and the temperature range to which the potential has been fitted. This holds for all model potentials mentioned in section 5.2 and shows that all these analytical forms are not flexible enough. They are therefore insufficient for an accurate determination of the pair potential curve. Attempts were made of a piecewise construction of the pair potential [26]. Recently [27] a numerically tabulated pair potential was used for those calculations. By a multivariational method it was possible to find the optimum potential function which fits all known dilute gas properties within experimental error. The obtained pair potential is compared with the 12/6- and the Kihara potential in fig. 5.5. It has a deeper minimum than the 12/6-potential, a wider bowl with a steeper attractive and a softer repulsive branch than

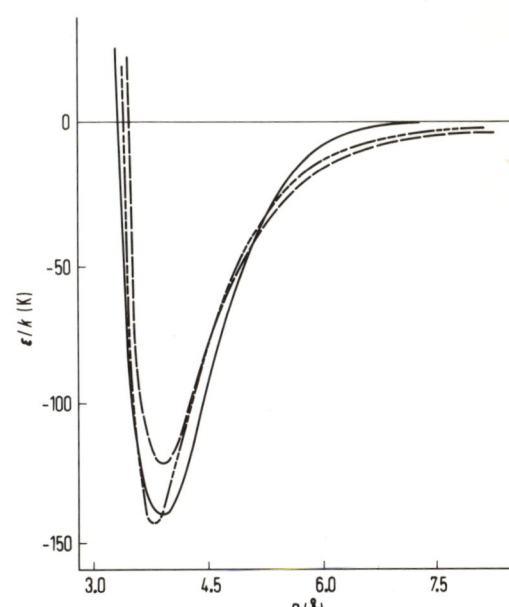

Fig. 5.5. The numerical potential of Dymond and Alder [27] compared with the 12/6-potential (dashed-) and the Kihara potential (dashed-dotted). From ref. [27].

5.3.1. Second Virial Coefficient

As shown in section 4.1, $g(r) = e^{-\varepsilon/kT}$ for low densities. For this case, eq. (4.20) becomes

$$\frac{PV}{N_A kT} = 1 + \frac{B_2}{V} = 1 - \frac{N_A}{6kTV}\int R\frac{\partial\varepsilon}{\partial R}e^{-\varepsilon/kT}d\vec{r} \qquad (5.32)$$

where B_2 is the second virial coefficient (referring to one mole). If there is no angular dependence in ε, $d\vec{r} = 4\pi R^2 dR$, and

$$B_2 = -\frac{4\pi N_A}{6kT}\int_0^\infty R^3 \frac{\partial\varepsilon}{\partial R}e^{-\varepsilon/kT}dR .$$

Integration by parts gives

$$B_2 = \frac{4\pi N_A}{6}\left\{\left| R^3 e^{-\varepsilon/kT}\right|_0^\infty - 3\int_0^\infty R^2 e^{-\varepsilon/kT}dR\right\} .$$

The lower limit of the first term is zero, and the higher is $R^3(R\to\infty) = 3\int_0^\infty R^2 dR$. Thus the final result is

$$B_2 = \frac{N_A}{2}\int_0^\infty (1 - e^{-\varepsilon/kT})4\pi R^2 dR \qquad (5.33)$$

Because of the simplicity of eq. (5.33) the second virial coefficient is very useful in providing information about the pair potential. But it is known that this property can be fitted by a variety of potentials. Even if $B_2(T)$ is known for all T, this determines only the positive part of the potential curve and (loosely speaking) the area between the negative part and the abscissa [28]. In fig. 5.6, $(e^{-\varepsilon/kT} - 1)$ is shown for $T = 0.5\varepsilon^*/k$ and for $T = 10\varepsilon^*/k$ (for comparison: ε^*/k for argon is

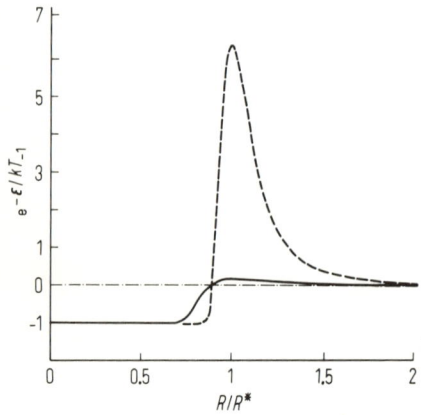

Fig. 5.6. The function $e^{-\varepsilon/kT} - 1$ according to a 12/6-potential for $kT/\varepsilon^* = 0.5$ (dotted) and for $kT/\varepsilon^* = 10$ (full line).

about 120 K). It is seen that the low temperature limit is mainly determined by the region of the potential around R^*, whereas for very high temperatures the repulsive branch becomes important. Unfortunately, the experiments have not been carried out yet to these extreme temperatures. At the lowest accessible temperatures (which are a little below the boiling point), the 12/6 potential gives B_2-values which are less negative than the experimental ones, which indicates that the minimum of the 12/6-potential is not deep enough.

5.3.2. Transport Properties of Gases

Gas viscosity, thermal conductivity and diffusion coefficient depend on the collisional cross-sections, which in turn are connected to the pair potential. If the kinetic energy of the colliding molecules is low (low temperature), the long range tail of the potential will suffice to deflect the particles. At very high temperatures the kinetic energy of the molecules will be so high that the attractive field can be almost neglected, and the collisional cross-section will be determined by the softness or hardness of the repulsive potential.

The low temperature limit of the transport properties can be used to obtain information on the coefficient of the R^{-6}-term [29]. For an attractive potential $\varepsilon = -\varepsilon'(R'/R)^6$ the collisional cross-section Ω is proportional to $T^{-1/3}$ for low temperatures. A plot of $\Omega\left(\dfrac{kT}{\varepsilon'}\right)^{1/3}$ vs. T yields in the limit of $T \to 0$ a quantity which is proportional to $(R')^2$. Such a plot is shown in fig. 5.7. The fact that the limit $T \to 0$ of $\Omega\left(\dfrac{kT}{\varepsilon'}\right)^{1/3}$ is finite is another proof for the R^{-6} dependence of the long range forces. Fig. 5.7 shows also the calculated values of $\Omega\left(\dfrac{kT}{\varepsilon'}\right)^{1/3}$ for some model potentials.

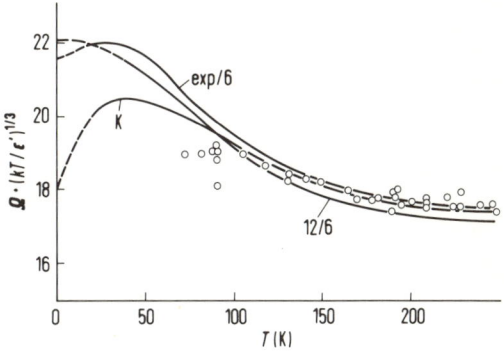

Fig. 5.7. A transformed collisional cross-section from transport properties of argon (in arbitrary units) versus temperature in comparison to model potentials (K means Kihara) (essentially after Rowlinson [29]).

The extrapolation of the experimental data is uncertain but it is seen that the exp/6- and the 12/6-potential extrapolate to a point, which is at least by a factor 1.2 too high. Considering that the coefficient of the R^{-6}-term is proportional to $(R')^6$ for given ε', the 12/6 potential gives a coefficient which is at least $1.2^3 = 1.7$ times too large. This is in agreement with semi-empirical calculations (cf. tables 5.1 and 5.3). The Kihara potential extrapolates to about the same point as the experimental data but shows at low temperatures a maximum which is too high. This

5. Pair Potential

is a confirmation of the fact that the model potentials can not be fitted to all data over the whole temperature range.

Table 5.3. Coefficients of the R^{-6}-term in units of 10^{60} erg cm^6 obtained from fitting model potentials to the properties of gaseous argon.

12-6 [23]	exp-6 [23]	Kihara [24]	Numerical [27]
−120.3	−73.7	−65.1	−66.9

The coefficients of the R^{-6} terms of the best model potentials are shown in table 5.3 together with the coefficient obtained by Dymond and Alder [27], which agrees reasonably well with the calculated ones (cf. table 5.1).

5.3.3. Scattering of Molecular Beams

The technique of molecular beams, known to all students from the experimental proof of the Maxwellian velocity distribution in a gas, has been revived and refined remarkably in the last years. Here we are interested only in elastic scattering between atoms. Elastic atom-molecule scattering is a promising tool for the study of anisotropy of molecular potentials.

It is necessary to distinguish between molecular beams at thermal energies, which are generated by effusion of the atoms out of an oven into the high vacuum, and high energy scattering, where ions are generated and accelerated, and then discharged.

There are many reasons why the evaluation of pair potentials by molecular beam techniques is only in the beginning. First, the theory of most of the phenomena involves complicated quantum mechanics, as simple approximations (e.g. the Born approximation, section 4.4.1) are usually not suitable in the atom-atom scattering case. Second, many of the phenomena have been found very recently, as the rigorous experimental requirements such as beam monochromaticity (i.e. uniform velocity) and angular resolution are not easy to meet. Third, most experimental results have been obtained on pairs of alkali atoms and heavy atoms, such as Li-Hg, Li-Xe, Li-Kr, K-Hg, K-Xe, K-Kr, for which comparison with other experimental investigations is not possible.

At present, several promising methods are worked out to evaluate details of pair potentials [30]. We will mention only the velocity dependence of total scattering cross-section $\sigma(v_r)$, where v_r is the relative velocity. The cross-section depends on the attractive branch of the pair potential for small v_r and on the repulsive branch for large v_r. Considering the small v_r-behaviour, and introducing $\varepsilon = -BR^{-m}$, it can be proved that m = 6, and, even more important, the coefficient B can be determined. For recent Ar-Ar scattering experiments, the experimental value of B is in agreement with the calculated values mentioned in paragraph 5.2.1 [31]. For reliable information on the repulsive part of the potential, it is necessary to leave the thermal range of v_r and to investigate the cross-section for high-energy beams. Series of such investigations have been carried out [32] and have led to our knowledge on the repulsive potential of a certain number of pairs. Fig. 5.3 and 5.4 show the experimental data together with some model potentials.

Other definite results on Ar-Ar interactions are not available yet, but will undoubtedly be obtained by this technique in the near future. Also information on many important details of the interaction between metal atoms in the vapour phase might be expected.

5.3.4. X-ray Determination of Pair Potentials

This is a recent development started by Pings and coworkers. The basic idea is that $g(r)$ should be equal to $e^{-\varepsilon/kT}$ at sufficiently low densities. Now, at such low densities meaningful diffraction data cannot be obtained. However, if $g(r)/e^{-\varepsilon/kT}$ is expanded into powers of density, the expansion coefficients are well-known cluster-integrals (discussed in section 7.3), of which the first is determined by the difference $h(r) - c(r)$ for low densities, accessible again from X-ray scattering data by eq. (4.56) and (4.57). A complication arises as the effect of three body potentials on the higher cluster integrals has also to be taken into account. Sufficient X-ray data in the low density region, allowing evaluation of pressure and temperature dependences, might ultimately lead to a direct experimental determination of pair potential. Presently, the existing data indicate only that the true pair potential of argon is a bit deeper and more symmetric than the 12/6 potential, in agreement with the evidence presented above [33].

5.4. Many Body Interactions

If more than two particles are near each other the sum of all pair contributions may differ from the total potential energy. The term which has to be added to the sum of the pair potentials to give the correct potential energy is called the many body potential. We do not know yet how important this term is. But it is sure that the assumption of the additivity of the pair potentials is not correct and that eq. (5.1) is only an approximation. All investigations hitherto made have suffered from the fact that the pair potential is not known accurately enough.

There are several reasons to believe that many body interactions are not negligible. One evidence is that the Lennard-Jones potential is better suited for calculations at higher densities than the Kihara model despite of the fact that the Kihara potential is closer to the true pair potential (see section 5.4.1). This has led to the concept of an "effective pair potential" applicable to states of high density, where many body contributions are approximately taken into account by a modified pair potential. For a clear distinction it is usual to refer to the potential between an isolated pair of particles as "true pair potential".

Another indication for the non-additivity of pair potentials is the third virial coefficient. Only simultaneous interactions of three particles (three body potentials) need to be considered. Calculations using Lennard-Jones potentials derived from second virial coefficients yield third virial coefficients which are too small. Corrections have been calculated for the non-additivity of the dispersion and of the repulsive forces. It has been shown that the latter is of opposite sign and of comparable magnitude to the dispersion non-additivity. The correction added to the additive contribution improves the agreement with experimental data. But details are sensitive to the used pair potential [34].

Three body forces are thought to be important for the calculation of certain properties of rare gas solids like zero point energy, Debye temperature, elastic constants and binding energy [35, 36]. The rare gases with the exception of helium crystallize in a f.c.c. packing. Calculations with a (n,6) Lennard-Jones and a (a,6) Buckingham potential for different values of n and a have shown that the h.c.p. structure is energetically more favourable than the f.c.c. structure. The relative difference in the energy is very small and constant with respect to the changes of the repulsive branch mentioned above. Jansen and coworkers [36] have shown that the inclusion of triple-dipole interactions makes the f.c.c. structure stable. This has to do with the fact that the h.c.p. structure has more equilateral triangles and less stretched configurations than the f.c.c. packing. In the case of the equilateral triangle ($R_{12} = R_{23} = R_{31}$) the three body potential has the opposite sign and amounts to ca. 20% of the sum of the pair energies. If, on the other hand, the particles 1, 2, and 3 form a linear configuration ($R_{31} = 2R_{12} = 2R_{23}$), then the three body potential has equal sign and amounts to ca. 5% of the sum of the pair energies. The dependence on distance is given by

$$\varepsilon_{123} = C \cdot (R_{12} R_{23} R_{31})^{-3} . \tag{5.34}$$

This discussion refers only to the triple-dipole interaction. The question concerning the importance of other contributions remains open [37].

Alder and Paulson [38] have shown that relatively small changes in the attractive part of the pair potential, e.g. like those proposed by Dymond and Alder (cf. fig. 5.5), can also account for the stability of the f.c.c. lattices of the heavy noble gases. A similar investigation has recently been performed on alkali halogenid crystals [39]. This shows how important an accurate knowledge of the pair potential is for the study of many body effects.

5.4.1. Effective Pair Potential

As already mentioned, it has become usual to take account of the three body effects by using an "effective pair potential", where an ensemble average over all three body configurations is divided into the pair contributions. The three body energies tend to make the minimum of the pair potential more shallow (equilateral triangles!) and to make the branch at larger R's more negative (linear configurations!), so that indeed a pair potential of the Kihara type is changed in the direction to a 12/6-potential. The effective pair potential depends on density, because many body effects contribute only if all corresponding distances are small. A further difficulty is that different effective pair potentials are needed to calculate the various thermodynamic functions or the disturbed pair distribution function $g(R)$ [40, 44].

An attempt to infer the effective pair potential from properties of solid argon has been started by O. K. Rice [41]. A similar work has been undertaken by Guggenheim and McGlashan [42]. The essential idea is to expand the pair energy around the minimum into a power series

$$\varepsilon = -\varepsilon^* + \kappa(R/R^* - 1)^2 - \alpha(R/R^* - 1)^3 + \beta(R/R^* - 1)^4 - \ldots \tag{5.35}$$

and to take care of the interaction energy due to the more distant neighbours by

$$\varepsilon = -\lambda(R^*/R)^6 \quad R \geq R^*\sqrt{2} . \tag{5.36}$$

Making a reasonable estimation of λ, the existing experimental data can be used to obtain values for R^*, κ, α, and β. These quantities can directly be related to the change of energy with volume and temperature expressed by a Debye or Einstein characteristic temperature and its volume derivative. Of course, the true frequency spectrum for a face-centered cubic crystal should be used to which the Debye model is a good approximation (which has been used by O. K. Rice).

The Einstein approximation (which has been used by Guggenheim and McGlashan) might lead to more serious errors. The value of ε^* can only be inferred from the energy of sublimation. Table 5.4 summarizes the values of the various coefficients, obtained on the basis of $\lambda/k = 143$ K, which agrees with theoretical calculations as well as with experimental evidence from the low temperature limit of transport properties of gaseous argon and from scattering of molecular beams. The most important difference between O. K. Rice and McGlashan is the value of κ, where the lower value of McGlashan is probably due to the inadequacy of the Einstein model

Table 5.4. Constants of eq. (5.35) [41, 42].

	Rice	McGlashan	12/6
$\kappa/k \cdot 10^{-3}$ (K)	53.28	48.15	43.6
$\alpha/k \cdot 10^{-3}$ (K)	25.16	24.70	30.5
$\beta/k \cdot 10^{-3}$ (K)	37.36	37.05	135
ε^*/k (K)	–	140.8	121
R^* (Å)	3.805	3.792	3.82

used. The higher value of O. K. Rice is better consistent with experimental values of sound velocities and compressibilities, which is very important, because volume derivatives offer usually a better test to pair potentials than temperature derivatives, which depend more on the statistical model. In table 5.4, the coefficients characteristic of the 12/6-potential are also given. It is seen that κ of the Lennard-Jones potential is too small, but this is compensated somewhat by the much larger values of α and β. The values of ε^* are not strictly comparable. The McGlashan-value depends on the choice of λ, which has been made to coincide with the long-range dependence of the true pair potential. If for the neighbours located at $R^*\sqrt{2}$ and $R^*\sqrt{3}$ a more attractive contribution would have been taken into account as might be well justified on the basis of three body effects, ε^* would come out correspondingly smaller.

It might be added that quite recently extensive Monte Carlo calculations have been presented for a liquid on the basis of the 12/6-potential with a remarkably good agreement to the experimental quantities of liquid argon. This indicates that the 12/6-potential is a good effective pair potential [25, 43]. This could also be confirmed using perturbation theory [44].

As shown in detail in chapter 7, statistical theories allow to calculate the pair distribution function from the pair potential. So the inverse procedure might be tried: To infer the form of pair potentials from $g(r)$-measurements by X-ray or neutron scattering plus employing a statistical theory. The attempts made have employed the Percus-Yevick theory. In the case of argon such an attempt gave an incredible density dependence of the effective pair potential [45], a clear sign of failure of the Percus-Yevick theory. But in the case of metals pseudopotentials derived by this procedure are regarded as relatively reliable [46].

Recently some progress has been made to derive accurate effective pair potentials from theory. Using third-order perturbation theory it could be shown that a nonpolar liquid environment can modify a pair potential derived from gas properties by as much as $10\% - 40\%$ [47]. Another method to calculate forces in dense media is the reaction field approach [48] which is based on the susceptibility method mentioned earlier.

5.5. Interaction Potentials in Ionic Melts

In some respects, the interaction between ions resembles the interaction between rare gas atoms. Neglecting transition elements, the electron configuration of the ions is similar to that of the rare gas atoms, leading to about the same repulsive and dispersion contribution of the interaction energy. Between an isolated pair, the dominant interaction is nevertheless the Coulomb interaction between the charged particles, attractive for unlike charges and repulsive for like charges. Another attractive term can be very important: the interaction between a charge and the dipole moments induced in the neighbouring molecule (of polarizability α) by the inhomogeneous electric field produced by the charge. It is clear that this induction effect is much more important than the induction effect dealt with in section 5.2.3, as the dipole moment induced by a charge is much larger than the one induced by a dipole, and as the charge-dipole interaction dies off slower than the dipole-dipole interaction. Summing up, the interaction energy between ions will be of the form [49]

$$\varepsilon_{ab} = C^{-aR} - BR^{-6} + q_a q_b R^{-1} - (q_a^2 \alpha_b + q_b^2 \alpha_a) 2^{-1} R^{-4} . \tag{5.37}$$

At first sight, one might be inclined to think of a pair potential in ionic melts in terms of a long range potential on account of the R^{-1} and R^{-4} term. However, the interaction between pairs of unlike charges is too strong to keep them isolated in pure components (as in solutions the additional effects of ion-solvent interaction and of the field-weakening due to the dielectric constant of the solvent must be taken into account). Even in the diluted gas the ions are associated to dipoles, quadrupoles or higher multipoles. The more we might expect the existence of multipole associates in the liquid, so that the R^{-1} and R^{-4} term effects only the next surroundings. The interaction between the multipole associates dies off rapidly with distance. This explains some striking similarities between ionic melts and molecular liquids.

5.6. Interaction Energy in Metals [50]

All pair potentials considered so far referred to a zero at infinite separation of the particles in vacuum. There have been already some difficulties in the last paragraph, as the gaseous state of an ionic melt is not realized by single ions at comparatively large distances but by ion pairs. The situation is even worse for metals. Here the gas consists of atoms or molecules (dimers, tetramers). If a certain number of these units are sufficiently close to each other, conversion to a metal seems to occur, where positive ions are embedded in a sea of common electrons [51]. The details of this conversion, including the energetic aspect, are by no means clear. At sufficiently high density and sufficiently low temperature the electronic levels of the common electrons might be considered to be almost the unperturbed levels of a free electron gas; for this limiting case a pair potential between the metallic ions might be given (often called the pseudo-pair potential), the energy zero being infinitely separated ions which are still embedded in an electronic gas of the same density. The density of the electrons is connected to the spacing ot their energy levels, i.e. to the highest filled level which is the Fermi energy $E_f = p_f^2 / 2 m_e$ (p_f being the magnitude of the corresponding momentum):

$$\frac{N}{v} = \frac{2}{h^3} \frac{4}{3} \pi p_f^3 = \frac{8\pi}{3h^3} (2 m_e E_f)^{3/2} . \tag{5.38}$$

The presence of a positive ion in this sea of free electrons perturbs the electron density in the neighbourhood of the ion. The electron density there will increase and screen the charge of the ion. A primitive argument shows that the electrostatic potential of the ion is changed by the screening from q/R to $(q/R)e^{-R/\xi}$, where ξ might be termed the radius of the extra electronic cloud. This result is in analogy to the Debye-Hückel treatment of very dilute ionic solutions, where each ion is surrounded by a cloud of counterions. The effect of the thermal energy in the theory of Debye-Hückel is here exerted by the kinetic energy of the electrons, ξ being related to p_f. In usual cases, ξ is of magnitude of 1 Å. If the above argument would be correct, the pair potential in a metal (referred to infinite separation of the ions at the same electronic density) would be solely repulsive:

$$\varepsilon_{ab} = (q_a q_b/R)e^{-R/\xi}. \tag{5.39}$$

Eq. (5.39) is in the same way a limiting case as is the Debye-Hückel theory. In both cases the counter-charge is treated as continuously distributed, and not as adherent to interacting particles. In the case of electrolyte solutions the Debye-Hückel picture has to be changed (at least at higher ionic densities and lower temperatures) to a lattice-like model, so that the monotonuous decrease of the electrostatic potential of a given ion is changed into a periodic function, as each shell of counterions is followed by a shell of like ions, and vice versa, this periodicity being damped out at large distances. In the case of the free electron gas, perturbed by the presence of a positive ion, there are some quantum mechanical restrictions which prevent a continuous decrease of extra electronic density, so that the higher electronic density in the neighbourhood of the ion is followed by a deficiency in electronic density, which in turn is followed by a surplus in electronic density and so on. Though these oscillations are vanishing at large distances, they are of much larger range than the exponential decrease of eq. (5.39) and lead to negative potentials if both ions are embedded in a surplus of electronic density. An approximate form for the potential at large R is the following:

$$\varepsilon_{ab} = q_a q_b \cos(2p_f R/h)/R^3 \quad \text{(large } R\text{)} \tag{5.40}$$

In reality it is not the interaction of point charges in the electronic sea but of ions with a finite size and a finite number of core electrons. This brings about additional corrections, but the

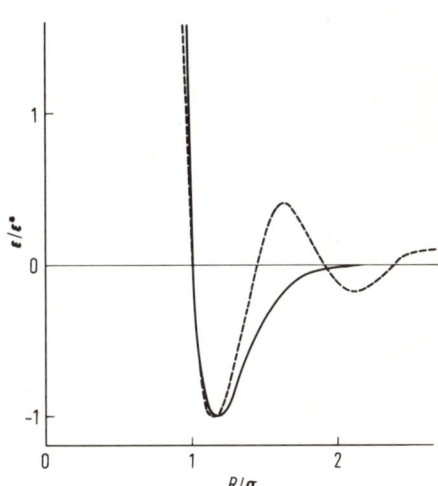

Fig. 5.8. The pair potential of Ar [27] (full line) and the pseudo-potential of Na (Paskin-Rahman potential (b)), after March [50] (dotted line) in reduced units.

important point remains valid that the pair potential of ions within the sea of electrons of constant density is oscillating and long-ranging, and has a softer repulsion than the pair potential of rare gas atoms. Fig. 5.8 shows a comparison between such a pseudo-pair potential and the pair potential of argon (but see also fig. 9.4a).

There are two important limitations on the use of metallic pseudo-pair potentials. First, they depend on electronic density, which enters, e.g., through p_f [cf. eq. (5.38)] into the argument of the cosine in eq. (5.40). That means that they are strongly dependent on external volume. This adds to the dependence of energy on volume which comes from the definition of the energy zero. Second, the oscillatory behaviour comes from quantum mechanical restrictions, which are completely valid only for a sharp Fermi energy. A distribution of electrons – at higher temperature or at higher disorder of the quasi-lattice – into various incompletely filled levels produces a trend towards eq. (5.39). This is illustrated by fig. 5.9.

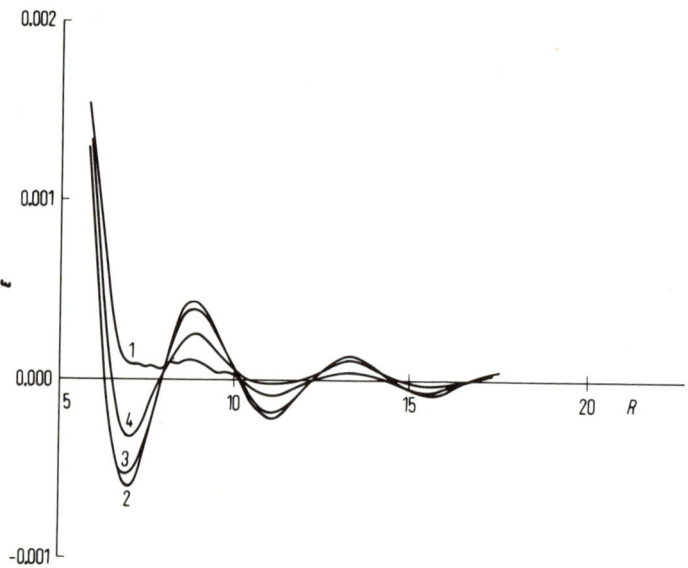

Fig. 5.9. The ion-ion interaction in a metal in dependence of the sharpness of the Fermi surface. Curve 1: $\Delta p_f/p_f = 0.2$; curve 2: $\Delta p_f/p_f = 0$ (sharp Fermi surface); curve 3: $\Delta p_f/p_f = 0.04$; curve 4: $\Delta p_f/p_f = 0.1$. After March [50], where details for the units of energy and distance are given.

5.7. References

1. cf. Margenau and N. R. Kestner, Theory of Intermolecular Forces, Pergamon Press, Oxford 1969; A. D. Buckingham and B. D. Utting, Ann. Rev. Physic. Chem. *21*, 287 (1970).
2. F. London, Trans. Faraday Soc. *33*, 8 (1937).
3. H. Eyring, J. Walter, and G. E. Kimball, Quantum Chemistry, Wiley, New York 1944, Chapter 8, 17.
4. For the bracket notation cf. P. A. M. Dirac, The Principles of Quantum Mechanics, 4th ed., Clarendon Press, Oxford 1958.
5. H. Margenau, Rev. Mod. Phys. *11*, 1 (1939).
6. A. E. Kingston, Phys. Rev. *A 135*, 1018 (1964); cf. also R. G. Gordon, J. Chem. Phys. *48*, 3929 (1968).

7. R. J. Bell, Proc. Phys. Soc. (London) *86*, 17 (1965).
8. L. Salem, Mol. Phys. *3*, 441 (1960).
9. J. C. Slater and J. G. Kirkwood, Phys. Rev. *37*, 682 (1931).
10. P. R. Fontana, Phys. Rev. *123*, 1865 (1961).
11. H. B. G. Casimir and D. Polder, Phys. Rev. *73*, 360 (1948).
12. B. V. Deryagin and I. I. Abrikosova, Phys. Chem. Solids *5*, 1 (1958); W. Black, J. G. V. de Jongh, J. T. G. Overbeek, and M. J. Sparnaay, Trans. Faraday Soc. *56*, 1597 (1960).
13. cf. H. C. Longuet-Higgins, Disc. Faraday Soc. *40*, 7 (1965); I. E. Dzyaloshinskii, E. M. Lifshits, and L. P. Pitaevskii, Adv. Physics *10*, 165 (1961); A. D. McLachlan, Proc. Roy. Soc. (London) *A 271*, 387 and *A 274*, 80 (1963), Mol. Phys. *7*, 381 (1964), Disc. Faraday Soc. *40*, 239 (1965).
14. H. C. Longuet-Higgins and L. Salem, Proc. Roy. Soc. (London) *A 259*, 433 (1961).
15. W. H. Keesom, Physik. Zeits. *22*, 129 (1921); M. Reinganum, Ann. Phys. *38*, 649 (1912); for a historical survey see ref. 5.
16. A. D. Buckingham, Adv. Chem. Phys. *12*, 107 (1967).
17. A. D. Buckingham, Quart. Rev. *13*, 183 (1959).
18. Gen. ref. 6.
19. C. A. Coulson: Valence. Clarendon Press, Oxford 1952, 93 ff.
20. W. B. Brown, Disc. Faraday Soc. *40*, 140 (1965); R. A. Buckingham, Trans. Faraday Soc. *54*, 453 (1958).
21. P. E. Phillipson, Phys. Rev. *125*, 1981 (1962); cf. also A. A. Abrahamson, Phys. Rev. *130*, 693 (1963).
22. L. Salem, Disc. Faraday Soc. *40*, 150 (1965); for a detailed discussion of the case of hydrogen see J. O. Hirschfelder and W. J. Meath, Adv. Chem. Phys. *12*, 3 (1967).
23. A. E. Sherwood and J. M. Prausnitz, J. Chem. Phys. *41*, 429 (1964).
24. J. C. Rossi and F. Danon, Disc. Faraday Soc. *40*, 97 (1965).
25. I. R. McDonald and K. Singer, J. Chem. Phys. *50*, 2308 (1969).
26. R. J. Munn, J. Chem. Phys. *40*, 1439 (1964); R. J. Munn and F. J. Smith, J. Chem. Phys. *43*, 3998 (1965).
27. J. H. Dymond and B. J. Alder, J. Chem. Phys. *51*, 309 (1969).
28. J. B. Keller and B. Zumino, J. Chem. Phys. *30*, 1351 (1959).
29. J. S. Rowlinson, Disc. Faraday Soc. *40*, 19 (1965); A. G. Clarke and E. B. Smith, J. Chem. Phys. *51*, 4156 (1969).
30. R. B. Bernstein, Science *144*, 141 (1964); M. R. C. McDowell (Ed.), Proceedings of the 3rd International Conference of Physics of Electronic and Atomic Collisions, North-Holland, Amsterdam 1964; C. Schlier, Ann. Rev. Physic. Chem. *20*, 191 (1969).
31. For references see: E. B. Smith, Ann. Rep. Progr. Chem. *63*, 13 (1966).
32. I. Amdur, J. E. Jordan, and R. R. Bertrand in: M. R. C. McDowell, ref. 30; I. Amdur and E. A. Mason, J. Chem. Phys. *22*, 670 (1954); H. W. Berry, Phys. Rev. *75*, 913 (1949) and *99*, 553 (1955); I. Amdur and R. R. Bertrand, J. Chem. Phys. *36*, 1078 (1962); I. Amdur, J. E. Jordan, and S. O. Colgate, J. Chem. Phys. *34*, 1525 (1961); I. Amdur and A. L. Harkness, J. Chem. Phys. *22*, 664 (1954); J. E. Jordan and I. Amdur, J. Chem. Phys. *46*, 165 (1967).
33. P. G. Mikolaj and C. J. Pings, Phys. Rev. Letters *16*, 4 (1966).
34. A. E. Sherwood, A. G. de Rocco, and E. A. Mason, J. Chem. Phys. *44*, 2984 (1966); but cf. also Rowlinson, ref. 29, and Barker and Pompe, ref. 40.
35. W. Götze and H. Schmidt, Z. f. Phys. *192*, 409 (1966); A. Hüller, W. Götze, and H. Schmidt, Z. f. Phys. *231*, 173 (1970).
36. L. Jansen and E. Lombardi, Disc. Faraday Soc. *40*, 78 (1965).
37. E. A. Mason and L. Monchick, Adv. Chem. Phys. *12*, 329 (1967), but cf. also D. A. Copeland and N. R. Kestner, J. Chem. Phys. *49*, 5214 (1968).
38. B. J. Alder and R. H. Paulson, J. Chem. Phys. *43*, 4172 (1965).
39. G. H. Kohlmaier, Ber. Bunsenges. *74*, 256 (1970).
40. I. R. McDonald and L. V. Woodcock, J. Phys. C, *3*, 722 (1970); G. Casanova, R. J. Dulla, D. A. Jonah, J. S. Rowlinson, and G. Saville, Mol. Phys. *18*, 589 (1970); J. A. Barker and A. Pompe, Austr. J. Chem. *21*, 1683 (1968).
41. O. K. Rice, J. Am. Chem. Soc. *63*, 3 (1941); O. K. Rice in: R. Smoluchowski, J. E. Mayer, W. A. Weyl (Eds.), Phase Transformations in Solids, Wiley, New York 1951; J. Elisha Mitchell Sci. Soc. *80*, 120 (1964); Disc. Faraday Soc. *40*, 118 (1965).

42. E. A. Guggenheim and M. L. McGlashan, Proc. Roy. Soc. (London) *A 255*, 456 (1960); M. L. McGlashan, Disc. Faraday Soc. *40*, 59 (1965).
43. J. P. Hansen and L. Verlet, Phys. Rev. *184*, 151 (1969).
44. J. A. Barker, D. Henderson, and W. R. Smith. Mol. Phys. *17*, 579 (1969).
45. S. C. Smelser and C. J. Pings, Acta Cyrst. *A 25*, S 3, 19 (1969).
46. M. D. Johnson, P. Hutchinson, and N. H. March, Proc. Roy. Soc. (London) *A 282*, 283 (1964); cf. also D. Schiff, Phys. Rev. *186*, 151 (1969).
47. T. Halicioglu and O. Sinanoglu, J. Chem. Phys. *49*, 996 (1968).
48. B. Linder, Adv. Chem. Phys. *12*, 225 (1967).
49. F. H. Stillinger in: M. Blander, Molten Salt Chemistry, Interscience, New York 1964.
50. N. H. March in gen. ref. 12; N. W. Ashcroft, Scient. Amer. *221*, 72 (1969); cf. also D. Schiff, Phys. Rev. *186*, 151 (1969).
51. cf. H. Renkert, F. Hensel, and E. U. Franck, Phys. Letters *A 30*, 494 (1969); F. Hensel, Phys. Letters *A 31*, 88 (1970).

6. Thermodynamic Properties of Liquids

6.1. The Theorem of Corresponding States

A comparison of the thermodynamic properties of various liquids can be carried out best on the basis of the theorem of corresponding states.

This theorem has been formulated and used on empirical grounds, but it can be derived statistically provided that the (effective) pair potential of all substances involved is a two-parameter potential of the same functional form [1, 2]:

$$\varepsilon = \varepsilon^* \varphi(R/R^*) \,, \tag{6.1}$$

where φ is a function common to all substances in question. The configurational partition function (1.25) can then be written

$$Q = \frac{1}{N!} \int \ldots \int e^{-\frac{\Sigma \varepsilon_{ik}(\vec{r}_i - \vec{r}_k)}{kT}} d\vec{r}_1 \ldots d\vec{r}_N$$

$$= \frac{R^{*3N}}{N!} \int \ldots \int e^{-\Sigma \varphi(\vec{r}_i^* - \vec{r}_k^*)/T^*} d\vec{r}_1^* \ldots d\vec{r}_N^* = R^{*3N} Q^* \,, \tag{6.2}$$

introducing the reduced variables $T^* = kT/\varepsilon^*$ and $\vec{r}_i^* = \vec{r}_i/R^*$. The integral Q^* of eq. (6.2) is then an universal function of T^* and $v^* = v/R^{*3}$. The configurational free energy is then

$$f' = -kT \ln Q = -3NkT \ln R^* - kT \ln Q^* \,. \tag{6.3}$$

All other configurational thermodynamic properties — i.e. properties which are not derived from the momentum integral $(2\pi mkT/h^2)^{3N/2}$ — are derived from f'. As the additive constant $(-3NkT \ln R^*)$ vanishes in the course of differentiation, all derived configurational quantities are again universal functions of T^* and v^*, times the appropriate factors in ε^* and R^*. For example, the pressure is given by

$$P = -\frac{\partial f'}{\partial v} = -\frac{1}{R^{*3}} \frac{\partial f'}{\partial v^*} = \frac{\varepsilon^*}{R^{*3}} T^* \frac{\partial \ln Q}{\partial v^*} = \frac{\varepsilon^*}{R^{*3}} P^*(T^*, v^*) \,. \tag{6.4}$$

As the argument is purely dimensional, each other set of characteristic temperature, volume, and pressure will do it than ε^*/k, R^{*3}, and ε^*/R^{*3}. In fact, the most used set for forming the reduced state variables is critical temperature, critical volume, and critical pressure. Another possibility would be the quantities at the triple point. However, the location of the triple point of saturated molecules is strongly influenced by deviations of the pair potential from spherical symmetry and the resulting question of mutual orientations of molecules at high density and order (cf. section 1.5.2 and chapter 10).

The best example for the obeyance of the theorem of corresponding states can be expected of the heavier rare gases; He and Ne show deviations due to quantum corrections which will not concern us here. Table 6.1 summarizes several important properties. First it gives the critical constants and the proportionality factors between the set ε^*/k, R^{*3} and T_c, V_c/N_A, P_c, where ε^* and R^* are evaluated for a 12/6-potential to fit the properties of condensed states. Then it gives the critical coefficient $P_c V_c/N_A k T_c$, which is the algebraic product of the proportionality factors. The boiling point would be a corresponding state only if the critical pressures of all substances would be the same. As corresponding point in the neighbourhood of the boiling point the one is taken for which the vapour pressure is $P_c/50$. The entropy change at vaporization at this point, the triple point, and the volume and entropy change at melting conform also very well to the theorem.

Table 6.1. The obeyance of the theorem of corresponding states by the heavier rare gases.

	$\varepsilon^*/k(K)$	$R^*(\text{Å})$	kT_c/ε^*	$V_c/N_A R^{*3}$	$P_c R^{*3}/\varepsilon^*$	$P_c V_c/N_A k T_c$
Ar	119.5	3.83 [3]	1.261 [4]	2.230 [5]	0.165 [4]	0.292 [4, 5]
Kr	165.2	4.11 [3]	1.267	2.206 [5]	0.167	0.291 [2]
Xe	229.8	4.49 [3]	1.261	2.179 [5]	0.168	0.290 [2]

	T/T_c (for $P_c/50$)	$\Delta S_{vap}/R$ (for $P_c/50$)	T_t/T_c	$\Delta S_m/R$	$\Delta V_m/V_s$
Ar	0.5769	9.017 [6]	0.5563	1.71	0.143 (cf. tab. 1.2)
Kr	0.5772	8.979 [7]	0.5528	1.70	0.142 [5, 7, 8]
Xe	0.580	9.057 [2]	0.5569	1.71	0.148 [2, 5, 8]

Many substances have critical pressures of very similar magnitudes. If the interaction energies are dispersion energies, ε^*/R^{*3} depends on the density and polarizability of the electrons in the electron clouds. Table 6.2 shows critical pressures of several substances. It is interesting to note

Table 6.2. Some critical pressures (in bar). References [4], [9] and [10], if not otherwise stated in the text.

Substance	P_c	Substance	P_c	Substance	P_c
Ar	48.55	CH_4	46.04	CF_4	37.4
Kr	55.0	C_2H_6	48.80	C_3F_8	26.80
Xe	58.8	C_3H_8	42.50	$n\text{-}C_6F_{14}$	19.0
CO	34.53	$n\text{-}C_6H_{14}$	29.7	$(CH_3)_2CO$	47.0
CO_2	72.85	C_2H_4	50.32	$(CH_3)_2O$	54
Cl_2	77.1	C_2H_2	61.39	CH_3OH	81.0
Br_2	103	$c\text{-}C_6H_{12}$	40.7	H_2O	221.2
SO_2	78.8	C_6H_6	48.98	KCl	232
HCl	82.6	C_2H_5Cl	53	Cs	110
SF_6	37.60	C_2H_5Br	62.3	Hg	1510

the low critical pressures of the hardly polarizable fluorocarbons, the influence of heavier atoms in organic compounds, and the effect of additional dipole-dipole interactions. Even strong quadrupole-quadrupole interactions (CO_2 and C_2H_2) are noticeable. For KCl only an extrapolated value can be given (which is based on the properties of liquid and vapour quite below the critical point [11]). It is remarkable that the interaction of metallic elements at the critical point [12] is also quite different from dispersion energy.

In the next section, we will discuss the thermodynamic properties along the saturation curve, which are most readily accessible. Section 6.3 will refer to some attempts to get hold of the volume dependence of thermodynamic quantities. In section 6.4 the thermodynamic characteristics of the melting point are summarized. Finally, section 6.5 deals with the phenomena around the gas-liquid critical point.

6.2. Thermodynamic Properties along the Saturation Curve

The densities of liquid and vapour along the saturation curve show two remarkable features: First they follow a cubic parabola, originating at the critical point; this point was emphasized first by Guggenheim [1]. Second, the arithmetic mean of liquid and vapour density follows a linear relationship; this is the well-known rule of the rectilinear diameter. To summarize, the density of rare gases and nearly spherical molecules can be very well approximated by Guggenheim's empirical formula:

$$\text{liquid: } \rho/\rho_c = 1 + 0.75(T_c - T)/T_c + 1.75[(T_c - T)/T_c]^{1/3}$$
$$\text{gas: } \rho/\rho_c = 1 + 0.75(T_c - T)/T_c - 1.75[(T_c - T)/T_c]^{1/3} . \tag{6.5}$$

The van der Waals equation gives a quadratic relation between temperature and density, and indeed a symmetrical relation would be expected. According to eq. (6.5), the coefficient of the cubic term changes sign at the critical point, i.e. there must be a singularity at that point. A similar

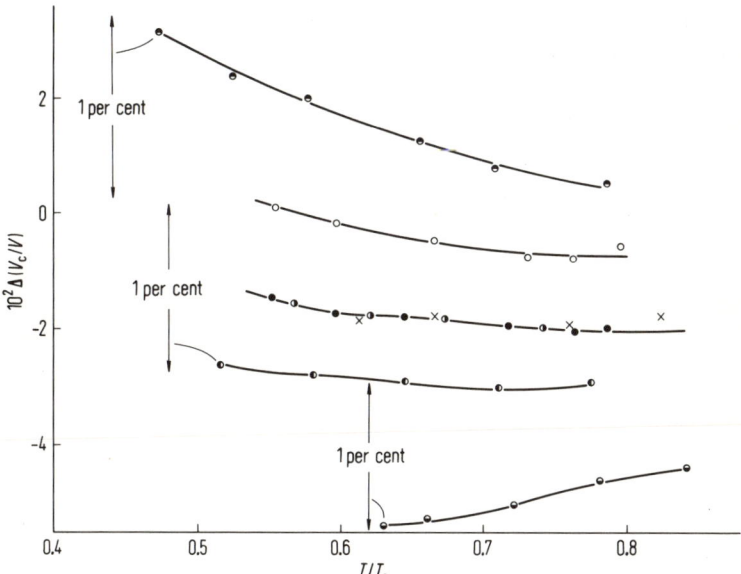

Fig. 6.1. The quantity $\Delta(V_c/V) = (\rho/\rho_c)_{\text{exp}} - (\rho/\rho_c)_{\text{eq. (6.5)}}$ for different liquids (◉ CH_4; ○ Ar; ● Kr; ◐ Xe; × N_2; ◓ O_2; ◔ CO). The arrows show the effect of an error of one percent in the critical volume on the points indicated. After ref. 5.

difficulty arises in the phase separation of liquid mixtures, where theory gives a quadratic behaviour and experiment a cubic one, even with almost the same coefficient as in the liquid-gas case. A test of eq. (6.5) [5] is shown in fig. 6.1 for a number of substances. The behaviour of different classes of liquids is indicated by fig. 6.2.

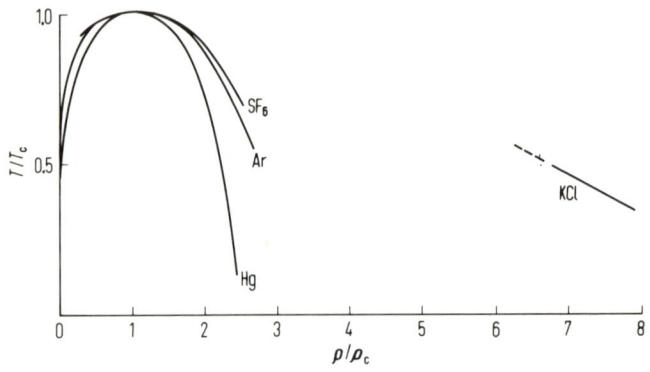

Fig. 6.2. The quantity ρ/ρ_c versus T/T_c for different classes of substances.

The dependence of vapour pressure on temperature is given by the Clausius-Clapeyron equation [eq. (1.24)], which might be rewritten

$$\frac{d \ln P}{d \ln T} = \frac{\Delta H_{vap}}{P \Delta V}. \tag{6.6}$$

The right hand side of eq. (6.6) is the ratio of the enthalpy of vaporization to the mechanical work done in the vaporization of one mole. It is noticeable that this ratio approaches a constant at the critical point [13] (5.81 for the rare gases). At temperatures below the boiling point, the Clausius-Clapeyron equation is usually written

$$\frac{d \ln P}{d T} = \frac{\Delta H_{vap}}{R T^2}, \tag{6.7}$$

approximating $\Delta V = V_{gas} - V_{liquid}$ by V_{gas}, and writing for $P V_{gas}$ the ideal value RT. Eq. (6.7), when integrated without considering the temperature dependence of ΔH_{vap}, gives

$$\ln P = -\frac{\Delta H_{vap}}{RT} + \text{const}. \tag{6.8}$$

An equation of this form which is exact below the boiling point has been discussed by Gottschal and Korvezee [14]:

$$\ln P = -\frac{\Delta H'_{vap}}{RT} + C - \frac{1}{RT}\int_{T'}^{T} \Delta C'_P dT + \frac{1}{R}\int_{T'}^{T} \frac{\Delta C'_P}{T} dT - \frac{(B_2 - V_{liq})(P - P')}{RT'} \tag{6.9}$$

with $$C = \frac{\Delta H'_{vap}}{RT'} + \ln P'. \tag{6.10}$$

In eq. (6.9) and (6.10) vaporization enthalpy and difference of heat capacity between gas and liquid are to be taken at a medium temperature T' (corresponding to the saturation pressure P') of the interval under consideration. Of the three correction terms in eq. (6.9), the first two form a parabola around T' plus higher terms, the first term being $\Delta C'_P(T - T')^2/(2RT)^2$. The last correction term is linear in $T - T'$ around T'. The correction terms cause a small curvature in the ln P vs. $1/T$ plot (e.g. for benzene, with $T' = 320$ K, the correction terms correspond at the boiling point of 353 K to 0.4% of the pressure). Above the boiling point, the terms with higher virial coefficients cannot be neglected. Nevertheless the linear plot of ln P vs. $1/T$ remains a reasonable approximation up to the critical point with a slope of $-\Delta H_{vap}/R$, where ΔH_{vap} is taken near the melting point. The deviation of the vapour pressure curve of argon from such a plot is shown in fig. 6.3. Fig. 6.4 shows plots of ln P vs. $1/T$ for a number of different substances in reduced units.

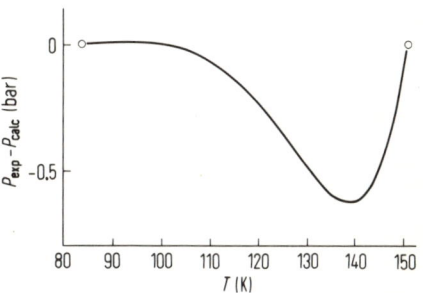

Fig. 6.3. The difference of the vapour pressure of argon [4, 6] to the equation $\ln P = -5.3355 \, (T_c/T) + 9.2182$, which fits both triple point and critical point.

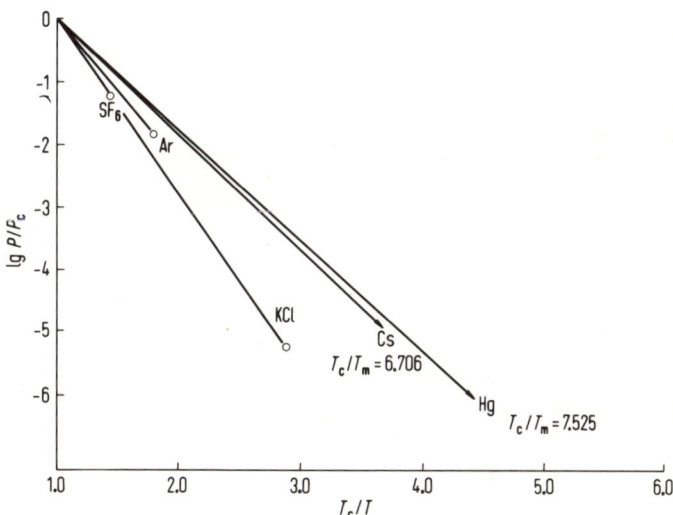

Fig. 6.4. A reduced plot of $\log(P/P_c)$ for different classes of substances. Circles mark the triple point.

It is an interesting problem, why metals on the one side and nearly spherical polyatomic molecules on the other side differ significantly from the behaviour of argon. In order to throw light on this question, a model calculation has been carried out [15] for the thermodynamic properties of a liquid with molecules interacting according to a 5/8-, 6/12-, and 7/28-Lennard-Jones potential,

6. Thermodynamic Properties of Liquids

and using the cell model and the disorder model of section 2.6.1. The vapour pressure curves could be approximated quite well, if 5/8-liquids are identified with metals, 12/6-liquids with rare gases, and 7/28-liquids with nearly spherical polyatomic molecules. But the order of the density curves came out contrary to fig. 6.2, the 5/8-liquid having the widest consolute curve and the 7/28-liquid the most narrow. Obviously the dependence of real pair potentials on volume (in the case of metals) and temperature (in the case of orientation-dependent molecules) cannot be neglected.

Though the deviations from the theorem of corresponding states are quite significant for several substances, it is interesting to compare to the reduced thermodynamic properties of argon. These are given in table 6.3 along the saturation curve, using new measurements of the ultrasonic sound velocity [16] and of the vapour pressure [4]*).

Table 6.3. The reduced thermodynamic properties of argon.

T/T_c	$\beta_s P_c$	$\beta_T P_c$	$\alpha_P T_c$	$\alpha_\sigma T_c$	$\gamma_v T_c/P_c$	$\Delta H_{vap}/R T_c$	C'_P/R	C'_v/R	C_P/C_v
$0.5563 = T_t/T_c$									
solid		0.00433	0.328		75.8	6.206	2.74	1.411	1.63
liquid	0.00463	0.00929	0.626	0.623	67.4	5.256	3.63	1.065	2.00
0.56	0.00468	0.00944	0.631	0.628	69.9	5.246	3.64	1.05	2.02
0.58	0.00498	0.01042	0.661	0.657	63.5	5.193	3.68	0.97	2.09
0.60	0.00530	0.01151	0.692	0.687	60.1	5.137	3.72	0.91	2.17
0.70	0.00770	0.0193	0.873	0.854	45.2	4.69	4.04	0.70	2.51
0.80	0.0125	0.0372	1.210	1.127	32.5	4.02	4.83	0.62	2.98
0.90	0.0270	0.111	2.30	1.90	20.7	3.06	7.26	0.63	4.11
1	∞?	∞	∞	∞	5.65	0	∞	∞?	∞

The importance of the sound velocity data, or the derived data of the adiabatic compressibility β_s [cf. eq. (1.17)]**), is twofold. First, β_s can be combined with PVT-data to furnish heat capacities, e.g. after eq. (1.21) and (1.23)

$$C_P = \frac{VT\alpha_P^2}{\beta_T - \beta_s}, \tag{6.11}$$

whereas PVT-data alone would need second derivatives, e.g.

$$\frac{\partial C_P}{\partial P} = -T\frac{\partial^2 V}{\partial T^2}, \tag{6.12}$$

which are hardly evaluable with sufficient precision. Determination of PVT-data combined with the use of β_s gives a very good access to the thermodynamic properties not too far from the critical point. At lower temperatures, heat capacity and expansion coefficient are usually not measured at constant pressure, but along the saturation curve (index σ). Again, β_s is useful to calculate the quantities at constant pressure, as follows from the set of equations [2]:

---

* cf. table 8.2 of ref. 2.
** Also the density data of ref. 16 have been used, as they deviate from ref. 4 about as much as the measurements of the British Oxygen Company, as indicated by ref. 4.

$$\alpha_\sigma = \alpha_P - \left(\frac{dP}{dT}\right)_\sigma \beta_T = \alpha_P - \gamma_\sigma \beta_T \tag{6.13}$$

$$C_\sigma = T\left(\frac{dS}{dT}\right)_\sigma = T\left(\frac{\partial S}{\partial T}\right)_P + T\left(\frac{\partial S}{\partial P}\right)_T \left(\frac{dP}{dT}\right)_\sigma = C_P - TV\alpha_P \gamma_\sigma \tag{6.14}$$

$$\beta_T - \beta_s = TV\alpha_P^2/C_P \tag{6.11}$$

These are three equations for the three unknown quantities C_P, α_P, and β_T, which give an explicit solution for β_T

$$\beta_T = \frac{\beta_s C_\sigma + TV\alpha_\sigma (\alpha_\sigma + \beta_s \gamma_\sigma)}{C_\sigma - TV\gamma_\sigma (\alpha_\sigma + \beta_s \gamma_\sigma)}. \tag{6.15}$$

The heat capacities of table 6.3 are based on an estimate of C_P, taking into account calculations from PVT-data at higher temperatures [16] and a calorimetric measurement at low temperatures [6]. Only the configurational part of the heat capacities is given in the table (i.e. for the full values $3R/2$ has to be added). The heats of vaporization are also based on measurements at low temperatures [6], calculations from PVT-data at high temperatures [17], and checked by the Clausius-Clapeyron equation (1.24), using coexistence curve [18] and vapour pressure curve [4]. Some properties of the solid are also listed in table 6.3 [19].

The relatively large values of α_P and C_P of the liquid can be explained qualitatively by the increase in disorder and the corresponding decrease in the number of nearest neighbours, but a more rigorous explanation is difficult. The heat capacity at constant volume has at low temperatures a value a little below the classical value of a solid. With increasing temperature and increasing volume, C_v becomes more translational than vibrational and decreases accordingly. The singularity at the critical point will be discussed later.

6.3. Volume Dependence of Thermodynamic Quantities

The dependence of energy on volume may be compared to the assumption of the van der Waals — or hard sphere — model, where the configurational energy U' is set equal to $-a/V$. A study by Wood [20] revealed that

$$\frac{\partial a}{\partial V} = -U' - V\frac{\partial U'}{\partial V} \tag{6.16}$$

is not negligible for most liquids, but covers a wide range from large positive to negative values. Examples are shown in table 6.4 for different classes of liquids, near the melting point, where in [cf. eq. (1.8), (1.10)]

$$\left(\frac{\partial U'}{\partial V}\right)_T = T\left(\frac{\partial S}{\partial V}\right)_T - P = T\gamma_v - P \tag{6.17}$$

the second term can be neglected. It is also shown in table 6.4 that the differences in γ_v, T_c/P_c are mainly caused by differences in the reduced compressibilities, with the exception of ionic melts, where the reduced expansion coefficient is very high. If the term $V\partial U'/\partial V$, reduced by RT_c, is compared with TC_v/RT_c, which can be easily formed from table 6.3, it is seen that the change of energy with volume is much more drastic than the change of energy with temperature.

Table 6.4. The volume dependence of configurational energy U' and of the van der Waals-parameter a (cf. also tables 1.2 – 1.4 and 6.1 – 6.3).

Substance	Ar	C_6H_6 [2]	KCl [21, 11]	Hg [22, 12, 23]
T/T_c	0.5563	0.496	0.348	0.1549
αT_c	0.626	0.675	1.15	0.320
$\beta_T P_c$	0.00929	0.00421	0.00832	0.0612
$\gamma_v T_c/P_c$	67.4	160	139	5.32
V/V_c	0.3745	0.3412	0.126	0.388
$P_c V_c/R T_c$	0.292	0.268	0.358	0.391
$U'/R T_c$	-4.700	-6.984	-6.99	-4.038
$\dfrac{V}{RT_c}\dfrac{\partial U'}{\partial V}$	4.100	7.257	2.18	0.123
$\dfrac{1}{RT_c}\dfrac{\partial a}{\partial V}$	$+0.60$	-0.27	$+4.81$	$+3.92$

It is also interesting to resolve the change of C_v along the saturation curve into the separated variations with volume or temperature. With the help of a Maxwell equation one has

$$\frac{1}{T}\frac{\partial C_v}{\partial V} = \frac{\partial \gamma_v}{\partial T} = \frac{1}{\beta_T}\left(\frac{\partial \alpha}{\partial T}\right)_v - \frac{\alpha}{\beta^2}\left(\frac{\partial \beta_T}{\partial T}\right)_v. \tag{6.18}$$

Converting the derivatives on the r.h.s. of eq. (6.18) to derivatives at constant (effectively zero) pressure

$$\left(\frac{\partial \alpha}{\partial T}\right)_v = \left(\frac{\partial \alpha}{\partial T}\right)_P + \frac{\partial \alpha}{\partial P}\gamma_v = \left(\frac{\partial \alpha}{\partial T}\right)_P - \left(\frac{\partial \beta_T}{\partial T}\right)_P \gamma_v, \tag{6.19}$$

yields
$$\left(\frac{\partial \beta_T}{\partial T}\right)_v = \left(\frac{\partial \beta_T}{\partial T}\right)_P + \frac{\partial \beta_T}{\partial P}\gamma_v \tag{6.20}$$

$$\frac{\partial C_v}{\partial V} = \frac{T}{\beta_T}\left\{\frac{\partial \alpha}{\partial T} - 2\gamma_v \frac{\partial \beta_T}{\partial T} - \frac{\alpha^2}{\beta_T^2}\frac{\partial \beta_T}{\partial P}\right\}. \tag{6.21}$$

The last term of eq. (6.21) can be evaluated by means of the Tait equation[*]. This is an empirical relation, which holds remarkably well at high densities, and which might be written [24]

$$\frac{1}{\beta_T} = \frac{1}{\beta_T^0} + 10.5 P. \tag{6.22}$$

Here β_T^0 is the compressibility at zero pressure. Eq. (6.22) is a rectangular hyperbola, β_T tending to infinity at a pressure of $P = -1/(10.5 \beta_T^0)$. The constant β_T^0 is the only quantity which is temperature-dependent. A test of eq. (6.22) for low pressures is shown in fig. 6.5. At very high pressures, the Tait equation fails as it would request the compressibility tending to zero, whereas fig. 1.2 has shown that β_T decreases very slowly at high pressures. Inserting the Tait equation, i.e. 10.5 for $-(\partial \beta_T/\partial P)/\beta_T^2$, eq. (6.21) can be evaluated by knowing the temperature dependence of α and β_T. Unfortunately, there arise large compensating terms, so that the errors

---

[*] For a discussion of this and similar equations, cf. also Rowlinson [2].

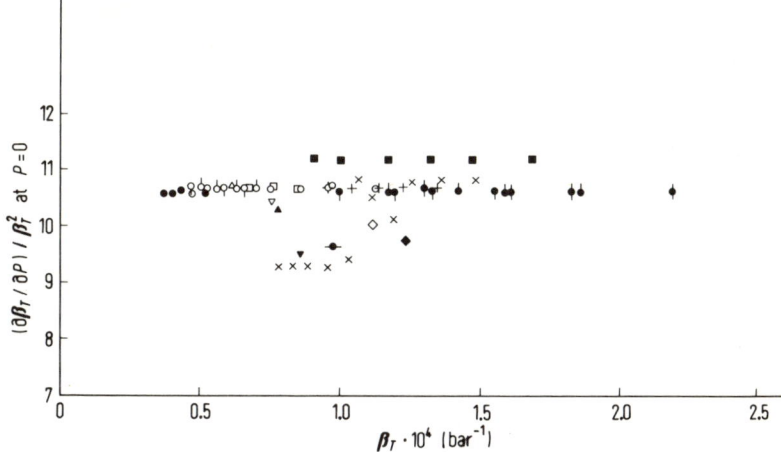

Fig. 6.5. The function $-(\partial \beta_T/\partial P)/\beta_T^2$ for zero pressure versus β_T. Data taken from the literature [22]: ● n-hexane; ◊ 1,2-dichloroethylene (trans); ● n-heptane; ▼ trichloroethylene; ● n-octane; ▽ tetrachlorethylene; ■ n-pentadecane; ● 1,2-ethanediol; + benzene; ♦ acetone; ● chloroform; ○ chlorobenzene; × carbon tetrachloride; □ bromobenzene; ▲ 1,2-dichloroethane; ⊙ nitrobenzene; △ 1,1,2,2-tetrachloroethane; ○ aniline. After ref. 24.

might be very large. An attempt of such an evaluation is shown in table 6.5 for argon. It seems that the change of C_v along the saturation curve is mostly caused by $\partial C_v/\partial V$ and not by $\partial C_v/\partial T$. Even for polyatomic molecules $\partial C_v/\partial V$ gives a substantial part of the change of C_v along the saturation curve [24], though there are some discrepancies to direct experimental observations of $\partial \gamma_v/\partial T$ [25].

Table 6.5. The volume dependence of C_v of argon at $T/T_c = 0.57$.

$\dfrac{\partial(\alpha T_c)}{\partial(T/T_c)}$	$-2\gamma_v \dfrac{T_c}{P_c} \dfrac{\partial(\beta_T P_c)}{\partial(T/T_c)}$	$10.5\,(\alpha T_c)^2$	$\dfrac{\partial(\gamma_v T_c/P_c)}{\partial(T/T_c)}$	$\dfrac{\partial(C_v/R)}{\partial(V/V_c)}$
1.50	-6.39	4.38	-51	-8.5

6.4. Melting [26]

Melting is a phase transformation of first order, as characterized by fig. 1.5. This figure implies that both solid and liquid state can be continued into a region of thermodynamic instability. This is true at least for the liquid, where undercooling is possible, as the difficulty in forming solid nuclei prevents the immediate transformation into the stable solid. It is of interest that such a kinetic hindrance does not seem to exist for the melting of solids.

The location of the melting point, and the change of several properties at melting is shown for different classes of substances in table 1.1 – 1.4, 6.1 and 6.3. Looking at the differences between metals and noble gases, which are partly caused by the different shape of the well of the pair potential, it is of interest to compare to another extrem, namely SF_6. This molecule may be regarded as almost spherical, but due to its numerous electrons on the periphery its pair potential has a very narrow well. That contributes [15] to the deviations of its melting properties from the properties of argon to the other side than metals: the reduced triple point is high ($T_t/T_c = 0.698$), and volume and entropy change at melting is large ($\Delta V/V_s = 0.357$, $\Delta S_m/R = 2.71$) [27]. As far as positional and not orientational disorder is concerned (cf. chapter 10), it is clear that entropy and volume change are in the main interrelated. This is shown in fig. 6.6. A separate consideration

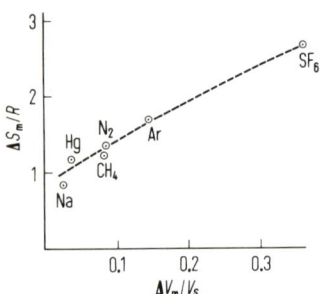

Fig. 6.6. Relation between entropy and volume change at melting for simple substances.

is necessary for ionic melts, where the entropy change at melting does not vary much in the series of alkalihalides, but where the volume change is strongly influenced by the size ratio of cations and anions and by their polarizability [28]. The entropy and volume changes of the alkalihalides are listed in table 6.6. The volume changes as function of the cation/anion

Table 6.6. Melting properties of some alkali halides. References are [29], [30] for melting points, [31] for the volume change, [32] for the entropy change (where the values of LiF, NaF, and KF are taken from older sources), and [30] for the slope of the melting curve, which is compared to $\Delta V/\Delta S_m$.

Substance	T_m(K) [29]	[30]	$\Delta V/V_s$ (%)	V_s (cm^3)	$\Delta S_m/R$	(dT/dP) (K bar^{-1})	$\Delta V/\Delta S_m$
LiF	1121		29.4	11.13	2.90		
LiCl	883		26.2	22.42	2.71		
LiBr	823		24.3	27.45	2.58		
NaF	1268	1265	27.4	16.91	3.19	0.0180	0.0175
NaCl	1073	1073.5	25.0	30.19	3.14	0.0241	0.0290
NaBr	1020	1014	22.4	36.01	3.08	0.0272	0.0315
NaI	933	928	18.6	46.16	3.04	0.0357	0.0340
KF	1131	1124	17.2	25.91	3.00	0.0226	0.0179
KCl	1043	1043	17.3	41.60	3.06	0.0291	0.0283
KBr	1007	1006	16.6	48.05	3.05	0.0379	0.0315
KI	954	957	15.9	58.46	3.03	0.0646	0.0369
RbCl	995	990.7	14.3	46.98	2.87	0.026	0.0282
RbBr	965	950	13.5	53.80	2.90	0.032	0.0301
CsCl	918		10.5	54.16	2.65		

size ratio are shown in fig. 6.7a, and this plot can be simplified by multiplying the relative volume changes by the average polarizabilities of the ion pairs $\bar{\alpha} = (\alpha_+ + \alpha_-)/2$ (fig. 6.7b). Going back now to simple molecules (fig. 6.6), it seems interesting to note that there is no lower limit for

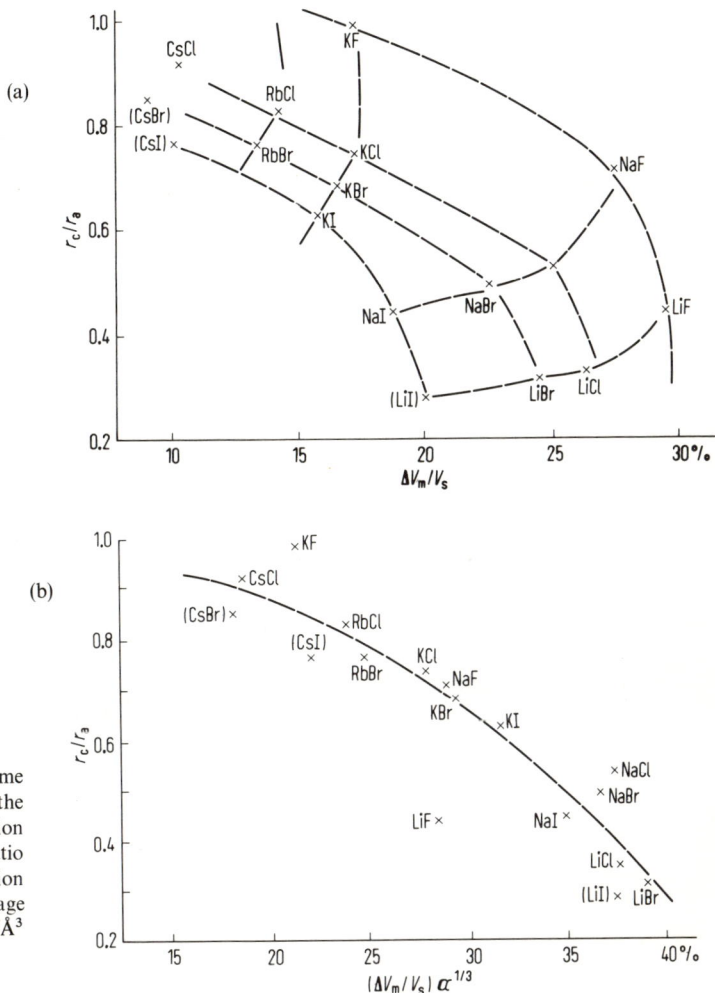

Fig. 6.7. The volume changes at melting of the alkalihalides as function of cation/anion size ratio (a), and with consideration of the (reduced) average polarizabilities $\alpha = \bar{\alpha}/\text{Å}^3$ (b). After ref. 28.

ΔV as is found in the hard sphere model of Bernal (chapter 2.4)*). Apparently also the amount of disorder can become relatively small in some liquids, especially in undercooled liquids [33]. Whatever the detailed picture of disorder in the liquid state, it seems to be natural to think of a disorder parameter, with respect to which the free energy can be minimized. At lower temperatures, the energy loss at disorder will be too big for the entropy gain, and the disorder

---

* cf. also the discussion of the melting curve at high pressures at the end of the section.

parameter will be small. In order that a clear intersection between the chemical potential of the solid and the liquid can occur, the chemical potential of the liquid for low disorder must be more positive than the chemical potential of the solid. This difference between solid and undercooled liquid at low disorder is brought about by coupling of the vibrations of the solid, leading to a frequency spectrum which can be approximated by a Debye spectrum. Fig. 6.8 shows the

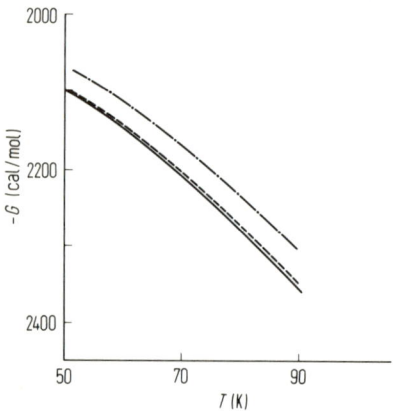

Fig. 6.8. The free enthalpy of solid argon, calculated with a cell model on the basis of an Einstein model (dashed-dotted), a Debye model (dashed), and the elaborate frequency spectrum of a f.c.c. crystal (full line).

chemical potential of solid argon [34], calculated by an uncorrelated cell model for single occupancy and the 12/6-potential, on the basis of the Einstein model, the Debye model, and the elaborate frequency spectrum [35]. Though the uncorrelated cell model may have some shortcomings (see chapter 7), even at these low temperatures and high densities, the relative behaviour of the three models expresses probably correctly the influence of coupling of the vibrations. Its significance for the melting process has been first stressed by O. K. Rice [36].

Another picture of melting has been given by Longuet-Higgins and Widom [37]. These authors took the gain in disorder from the phase change of hard spheres (calculated by Monte Carlo or molecular dynamics, see chapter 3), and added an attractive potential as in the hard sphere

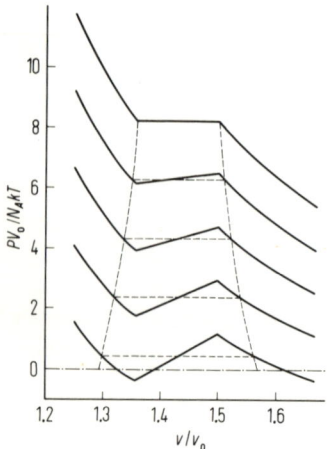

Fig. 6.9. The isotherms according to the model of Longuet-Higgins and Widom, for the values 0, 4, 8, 12, and 16 of the parameter $a/V_0 N_A k T$. The heterogeneous region is constructed on the basis of an equal area criterion.

model (section 2.3), namely $U = -a/V$. There are two parameters in this model, the hard sphere diameter, which determines the close packed volume V_0, and the van der Waals parameter a. But in order to bring $PV_0/N_A kT$ vs. V/V_0 down to effectively zero at the triple point, $a/V_0 N_A kT$ must have a value of 16.9 [cf. eq. (2.2) and fig. 6.9]. This enables the calculation of the dimensionless quantities given in table 6.7 in comparison to several liquids*). The agreement cannot be complete, as real liquids do not quite conform to the underlying assumption that $U = -a/V$ (cf. table 6.4).

Table 6.7. Predictions of the model for melting of Longuet-Higgins and Widom in comparison to experimental values of several substances (configurational energy, and PV refer to the liquid at the triple point).

	$\Delta V_m/V_s$	$\Delta S_m/R$	U'/RT_t	$\ln(PV/RT_t)$
model	0.214	2.29	−10.1	−5.7
Ar	0.143	1.71	−8.5	−5.9
SF_6	0.357	2.71	−8.2	−4.65
Hg	0.036	1.18	−30.7	−26.7
KCl	0.173	1.53[a]	−20.1[b]	−15.1

a) for half a mole KCl or one mole of ions.
b) for the assumption of monomeric units of KCl in the vapour.

In the hard sphere fluid, the enthalpy change at melting is given by

$$\Delta H = P\Delta V, \tag{6.23}$$

and the melting curve has the form [cf. eq. (6.6)]

$$d\ln P/d\ln T = 1. \tag{6.24}$$

In a real liquid, the high density needed for crystallization is upheld at low temperatures by the internal pressure. Up to this pressure the melting curve is very steep and is mainly determined by the attractive interaction. At much higher pressures, where the external pressure alone is capable of maintaining the high density needed for crystallization, the melting curve bends over in a $\ln P/\ln T$ plot (cf. fig. 1.4). According to Rowlinson [2, 39], the melting pressure P_M is then the melting pressure of a hard sphere fluid P_M^0 corrected for the finite steepness of the repulsive potential:

$$P_M = P_M^0/g^3 \tag{6.25}$$

where g is a characteristic length decreasing with temperature [cf. eq. (6.4)]. If the repulsive potential goes with R^{-n}, the high pressure limit should be given by

$$\frac{d\ln P_M}{d\ln T} = 1 + \frac{3}{n}, \tag{6.26}$$

---

* The values given here are recalculated on the basis of the hard sphere isotherm derived by Hoover and Ree [38].

which is 5/4 for the 12/6-potential. In reality the limit appears to be somewhat higher [Ebert, ref. 13].

If the melting curve would end on a critical point, volume change and entropy change at melting would become zero simultaneously. With $d \ln P_M/d \ln T$ approaching a constant, say C, ΔS becomes $CP\Delta V/T$. So the problem reduces to the question if ΔV can become zero at high pressures. Recent Monte Carlo calculations on soft spheres [40] (particles interacting with an inverse power repulsive potential) showed that ΔV at the melting transition of soft spheres is much lower than in the case of hard spheres. Apparently the arrangement in the liquid is much influenced by the possibility that some particles can be squeezed. It is interesting to note that ΔV of argon is still more reduced at higher pressures than in the case of soft spheres, as comparison of experimental values [41] with a Monte-Carlo calculation on the basis of the 12/6 potential [42] shows (fig. 6.9). Nevertheless, it seems that no critical point can be expected.

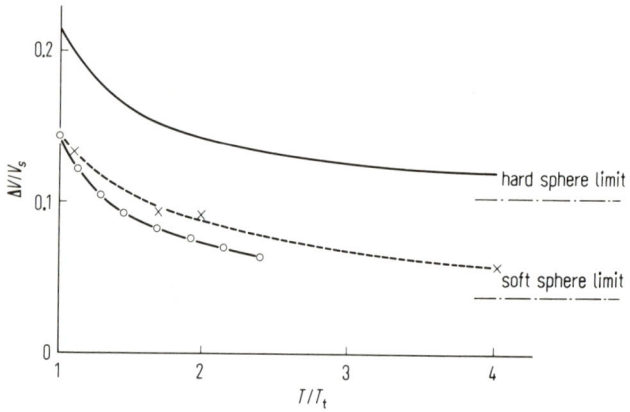

Fig. 6.10. The relative volume change along the melting line. The full curve is for the model of Longuet-Higgins and Widom, the dashed curve represents Monte Carlo calculations for a 12/6 potential, circles are experimental results on argon.

6.5. The Critical Point [43]

At the critical point the coexistent liquid and gas phases become identical and the visible meniscus between them vanishes. Thermodynamically, the four points A, B, C, and D of fig. 1.6 become identical, i.e. the critical isotherm has a horizontal tangent with inflection in the P-v diagram. This makes necessary the following conditions for the derivatives of the Helmholtz free energy (or, at least, for their limiting values when going with v, T to v_c, T_c):

$$\frac{\partial f}{\partial v} < 0, \quad \frac{\partial^2 f}{\partial v^2} = 0, \quad \frac{\partial^3 f}{\partial v^3} = 0, \tag{6.27}$$

which is equivalent to

$$P > 0, \quad \frac{\partial P}{\partial v} = 0, \quad \frac{\partial^2 P}{\partial v^2} = 0. \tag{6.28}$$

The first non-vanishing derivative of f, if existing, must be of even order and positive, since mechanical stability requires that pressure must fall with increasing volume in the neighbourhood of the critical point. But no further general statement can be made. Below the critical point,

6.5. The Critical Point

the free energy of real systems is a continuous function of v and T, but it is non-analytic[*] across the coexistence curve. So it is perhaps not too surprising, that the free energy is also non-analytic at the critical point, as we will show below. Nevertheless, it should be pointed out that the singularity at the critical point (which has remarkable similarities to the behaviour at the critical point of mixtures and at the transition point to ordered states [44]) is in contradiction to the assumption of a continuous, though at lower temperatures partly instable, sequence of homogeneous states between liquid and gas, first formulated by van der Waals [45].

Before showing the non-analytic behaviour of the free energy at the critical point, we will review the general behaviour of the thermodynamic quantities there. As the critical isotherm (P vs. v) has a horizontal tangent at the critical point [eq. (6.28)], one may think of a series of states of different volume and energy at virtually the same pressure and the same temperature. Therefore, the corresponding differentials $(\partial v/\partial P)_T$, $(\partial v/\partial T)_P$, and $(\partial u/\partial T)_P$ diverge, and so do β_T, α, and c_P. Next we consider the thermal pressure coefficient $\gamma_v = \alpha/\beta_T$. The general behaviour of this coefficient is shown in fig. 6.11. At the critical point, γ_v is identical with the limiting slope of the

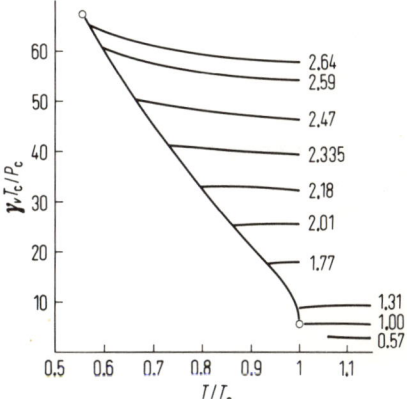

Fig. 6.11. The reduced thermal pressure coefficient of argon. The curves beyond the saturation line are estimations based on the results of ref. 16 and 59. Parameters give the values of ρ/ρ_c.

vapour pressure curve γ_σ. This follows from eq. (6.13) and (1.18), combining to $\alpha_\sigma = \beta_T(\gamma_v - \gamma_\sigma)$, together with the experimental evidence that α_σ goes slower to infinity than α_P or β_T at the critical point (cf. table 6.3). The well-behaving of γ_v is one important reason for the emphasis placed by Rowlinson [2] on this coefficient. The derivative $\partial \gamma_v/\partial T$ is related to the volume derivative of c_v [see eq. (6.18)]. The fact that $\partial \gamma_v/\partial T$ changes sign at or near the critical isochore (positive for smaller and negative for larger volumes) means a maximum in c_v. In CO_2, this maximum is still noticeable 100 K above the critical point [46]. Calorimetric work shows that this maximum increases probably to a logarithmic or slightly stronger infinity at the critical point (cf. fig. 6.12) [47].

The way in which c_v goes to the critical point is intimately connected to the adiabatic compressibility or the sound velocity [cf. eq. (1.21) and (1.23)]:

$$\frac{1}{\beta_s} = \frac{1}{\beta_T}\left(1 + \frac{Tv\alpha^2}{c_v\beta_T}\right) = \frac{1}{\beta_T} + \frac{Tv}{c_v}\gamma_v^2 . \tag{6.30}$$

---

[*] An "analytic" function can be expanded into a Taylor series at a given point.

6. Thermodynamic Properties of Liquids

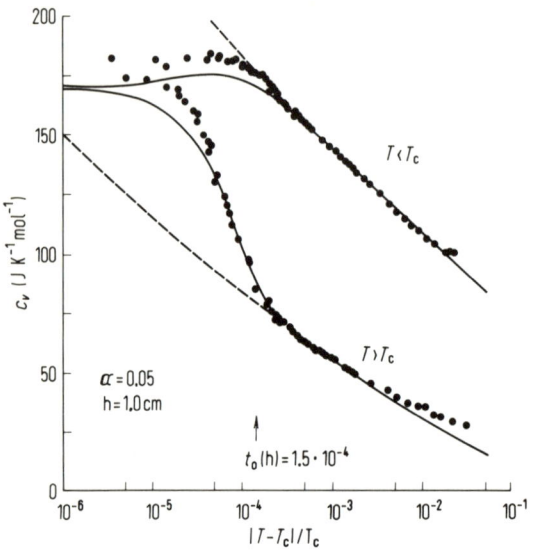

Fig. 6.12. Heat capacity C_v of Xe [47] versus the logarithm of the reduced deviation from critical temperature. The dashed lines represent the curves corrected for absence of gravity, calculated for a sample height h = 1.0 cm. Outside the temperature interval $t_0(h)$ gravity effects are negligible. The plot leads to a critical index [cf. eq. (6.37)] $\alpha = 0.05$. After ref. 49.

At the critical point $1/\beta_T$ vanishes, and the behaviour of $1/\beta_s$ or c^2 [cf. eq. (1.17)] depends on $1/c_v$, as all other quantities remain certainly finite. Now the observed ultrasonic velocities have a minimum but non-zero value at the critical point (fig. 6.13). At some distance from the critical point the decrease of the sound velocity corresponds to the increase in c_v [48]. This "rounding"

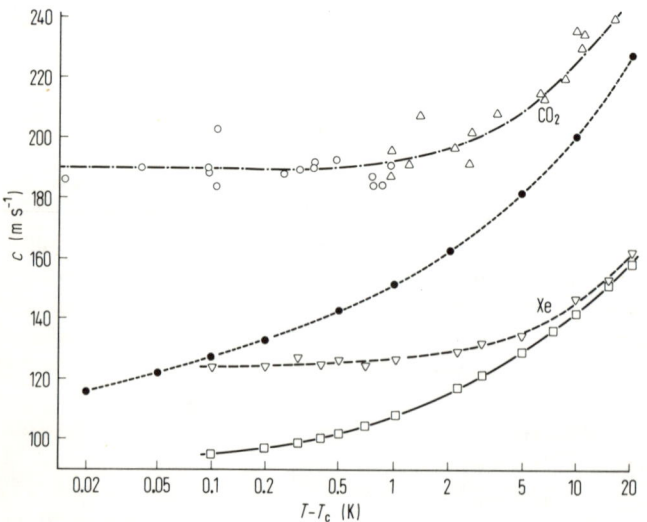

Fig. 6.13. The sound velocities of CO_2 (hypersound by Brillouin scattering —·—·— [52], ultrasound ····· [48]) and Xe (hypersound ---- [52], ultrasound ——— [48]) versus the logarithm of $T - T_c$.

of the transition occurs in all practical measurements [49, 50]. Essentially two explanations have been advanced for the apparant non-zero value at the critical point: a) Relaxation processes, which cause a higher value of c than the equilibrium value at the applied frequency [51]. The difference between the hypersonic sound velocity (determined by Brillouin scattering [52]) and the ultrasonic sound velocity of Xe and CO_2 given in fig. 6.13 may be explained in this way. However, no clear determination of the relaxation frequency has been done yet which would allow an extrapolation of the sound velocity to zero frequency. b) Gravitational effects producing density gradients. By consideration of this effect the non-zero value of the ultrasonic sound velocity in ^4He could be explained to a good deal [49].

6.5.1. Classical Theory, Based on Analycity of Free Energy Around the Critical Point

If the free energy is an analytic function around the critical point, it may be expanded by a double Taylor series in terms of $\Delta T = T - T_c$ and $\Delta v = v - v_c$:

$$f = f_c + \sum_{n=1}^{\infty} \sum_{m=0}^{n} \frac{f_{m,n-m}}{m!(n-m)!} \Delta v^m \Delta T^{n-m} \tag{6.31}$$

where the derivative $f_{m,n-m} = \partial^n f / \partial v^m \partial T^{n-m}$, taken at the critical point. A typical model leading to such an expandable free energy is the van der Waals or hard sphere model introduced in section 2.3. If the pressure of the hard sphere assembly is approximated by

$$P_{h.s} \approx \frac{NkT}{v-b} \quad \text{with} \quad b = \frac{2\pi}{3} N\sigma^3 \tag{6.32}$$

then the old van der Waals equation results [53]. A much better equation for $P_{h.s}$ will be given in chapter 7. With eq. (7.67) the van der Waals model furnishes the following relations at the critical point: $V_c/b = 1.906$ (this corresponds to $V_c/V_0 = 5.64$ compared to a value of 4.46 following for argon from table 6.1), $a/V_0 N_A k T_c = 7.88$ (which gives combined with the model for melting of Longuet-Higgins and Widom $T_m/T_c = 0.467$), and $P_c V_c/N_A k T_c = 0.357$.

On account of the assumption of the uniform background potential the van der Waals model is also called mean field or molecular field approximation. Besides the possibility of expanding f in the form of eq. (6.31) it is characterized by the condition $f_{4,0} \neq 0$. The general theory based on eq. (6.31) with $f_{2n,0} \neq 0$ ($n > 2$) is due to Landau [54]. From eq. (6.31) and (6.27) the leading terms in expansions of derivatives of f are easy to find, e.g.:

$$P = -f_{1,0} - f_{1,1}\Delta T - f_{2,1}\Delta v \Delta T - \frac{1}{6}f_{4,0}\Delta v^3 - \cdots \tag{6.33}$$

$$c_v = -T(f_{0,2} + f_{0,3}\Delta T + \cdots) = -T_c f_{0,2} - (f_{0,2} + T_c f_{0,3})\Delta T - \cdots. \tag{6.34}$$

It has become customary to define critical indices [44, 55] for the way in which thermodynamic properties approach the critical point. From eq. (6.33) it follows that in

$$|\Delta P| \sim |\Delta v|^\delta \quad (T = T_c) \tag{6.35}$$

the index δ is equal to $\delta = 3$. As $\partial P/\partial v = -f_{2,1}\Delta T - \cdots$, in

$$\beta_T \sim |\Delta T|^{-\gamma} \tag{6.36}$$

the index γ becomes $\gamma = 1$ along the critical isochore (for $T > T_c$) or along the coexistence curve *) (for $T < T_c$, index γ'). As γ_v has no singularity at the critical point, the thermal expansion coefficient α_P has the same singularity as β_T. As c_v has no singularity when approaching the critical point either along the critical isochore from above ($T > T_c$, index α) or along the coexistence curve from below ($T < T_c$, index α'), c_P shows the same singularity as β_T. The critical index for c_v is accordingly given by

$$c_v \sim |\Delta T|^{-\alpha} \tag{6.37}$$

with $\alpha = \alpha' = 0$. It is more complicated to state the behaviour of c_v when the critical point is approached from the two-phase region along the critical isochore ($T < T_c$), as liquid is transformed into vapour along this path. Here it should only be mentioned that in the van der Waals model c_v is larger along this path than above the critical point, but not infinite. So c_v has a discontinuity, but no singularity at the critical point. Again

$$c_v \sim |\Delta T|^{-\alpha''} \tag{6.38}$$

with $\alpha'' = 0$ where α'' is for an approach along the critical isochore from below ($T < T_c$)**). The shape of the coexistence curve is found by applying the expansion (6.33) on both the liquid and the gas branch:

$$\Delta P = -f_{1,1}\Delta T - f_{2,1}\Delta T \Delta v_l - \frac{1}{6}f_{4,0}\Delta v_l^3$$

$$\Delta P = -f_{1,1}\Delta T - f_{2,1}\Delta T \Delta v_g - \frac{1}{6}f_{4,0}\Delta v_g^3 .$$

Subtraction and insertion of the law of the rectilinear diameter $\Delta v_g = -\Delta v_l$ (which can be derived independently from the equality of chemical potentials [43]) yields

$$f_{2,1}\Delta T + \frac{1}{6}f_{4,0}\Delta v^2 = 0, \tag{6.39}$$

where Δv can be either Δv_g or Δv_l. Therefore, we have for the coexistence curve, which is described by

$$|\Delta v| \sim |\Delta T|^\beta , \tag{6.40}$$

the relation $\beta = 1/2$.
The critical indices according to the van der Waals model or to the more general Landau theory are summarized in table 6.8, together with the experimental evidence.
The experimental basis [50] of determination of the critical indices are classical P-v-T measurements [58, 59], observations of the index of refraction as function of sample height in the neighbourhood of the critical point [60], calorimetric measurements [47], ultrasonic experiments [48, 49], NMR-techniques [61], and scattering of radiation. Scattering of laser beams [62, 63] gives information on the compressibility (by total scattered intensity), on the ratio $(C_P - C_v)/C_v$ (by the ratio of intensity of Rayleigh-line to Brillouin-lines), on the hypersonic sound velocity (by the distance between Rayleigh and Brillouin-line), on the thermal diffusivity D_T (by the

---

* assuming $|\Delta v| \sim |\Delta T|^\beta$ with $\beta > 0$ for the coexistence curve [cf. eq. (6.40)].
** The index zero describes both a simple discontinuity and a logarithmic singularity. This is the reason for the indications in brackets in table 6.8. For the logarithmic or inverse power singularity eq. (6.37) and (6.38) are usually written [56, 57] $c_v = A(|\Delta T/T_c|^{-\alpha} - 1)/\alpha + B$.

half width of the Rayleigh line, for the formulation see section 9.4.3), and on the sound absorption (by the half width of Brillouin lines). Small angle scattering of X-rays as well as Rayleigh scattering is a possibility to determine the correlation length of critical fluctuations [63, 64, 65] [section 6.5.3, esp. eq. (6.46) to (6.49)]. Critical neutron scattering has only been applied to critical points of liquid mixtures [66].

Table 6.8. Critical indices and model predictions. The values for Xe [81] are fits to experimental data forced to fulfill the scaled equation of state.

Index	Definition	Way of approach to critical point	Classical (general)
α''	$c_v \sim \|\Delta T\|^{-\alpha''}$	$\rho = \rho_c \quad \Delta T \leq 0$	$(n-2)/(n-1)$
β	$\|\Delta \rho\| \sim \|\Delta T\|^{\beta}$	coex. curve	$1/(2n-2)$
γ'	$\beta_T \sim \|\Delta T\|^{-\gamma'}$	coex. curve	1
ν'	$\xi \sim \|\Delta T\|^{-\nu'}$	coex. curve	0.5
δ	$\|\Delta P\| \sim \|\Delta \rho\|^{\delta}$	$\Delta T = 0$	$2n - 1$
η	$h(R) \sim R^{-(1+\eta)}$	$\Delta T = 0$	0
α	$c_v \sim \|\Delta T\|^{-\alpha}$	$\rho = \rho_c \quad \Delta T \geq 0$	0
γ	$\beta_T \sim \|\Delta T\|^{-\gamma}$	$\rho = \rho_c \quad \Delta T \geq 0$	1
ν	$\xi \sim \|\Delta T\|^{-\nu}$	$\rho = \rho_c \quad \Delta T \geq 0$	0.5

Index	Classical v. d. Waals	Ising 2d	Ising 3d	Exp. values (Xe, fitted)	Exp. values for various fluids [58]
α''	0 (discont.)	0 (log.)	0.063	0.04	≤ 0.1
β	0.5	0.125	0.313	0.35	0.35
γ'	1	1.75	1.31	1.26	≥ 1.2
ν'	0.5	1	0.67	0.65	
δ	3	15	5.2	4.6	4.1
η	0	0.25	0.06	0.06	≈ 0
α	0 (discont.)	0 (log.)	0.12	0.04	≤ 0.1
γ	1	1.75	1.25	1.26	≥ 1.1
ν	0.5	1	0.65	0.65	≈ 0.5

As table 6.8 shows no agreement between experimentally determined critical indices and those derived from classical theory, we must conclude that the main assumption is not fulfilled and that the free energy is non-analytic at the critical point. The details of singularity are still unknown.

6.5.2. The Lattice Gas (Ising Model)

Only for two models one has succeeded to calculate an exact partition function: the one-dimensional van der Waals gas [67] and the plane lattice gas. The latter is equivalent to the two-dimensional Ising model of ferromagnetism [68, 55]. A short characterization has been given in section 2.2. From the exact solution of the two-dimensional problem first given by Onsager [34], we only note that the free energy around the critical point is given by

$$f = f_c - a(\Delta T) - b(\Delta T)^2 \ln |\Delta T| + \cdots \tag{6.41}$$

which includes the function $\ln |\Delta T|$ demonstrating its non-analyticity at the critical point. From eq. (6.41) a symmetric logarithmic singularity follows for c_v

$$c_v = -T \frac{\partial^2 f}{\partial T^2} = 3bT + 2bT \ln |\Delta T| + \cdots \tag{6.42}$$

corresponding to the critical indices $\alpha = \alpha' = 0$ (logarithmic) in table 6.8.

Naturally the two-dimensional (2d) model is too artificial to give agreement with experiments on three-dimensional fluids. Indeed, table 6.3 shows large discrepancies. Nevertheless this model is important because its exact solution may serve as test case for approximate methods like power series expansions which have been introduced to handle the three-dimensional (3d) Ising model [55, 70]. Estimated critical coefficients from the 3d-lattice gas are included in table 6.8. The singularity for c_v is a bit stronger than a logarithmic singularity and non-symmetric. The good agreement with experiment confirms that the 3d-Ising model contains the essential characteristics of the critical point. Since only interactions between nearest neighbours are considered in this model one may state that the critical phenomena are not so much influenced by long range interactions (uniform background potential in the van der Waals model), but rather by the long-ranging correlations of many particles built up by the attractive part of the pair potential, which may be of relatively short range. This is confirmed by the fact that critical phenomena in systems with very different interactions (ferromagnetic and antiferromagnetic systems, liquid mixtures) exhibit the same critical indices as in the liquid-gas case [44, 50, 55].

6.5.3. Correlations in the Critical Region

The importance of the correlations in the critical region is obvious from the phenomenon of a critical opalescence. Theoretically, it is contained in the growing influence of the long range tail of the total correlation function $h(R)$, which makes ultimately the integral $\int h(R) d\vec{r}$ divergent (cf. section 4.2.1). Consequently, $h(R)$ must decay more slowly than $1/R^3$ near the critical point. A more detailed statement about the decay of $h(R)$ can be made by means of the direct correlation function $c(R)$ and the Ornstein-Zernike eq. (4.30) [or eq. (4.32)].

At the critical point, there is $\int c(R) d\vec{r} = \tilde{c}(0) = 1/\rho_c$ which is finite [eq. (4.35)]. This suggests an expansion of $\tilde{c}(Q)$ for small Q. It can be shown that such an expansion which assumes an analytic behaviour of $\tilde{c}(Q)$ around $Q = 0$ is somewhat equivalent to an expansion of the free energy around the critical point [56, 71]. Such an expansion of $\rho \tilde{c}(\vec{Q}) = \rho \int c(\vec{r}) e^{-i\vec{Q}\cdot\vec{r}} d\vec{r}$ yields, after averaging over all directions,

$$\rho \tilde{c}(Q) = \rho \tilde{c}(0) - Q^2 \langle R^2 \rangle + \cdots \quad \text{with} \quad \langle R^2 \rangle = \frac{1}{6} \rho \int R^2 c(R) 4\pi R^2 dR, \tag{6.43}$$

where the relation $\langle \cos^2 \vartheta \rangle = 1/3$ has been used. Inserting eq. (6.43) into the Ornstein-Zernike eq. (4.32), the result is

$$\tilde{h}(Q) = \tilde{c}(Q) + \tilde{h}(Q) \rho \tilde{c}(0) - \langle R^2 \rangle Q^2, \tag{6.44}$$

or, taking transforms

$$h(R)[1 - \rho\tilde{c}(0)] = c(R) - \langle R^2 \rangle \frac{1}{(2\pi)^3} \int Q^2 \tilde{h}(Q) e^{i\vec{Q}\cdot\vec{r}} \mathrm{d}\vec{Q} = c(R) + \langle R^2 \rangle \nabla^2 h(R). \tag{6.45}$$

For large R, $c(R)$ decays much faster than $h(R)$ and can be ignored, so that eq. (6.45) yields the differential equation

$$\frac{1}{\xi^2} h(R) = \nabla^2 h(R) \quad \text{with} \quad \xi^2 = kT\langle R^2 \rangle \frac{\partial \rho}{\partial P}. \tag{6.46}$$

Here eq. (4.34) has been used. The meaning of ξ is that of a characteristic correlation length. It behaves at the critical point as $\beta_T^{1/2}$, if the expansion (6.43) is possible ($\langle R^2 \rangle$ finite). Integration of eq. (6.46) yields

$$h(R) = \frac{A e^{-R/\xi}}{R}, \tag{6.47}$$

so that $h(R)$ goes with R^{-1} at the critical point. Eq. (6.47) has been first obtained by Ornstein and Zernike [72] and has been rederived on the basis of different arguments [54, 56, 73]. There is one important consequence which can be checked experimentally. Considering the connection between scattering function $i(Q)$, or structure factor $S(Q) = 1 + i(Q)$, to $\tilde{h}(Q)$ [eq. (4.55)], the Ornstein-Zernike eq. (4.33) combined with the expansion (6.43) gives the relation

$$1 + \rho\tilde{h}(Q) = S(Q) = \frac{1}{1 - \rho\tilde{c}(0) + \langle R^2 \rangle Q^2}. \tag{6.48}$$

The inverse of $S(Q)$ should be proportional to Q^2. The intercept in such a plot is given by $1 - \rho\tilde{c}(0)$, which is proportional to $1/\beta_T$. Such a plot is shown in fig. 6.14.

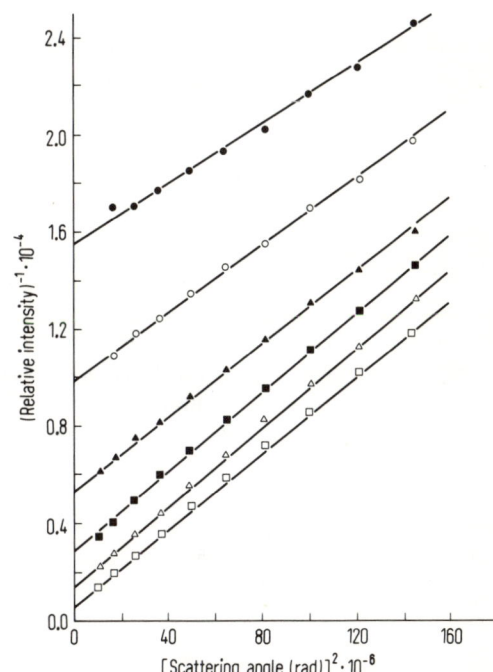

Fig. 6.14. The reciprocal intensity of light scattered by argon near T_c as function of Q^2. The lines are for the following deviations from the temperature of maximal scattering ($\approx T_c$): 4° (●), 2° (○), 1° (▲), 0.4° (■), 0.2° (△), 0° (□).

For real fluids, the expansion (6.43) is open to question. It is not possible in the two-dimensional Ising model. On the other hand, fig. 6.14 has shown that one cannot be far away from eq. (6.47) or (6.48). Accordingly, the following generalization of eq. (6.47) has been suggested [56]:

$$h(R) = \frac{A e^{-R/\xi}}{R^{1+\eta}}, \tag{6.49}$$

where η is treated like a critical index and should be 0.06 according to the 3d-Ising model. If an expansion of $\tilde{c}(Q)$ around $\tilde{c}(0)$ is not possible, eq. (6.49) may be taken as the defining equation for ξ. Proportionality would then exist between $1/S(Q)$ and $Q^{2-\eta}$, i.e. the lines of fig. 6.14 should be curved downward near the critical point for positive values of η. The experimental evidence is not yet convincing for gas-liquid critical points but quite satisfactory for liquid-liquid critical points [58]. The constant η could be evaluated from a relation between $c(R)$ and $h(R)$ for large R. The hypernetted chain theory, dealt with in the next chapter, gives [taking the logarithm of eq. (7.48) and expanding $\ln(1+h)$]

$$c(R) = -\varepsilon(R)/kT + h^2(R)/2 + \text{terms of higher order}. \tag{6.50}$$

A more detailed analysis by Stell [74] suggests rather

$$c(R) = -\varepsilon(R)/kT + Bh^\delta(R) + \cdots, \tag{6.51}$$

where δ is the index for the critical isotherm. If the term $\varepsilon(R)/kT$ can be neglected for large R, it can be shown that eq. (6.51) leads to

$$\eta = \frac{5-\delta}{\delta+1} \quad \text{if} \quad \delta < 5$$
$$\eta = 0 \quad \text{if} \quad \delta \geq 5. \tag{6.52}$$

6.5.4. Relations Between the Critical Indices. Scaling Laws

From purely thermodynamic arguments inequalities can be derived which connect the critical indices. Their derivation is not based on the assumption that $f(v,T)$ is an analytic function at and around the critical point, but uses only continuity properties and criteria of stability [75, 43]. We only list the results:

$$\begin{aligned} &\alpha' + \beta(\delta+1) \geq 2 \\ &\alpha' + 2\beta + \gamma' \geq 2 \\ &\gamma' - \beta(\delta-1) \geq 0 \,. \end{aligned} \tag{6.53}$$

It is implied that the indices characteristic for the coexistence curve are the same for both branches $\rho < \rho_c$ and $\rho > \rho_c$, which seems to be in accord with experiments. Classical theory and the plane lattice gas demands that the equality signs are used in eq. (6.53) (cf. table 6.8). Also experiments make it probable that eq. (6.53) hold as equalities. They can be derived from an "ad hoc" assumption on the behaviour of the chemical potential [76, 57], which leads to further interesting relations. The "ad hoc" assumption, which will be described below, has been made plausible by Kadanoff [77, 44] on the basis of an extension of the Ising model. The following relations have been formulated in addition to eq. (6.53) read with the equality sign:

$$\begin{aligned} &\gamma = \gamma' = v(2-\eta) \\ &\alpha = \alpha' \\ &v = v' = (2-\alpha)/3 \,. \end{aligned} \tag{6.54}$$

All these equalities (seven independent equations with nine variables) are called scaling laws. Two of the indices can be assigned independently. Combining the proper equations the relation (6.52) is recovered, which has been derived by quite independent arguments. It is seen from table 6.8 that the "best" approximations for the 3d-Ising model are neither in full agreement with the scaling laws, nor with experiment. The remaining discrepancies are believed to be caused by oversimplifications of the Ising model, which cannot account for the complex nature of real surroundings of a given molecule [78].

Starting point of the above mentioned assumption on the behaviour of the chemical potential [76] is a consideration of $\mu(\rho, T)$ as a function of T which is given in fig. 6.15. Denoting the tem-

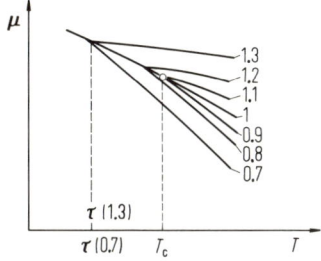

Fig. 6.15. The schematical behaviour of the chemical potential $\mu(T)$ for the various parameters ρ/ρ_c, which are indicated in the figure.

perature, where the isochores cross the coexistence curve, by $\tau(\rho)$, the analytic formulation corresponding to fig. 6.15 is

$$\mu(\rho, T) - \mu(\rho_c, T) = \Delta\mu(T) = \Delta\rho[T - \tau(\rho)]\phi, \tag{6.55}$$

where $\Delta\rho$ means $\rho - \rho_c$ and ϕ is in general a function of T and ρ. In classical theory, ϕ becomes a constant in the immediate neighbourhood of the critical point. If $\Delta T = T - T_c$ and $T_c - \tau(\rho) = B|\Delta\rho|^{1/\beta}$ are used as variables instead of T and ρ, then the assumption leading to the scaling laws is that $\phi(\Delta T, |\Delta\rho|^{1/\beta})$ is a homogeneous function of degree $\gamma - 1$, where γ turns out to be identical with the critical index for the compressibility [eq. (6.36)]. Eq. (6.55) can then be reformulated as follows:

$$\Delta\mu(T) = \Delta\rho(\Delta T + B|\Delta\rho|^{1/\beta})|\Delta\rho|^{(\gamma-1)/\beta}\phi(\Delta T/|\Delta\rho|^{1/\beta}, 1) = \Delta\rho|\Delta\rho|^{\gamma/\beta}\Psi(\Delta T/|\Delta\rho|^{1/\beta}). \tag{6.56}$$

Now $\Delta\mu(T)$ has been expressed as a function Ψ of the single variable $\Delta T/|\Delta\rho|^{1/\beta}$; Ψ is an analytic function in the region of a homogeneous fluid. For $T = T_c$, $\Delta T = 0$, it follows immediately that Ψ has a fixed value and $|\Delta\mu| = |\Delta\rho|^{1+\gamma/\beta}$, whence the last equality of (6.53) is recovered. Writing now $\delta = 1 + \gamma/\beta$, and dividing both sides of eq. (6.56) by $|\Delta T|^{\beta\delta}$, we have

$$\frac{|\Delta\mu(T)|}{|\Delta T|^{\beta\delta}} = \left(\frac{|\Delta\rho|^{1/\beta}}{|\Delta T|}\right)^{\beta\delta}\Psi\left(\frac{\Delta T}{|\Delta\rho|^{1/\beta}}\right). \tag{6.57}$$

Therefore, $|\Delta\mu(T)|/|\Delta T|^{\beta\delta}$ is an universal function of $\Delta T/|\Delta\rho|^{1/\beta}$ in the neighbourhood of the critical point [79]. It has two branches for $\Delta T > 0$ and $\Delta T < 0$, resp., which become asymptotically identical for $\Delta T \to 0$. Such a scaled plot is shown in fig. 6.16[*]. It is believed that the

---

[*] For other tests of "scaled" equations of state cf. Vicentini-Missoni et al. [59] and ref. 80.

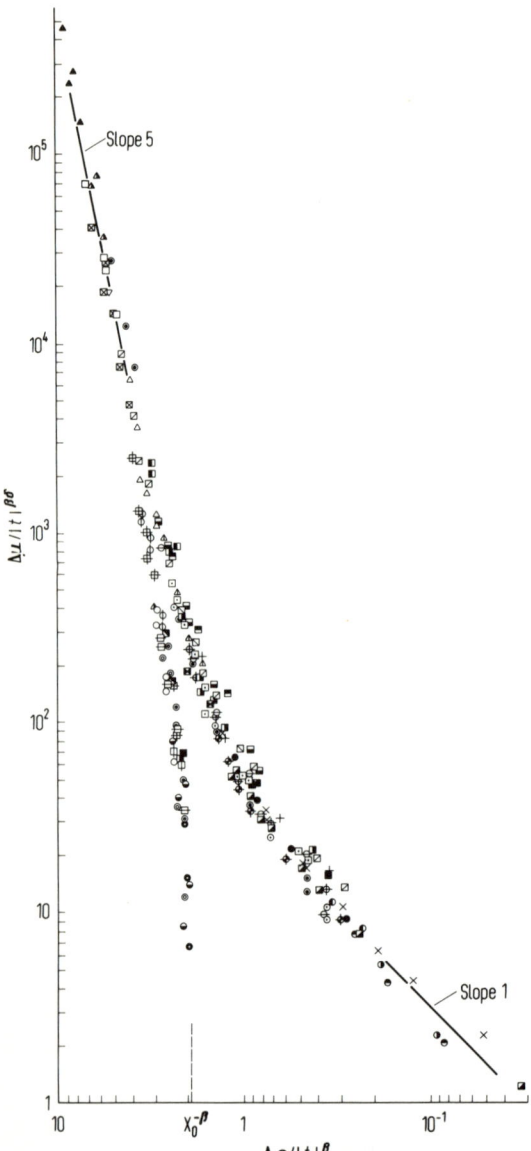

Fig. 6.16. A scaled plot of $|\Delta\mu|/|t|^{\beta\delta}$ versus $|\Delta\rho|/|t|^{\beta}$ in the critical region of N_2O, $CClF_3$, CO_2, SF_6, and Xe. The quantity t is defined as $\Delta T/T_c$. For β and δ the values $\beta = 0.35$ and $\delta = 5$ have been used. After ref. 79.

scaling laws describe the data of all non-electrolytic substances within the experimental precision and within a range of 30% in $\Delta\rho$ and 1–3% in ΔT. Another consequence of eq. (6.55) is the $(\rho - \rho_c)^{3/2}$ behaviour of the locus of the inversion points of the P-ρ-isotherms above the critical (fig. 6.17). Though this dependence has not yet been proved in detail experimentally, it seems that the locus of the inversion points is not located on the continuation of the rectilinear diameter [76].

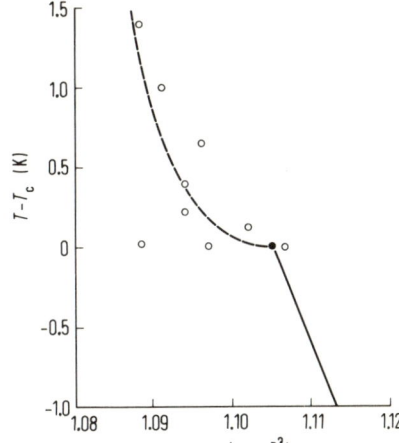

Fig. 6.17. The locus of inversion points of the $P\text{-}v$-isotherms above T_c (circles are experimental values for Xe) in comparison to the scaling model of Widom (dashed curve). Full line is the rectilinear diameter below T_c. After ref. 76.

6.6. References

1. J. de Boer and A. Michels, Physica 5, 945 (1938); K. S. Pitzer, J. Chem. Phys. 7, 583 (1939); E. A. Guggenheim, J. Chem. Phys. 13, 253 (1945).
2. Rowlinson, gen. ref. 11, ch. 8 and ch. 2.
3. A. J. Leadbetter and H. E. Thomas, Trans. Faraday Soc. 61, 10 (1965), mainly following G. Boato and G. Casanova, Physica 27, 571 (1961); cf. also J. C. Rossi and F. Danon, Disc. Faraday Soc. 40, 97 (1965), and (for Ar and Kr) D. D. Fitts, Ann. Rev. Physic. Chem. 17, 59 (1966).
4. W. D. McCain, Jr. and W. T. Ziegler, J. Chem. Eng. Data 12, 199 (1967).
5. M. J. Terry, J. T. Lynch, M. Bunclark, K. R. Mansell, and L. A. K. Staveley, J. Chem. Thermodyn. 1969, 413.
6. P. Flubacher, A. J. Leadbetter, and J. A. Morrison, Proc. Phys. Soc. (London) 78, 1449 (1961).
7. R. H. Beaumont, H. Chihara, and J. A. Morrison, Proc. Phys. Soc. (London) 78, 1462 (1961).
8. A. Michels and C. Prins, Physica 28, 101 (1962).
9. K. A. Kobe and R. E. Lynn, Chem. Rev. 52, 117 (1953).
10. A. P. Kudchadker, G. H. Alani, and B. J. Zwolinski, Chem. Rev. 68, 659 (1968).
11. F. Kohler, Monatsh. Chem. 103 (1972), in press.
12. F. Hensel and E. U. Franck, Rev. Mod. Phys. 40, 697 (1968).
13. L. Ebert, Österr. Chem. Zeitung 55, 1 (1954); R. Plank and L. Riedel, Ing. Archiv 16, 255 (1948).
14. A. J. Gottschal and A. E. Korvezee, Rec. Trav. Chim. Pays-Bas 72, 465 (1953).
15. F. Kohler, Monatsh. Chem. 101, 1493 (1970).
16. J. Thoen, E. Vangeel, and W. van Dael, Physica 45, 339 (1969).
17. A. Michels, J. M. Levelt, and G. J. Wolkers, Physica 24, 769 (1958).
18. A. Michels, J. M. Levelt, and W. de Graaff, Physica 24, 659 (1958).
19. O. G. Peterson, D. N. Batchelder, and R. O. Simmons, Phys. Rev. 150, 703 (1966).
20. S. E. Wood, O. Sandus, and S. Weissman, J. Am. Chem. Soc. 79, 1777 (1957).
21. H. Bloom and J. O'M. Bockris in J. O'M. Bockris, Modern Aspects of Electrochemistry, Vol. II, Butterworth, London 1959.
22. Handbook of Chemistry and Physics, The Chemical Rubber Co., Cleveland 1964.
23. R. H. Busey and W. F. Giauque, J. Am. Chem. Soc. 75, 806 (1953).
24. E. Wilhelm, R. Schano, G. Becker, G. H. Findenegg, and F. Kohler, Trans. Faraday Soc. 65, 1443 (1969).
25. U. Bianchi, G. Agabio, and A. Turturro, J. Phys. Chem. 69, 4392 (1965).
26. A. R. Ubbelohde, Melting and Crystal Structure, Clarendon Press, Oxford 1965.

27. H. P. Clegg, J. S. Rowlinson, and J. R. Sutton, Trans. Faraday Soc. *51*, 1327 (1955); J. Otto and W. Thomas, Z. Phys. Chem. NF *23*, 84 (1960); L. A. Makarevich, E. S. Sokolova, and G. A. Sorina, Zh. Fiz. Khim. *42* (1), 22 (1968); Gmelins Handbuch der anorgan. Chemie 8. Auflage, Nr. 9, B 3, Verlag Chemie, Weinheim 1963, S 1714 ff.; Selected Values of Chemical Thermodynamic Properties, Circular 500 Natl. Bur. Stand., Washington 1952.
28. K. Furukawa, Disc. Faraday Soc. *32*, 53 (1961).
29. A. Klemm in: M. Blander, Molten Salt Chemistry, Interscience, New York 1964.
30. C. W. F. T. Pistorius, J. Phys. Chem. Solids *26*, 1543 (1965), J. Chem. Phys. *43*, 1557 (1965) and *45*, 3513 (1966).
31. H. Schinke and F. Sauerwald, Z. anorg. und allgem. Chem. *287*, 313 (1956).
32. A. S. Dworkin and M. A. Bredig, J. Phys. Chem. *64*, 269 (1960).
33. F. Klein and H. Ruppersberg, Adv. Phys. *16*, 271 (1967), H. Ruppersberg, Z. Naturforsch. *24a*, 1034 (1969).
34. F. Kohler and F. Weissenböck, unpublished.
35. R. B. Leighton, Rev. Mod. Phys. *20*, 165 (1948).
36. O. K. Rice in: R. Smoluchowski, J. E. Mayer, and W. A. Weyl, Phase Transformations in Solids, Wiley, New York 1951.
37. H. C. Longuet-Higgins and B. Widom, Mol. Phys. *8*, 549 (1964).
38. W. G. Hoover and F. H. Ree, J. Chem. Phys. *49*, 3609 (1968).
39. J. S. Rowlinson, Mol. Phys. *8*, 107 (1964); Proceedings of the First International Conference on Calorimetry and Thermodynamics, Warsaw, Sept. 1969, Polish Scientific Publ.
40. W. G. Hoover, M. Ross, K. W. Johnson, D. Henderson, J. A. Barker, and B. C. Brown, J. Chem. Phys. *52*, 4931 (1970).
41. W. van Witzenburg and J. C. Stryland, Can. J. Phys. *46*, 811 (1968); R. K. Crawford and W. B. Daniels, Phys. Rev. Letters *21*, 367 (1968).
42. J. P. Hansen and L. Verlet, Phys. Rev. *184*, 151 (1969).
43. cf. J. S. Rowlinson, gen. ref. 11, ch. 3; Discussion Meeting of the Deutsche Bunsengesellschaft at Lindau, Sept. 1971, to be published in Ber. Bunsenges. *76* (1972).
44. L. P. Kadanoff, W. Götze, D. Hamblen, R. Hecht, E. A. S. Lewis, V. V. Palciauskas, M. Rayl, and J. Swift, Rev. Mod. Phys. *39*, 395 (1967).
45. J. D. van der Waals, Die Kontinuität des gasförmigen und flüssigen Zustands (2nd ed., vol. 1), Barth, Leipzig 1899.
46. A. Michels, A. Bijl, and C. Michels, Proc. Roy. Soc. (London) *A 160*, 376 (1937).
47. M. I. Bagatskii, A. V. Voronel, and V. G. Gusak, Sov. Phys.-JETP *16*, 517 (1963) (Ar); A. V. Voronel, Ya. R. Chaskin, V. A. Popov, and V. G. Simkin, Soc. Phys.-JETP *18*, 568 (1964) (CO_2); M. R. Moldover and W. A. Little in M. S. Green and J. V. Sengers, Critical Phenomena, proceedings of a conference at Washington, D. C., 1965, Natl. Bur. Stand. Miscell. Publ. 273 (1966) (He); E. Edwards, J. A. Lipa, and M. J. Buckingham, Phys. Rev. Letters *20*, 496 (1968) (Xe); J. A. Lipa, C. Edwards, and M. J. Buckingham, Phys. Rev. Letters *25*, 1086 (1970) (CO_2); M. R. Moldover, Phys. Rev. *182*, 342 (1969) (He).
48. R. C. Williamson and C. E. Chase, Phys. Rev. *176*, 285 (1968) (He); G. T. Feke, K. Fritsch, and E. F. Carome, Phys. Rev. Letters *23*, 1282 (1969) (CO_2); M. Barmatz, Phys. Rev. Letters *24*, 651 (1970) (He); H. Z. Cummings and H. L. Swinney, Phys. Rev. Letters, *25*, 1165 (1970) (Xe).
49. M. Barmatz and P. C. Hohenberg, Phys. Rev. Letters, *24* 1225 (1970).
50. P. Heller, Rep. Progr. Phys. *30*, 731 (1967).
51. K. F. Herzfeld and T. A. Litovitz, Absorption and Dispersion of Ultrasonic Waves, Academic Press, New York 1959.
52. N. C. Ford, Jr., K. H. Langley, and V. G. Puglielli, Phys. Rev. Letters *21*, 9 (1968) (CO_2); D. S. Cannell and G. B. Benedek, Phys. Rev. Letters *25*, 1157 (1970) (Xe).
53. M. Rigby, Quart. Rev. *24*, 416 (1970).
54. L. D. Landau and E. M. Lifshitz, Statistical Physics, Pergamon Press, London 1958.
55. M. E. Fisher, Rep. Progr. Phys. *30*, 615 (1967).
56. M. E. Fisher, J. Math. Phys. *5*, 944 (1964).
57. R. B. Griffiths, Phys. Rev. *158*, 176 (1967).
58. P. A. Egelstaff and J. W. Ring in gen. ref. 12; J. E. Thomas and P. W. Schmidt, J. Chem. Phys. *39*, 2506 (1963).

59. O. B. Verbeke, V. Jansoone, R. Gielen, and J. de Boelpaep, J. Phys. Chem. *73*, 4076 (1969); M. Vicentini-Missoni, R. I. Joseph, M. S. Green, and J. M. H. Levelt-Sengers, Phys. Rev. *B 1*, 2312 (1970).
60. E. H. Schmidt in M. S. Green and J. V. Sengers, quoted in ref. 47.
61. L. M. Stacey, B. Pass, and H. Y. Carr, Phys. Rev. Letters *23*, 1424 (1969).
62. R. D. Mountain, Rev. Mod. Phys. *38*, 205 (1966).
63. D. McIntyre and J. V. Sengers in gen. ref. 12.
64. Proceedings of the Meeting on Small-Angle X-ray Scattering at Graz (1970), to be published.
65. B. Chu, J. S. Lin, and J. A. Duisman, Phys. Letters *32 A*, 95 (1970).
66. B. Jacrot in gen. ref. 14.
67. M. Kac, G. E. Uhlenbeck, and P. C. Hemmer, J. Math. Phys. *4*, 216 (1963).
68. C. N. Lang and T. D. Lee, Phys. Rev. *87*, 404, 410 (1952); M. E. Fisher, Lectures in Theoretical Physics Vol. VII C, University of Colorado Press, Boulder 1965.
69. L. Onsager, Phys. Rev. *65*, 117 (1944).
70. C. Domb in G. T. Rado and H. Suhl, Magnetism, Vol. II A, Academic Press, New York 1963.
71. A. Münster, Statistical Thermodynamics, Vol. I, Engl. ed., Springer and Academic Press, Berlin and New York 1969.
72. L. S. Ornstein and F. Zernike, Proc. Acad. Sci. Amsterdam *17*, 793 (1914); Physik. Zeits. *19*, 134 (1918) and *27*, 761 (1926); F. Zernike, Proc. Acad. Sci. Amsterdam *18*, 1520 (1916).
73. M. Fixman in Study Week on Molecular Forces, Pontificiae Academiae Scientiarum Scripta Varia. North-Holland, Amsterdam 1967.
74. G. Stell, Phys. Rev. Letters *20*, 533 (1968), and Phys. Rev. *B 1*, 2265 (1970); for another extension of the Ornstein-Zernike theory cf. M. S. Green, J. Math. Phys. *9*, 875 (1968).
75. G. S. Rushbrooke, J. Chem. Phys. *39*, 842 (1963); R. B. Griffiths, J. Chem. Phys. *43*, 1958 (1965).
76. B. Widom, J. Chem. Phys. *43*, 3898 (1965).
77. L. P. Kadanoff, Physics *2*, 263 (1966).
78. M. E. Fisher, Phys. Rev. *176*, 257 (1968); M. E. Fisher and P. E. Scesney, Phys. Rev. *A 2*, 825 (1970); N. E. Frankel, Progr. Theor. Phys. (Kyoto) *43*, 1148 (1970).
79. M. S. Green, M. Vicentini-Missoni, and J. M. H. Levelt-Sengers, Phys. Rev. Letters *18*, 1113 (1967).
80. P. Schofield, J. D. Litster, and J. T. Ho, Phys. Rev. Letters *23*, 1098 (1969).
81. M. Vicentini-Missoni, J. M. H. Levelt-Sengers, and M. S. Green, Phys. Rev. Letters *22*, 389 (1969).

7. Equilibrium Theories of the Liquid State

7.1. Introduction

A fully satisfying equilibrium theory of the liquid state does not yet exist. Attempts go in two different directions. One tries to evaluate the canonical partition function directly by some model-like approximations. One of these models — the cell model — will be described in section 7.2.

The other direction aims at a calculation of the pair distribution function $g(R)$. As was shown in chapter 4, $g(R)$ is closely related to the result of scattering experiments and to thermodynamic quantities. Pair distribution function theories, which will be discussed in section 7.3, use much of the cluster expansion technique, which is most naturally done by a grandcanonical formalism.

Cluster expansion proved to be extremly useful in the case of a real gas, where it lead to the virial equation [1, 2, 3, 4]

$$\frac{Pv}{NkT} = 1 - \sum_{k=1}^{\infty} \frac{k}{k+1} \beta_k \rho^k . \qquad (7.1)$$

Here the well defined irreducible cluster integrals β_k consider the interaction of $(k+1)$ particles and it is readily observed, that the quantity $-k\beta_k/(k+1)$ is identical with the $(k+1)$-th virial coefficient B_{k+1} (per molecule). As the calculation of β_k requires the evaluation of a $3k$ dimensional integral, even with the aid of computers only the first coefficients have been determined for some simple potentials [5, 6]. So the above expansion cannot be used for densities as high as those of a liquid. Nevertheless it is of value in testing the low density behaviour of other theories.

The major difficulty in testing the theories by experiment is caused by the limitation of our knowledge of the intermolecular interaction. Luckily enough, "exact" results for an assumed pair potential can be provided by machine calculations, and in the low density region also by eq. (7.1). The accumulation of data for the hard sphere system by machine calculations has stimulated the development of perturbation methods, which will be dealt with in section 7.4.

As the theories are in every case rather complicated we consider only the simplest case of additivity of the pair potentials [eq. (1.44)] and spherical symmetry of the particles.

7.2. Cell Model

The cell model tries to evaluate the configurational partition function $Q(N,v,T)$ by splitting it into the partition function for single occupancy $Q^{(1)}(N,v,T)$ and the parameter σ^N, which takes account of fluid disorder (cf. section 2.6). Here we are left with the problem of calculating

$$Q^{(1)} = \int_{W_1} \int_{W_2} \cdots \int_{W_N} \exp\left\{-\frac{1}{kT} E(\vec{r}_1, \ldots, \vec{r}_N)\right\} d\vec{r}_1 \ldots d\vec{r}_N , \qquad (7.2)$$

where particle 1 is confined to the cell W_1, particle 2 to the cell W_2, and so on.

The probability density for a certain configuration in the single occupancy model is given by its Boltzmann factor

$$p(\vec{r}_1, \ldots, \vec{r}_N) = \frac{1}{Q^{(1)}} \exp\left\{-\frac{1}{kT} E(\vec{r}_1, \ldots, \vec{r}_N)\right\}, \qquad (7.3)$$

where the position coordinates refer to the dislocations of the particles from the centers of their cells. Two different ways have been tried to simplify eq. (7.3). The results can be checked with Monte Carlo calculations on the single occupancy model [7, 8].

7.2.1. The Lennard-Jones and Devonshire Model [9]

From the historical point of view, the model of Lennard-Jones and Devonshire (LJD) was the first molecular theory of liquids. It starts from the assumption that every molecule moves in the field that would be produced by its neighbours if they are fixed at the centers of their cells. Let ε_{i0j0} denote the interaction energy between particles i and j with both particles at their cell centers, and ε_{ij0} the interaction with particle i in any position $\vec{r}_i$ and particle j at its cell center. If all particles are located at their cell centers the energy E_s of the whole system is given by

$$E_s = \sum_{i<j} \varepsilon_{i0j0}. \tag{7.4}$$

If particle i is placed at $\vec{r}_i$ and all the other particles are fixed at their cell centers, the additional energy is given by

$$\psi(\vec{r}_i) - \psi(0) = \sum_j (\varepsilon_{ij0} - \varepsilon_{i0j0}). \tag{7.5}$$

Making the inconsistent assumption that this happens for every particle, we get as energy for the whole system

$$E = E_s + \sum_i [\psi(\vec{r}_i) - \psi(0)]. \tag{7.6}$$

Inserting this in eq. (7.2), the integrand can be written as a product where each factor depends only on the position of one particle, and hence can be easily evaluated:

$$Q_1^{(1)} = \exp\left\{-\frac{E_s}{kT}\right\} \left[\int_{W_1} \exp\left\{-\frac{1}{kT}[\psi(\vec{r}_1) - \psi(0)]\right\} d\vec{r}_1\right]^N. \tag{7.7}$$

This result is similar to that obtained for the Einstein solid, eq. (1.40). Thus the LJD theory is essentially an anharmonic Einstein model. The integral gives again the mean volume accessible by each particle and hence is called "free volume".

The LJD model has been studied very extensively for the 12/6 potential [10], usually under the assumption of spherical cells and a sphericalized ψ-function (for the effect of this "smearing approximation", cf. [11]). In fig. 7.1 the result for a temperature somewhat below the critical is checked with a Monte Carlo calculation. As Einstein model the LJD model gives reasonable results rather at high densities. In fact it is regarded nowadays as the best simple model for the solid state [12]. At lower densities correlations between positions of neighbouring molecules become increasingly important, and the LJD model breaks down.

It is possible, however, to give an expansion of $Q^{(1)}$ in terms of correlated motions of molecules, which contains the LJD model as first approximation [13]. With the correlation term

$$\Delta_{ij} = \varepsilon_{ij} - \varepsilon_{ij0} - \varepsilon_{i0j} + \varepsilon_{i0j0}, \tag{7.8}$$

the energy of the whole system can be expressed exactly by

$$E = E_s + \sum_i [\psi(\vec{r}_i) - \psi(0)] + \sum_{i<j} \Delta_{ij}. \tag{7.9}$$

The first approximation is $\Delta_{ij} = 0$, i.e. the LJD model. As a next step the motion of two neighbouring molecules, correlated by Δ_{12}, is considered in the field of the other particles which are fixed at their cell centers. This leads to the second approximation

$$Q_2^{(1)} = Q_1^{(1)} \left[\frac{\iint\limits_{W_1\,W_2} \exp\left\{-\frac{1}{kT}[\psi(\vec{r}_1) + \psi(\vec{r}_2) - 2\psi(0) + \Delta_{12}]\right\} d\vec{r}_1 d\vec{r}_2}{\left[\int\limits_{W_1} \exp\left\{-\frac{1}{kT}[\psi(\vec{r}_1) - \psi(0)]\right\} d\vec{r}_1\right]^2} \right]^{\frac{zN}{2}} \quad (7.10)$$

where $zN/2$ is the number of neighbouring pairs. Unfortunately, the computation of this approximation is already very difficult. Only a few results are known, calculated with a low and a high density approximation to the integral of eq. (7.10). As fig. 7.1 shows, this brings a considerable improvement.

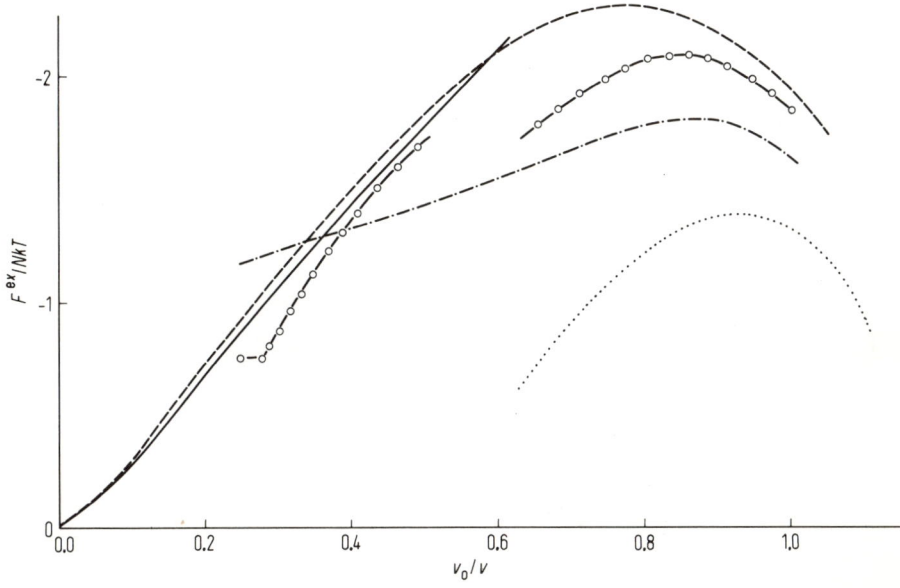

Fig. 7.1. Results from different theories for the residual free energy F^{ex} of a single occupancy model with 12/6 potential as a function of density at a temperature $kT/\varepsilon^* = 1.2$. The residual free energy is given by $F^{ex} = F^{(1)} - NkT \ln(v/N)$ and the volume v_0 is obtained for f.c.c. packing with nearest neighbour distance R^*. ----- shows the result of a Monte Carlo calculation [8], –·–·–·– the L.J.D. model [10], –○–○–○– the correlated L.J.D. model of Barker [13], ····· the uncorrelated self-consistent model [16] and ——— the correlated self-consistent model [17].

Nevertheless both models are inconsistent in the sense, that they calculate on the one hand the motion of one or two molecules in the field of all other molecules at fixed positions, and they consider on the other hand all molecules moving around. This inconsistency is avoided by the self-consistent ansatz.

7.2.2. The Self-Consistent Model [14]

This model assumes self-consistent simultaneous motions of the molecules. As first approximation, the probability to find a molecule at a certain point of its cell is the same for all molecules and is independent from the instantaneous positions of the other molecules; it shall be denoted by $\varphi(\vec{r}_i)$. Thus the exact expression for $p(\vec{r}_1, \ldots, \vec{r}_N)$ given by eq. (7.3) is replaced by the approximation

$$p(\vec{r}_1, \ldots, \vec{r}_N) = \prod_i \varphi(\vec{r}_i). \tag{7.11}$$

The function φ is determined such that the configurational free energy $F^{(1)}$ of the single occupancy model becomes a minimum. For this reason we first express $F^{(1)}$ by p, then insert the above approximation and finally use the variational method to determine φ.

Taking the logarithm of eq. (7.3) and multiplying it with kTp we get

$$kTp \ln p = kTp\left(-\frac{E}{kT}\right) - kTp \ln Q^{(1)}. \tag{7.12}$$

As the integral over the probability density must be unity

$$\int p \, d\vec{r}_1 \ldots d\vec{r}_N = 1, \tag{7.13}$$

integration of eq. (7.12) over all cells gives

$$kT \int p \ln p \, d\vec{r}_1 \ldots d\vec{r}_N = -\int pE \, d\vec{r}_1 \ldots d\vec{r}_N - kT \ln Q^{(1)}. \tag{7.14}$$

Hence we get for the free energy

$$F^{(1)} = -kT \ln Q^{(1)} = E^{(1)} - TS^{(1)} \tag{7.15}$$

with

$$E^{(1)} = \int pE \, d\vec{r}_1 \ldots d\vec{r}_N \tag{7.16}$$

and

$$S^{(1)} = -k \int p \ln p \, d\vec{r}_1 \ldots d\vec{r}_N. \tag{7.17}$$

Inserting the approximation (7.11) and using the abbreviation $\varphi_i = \varphi(\vec{r}_i)$, we obtain for eq. (7.13) and (7.15)

$$G \equiv \int \varphi_1 \, d\vec{r}_1 - 1 = 0 \tag{7.18}$$

$$F^{(1)} = N \left\{ \frac{1}{2} \sum_{i=2}^{N} \int \varepsilon_{1i} \varphi_1 \varphi_i \, d\vec{r}_1 \, d\vec{r}_i + kT \int \varphi_1 \ln \varphi_1 \, d\vec{r}_1 \right\}. \tag{7.19}$$

Thus we have found $F^{(1)}$ as functional of φ. The minimum of $F^{(1)}$ is obtained by functional differentiation, whereby the normalizing condition eq. (7.18) has to be taken as boundary condition. Straightforward calculation of

$$\frac{\delta F^{(1)}}{\delta \varphi} + \lambda NkT \frac{\delta G}{\delta \varphi} = 0, \tag{7.20}$$

where λNkT is the Lagrange multiplier, leads to

$$\varphi_1 = \exp\{-\lambda - 1\} \exp\left\{-\frac{1}{kT} \sum_{i=2}^{N} \int \varepsilon_{1i} \varphi_i \, d\vec{r}_i\right\}. \tag{7.21}$$

Now eq. (7.21) and (7.18) can be solved for φ and λ. Insertion of φ into eq. (7.19) gives the free energy for that model. Eq. (7.21) states that the probability to find particle 1 at $\vec{r}_1$ is proportional to the Boltzmann factor of the mean potential energy at this place.

The self-consistent model has also been studied for the hard sphere [15] and the 12/6 potential [16]. A result for the latter, which is typical for the model, is given in fig. 7.1. It shows that the free energy becomes much too positive. For an explanation consider e.g. the case of hard discs in two dimensions. The self-consistent assumption allows the centers of the particles only to be in the white section of the cells of fig. 7.2 a. But in reality the center of particle 1 could enter also the dashed section of its cell if particle 2 is kept at the center of its cell. That means that eq. (7.21) gives too small a volume accessible by each particle, and hence also too small an entropy of the whole system.

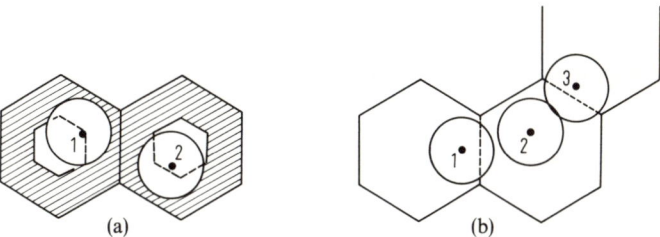

Fig. 7.2. The self-consistent cell model for hard discs
a) The uncorrelated model. The center of the discs must not enter the dashed section, which is as broad as the discs radius.
b) The correlated model which accounts only for binary correlations. If the correlated motion of particles 1 and 2 is considered, the motions of particles 2 and 3 relative to each other are not restricted, so that a forbidden overlapping can occur.

On the basis of this explanation the self-consistent model should be improved allowing for correlated motions [17]. Thus eq. (7.11) is replaced by

$$p(\vec{r}_1,\ldots,\vec{r}_N) = \prod_i \varphi(\vec{r}_i) \prod_{i<k} (1 + \chi(\vec{r}_i,\vec{r}_k)), \tag{7.22}$$

where the function $\chi(\vec{r}_i,\vec{r}_k)$ is the still unknown correlation term. Insertion into eq. (7.15) to (7.17) leads to an expansion for $F^{(1)}$ in terms of the correlated motion of two molecules, three molecules and so on. If we take as a first approximation only the binary correlations between neighbouring molecules and omit all higher terms we get for $F^{(1)}$

$$F^{(1)} = N \left\{ \frac{z}{2} \frac{\iint \varphi_1 \varphi_n (1 + \chi_{1n})[\varepsilon_{1n} + kT\ln \varphi_1 \varphi_n (1+\chi_{1n})]\,d\vec{r}_1\,d\vec{r}_n}{\iint \varphi_1 \varphi_n (1 + \chi_{1n})\,d\vec{r}_1\,d\vec{r}_n} + \right.$$

$$\left. + (1-z)kT \frac{\int \varphi_1 \ln \varphi_1\,d\vec{r}_1}{\int \varphi_1\,d\vec{r}_1} + \frac{1}{2}\sum_k \frac{\iint \varphi_1 \varphi_k \varepsilon_{1k}\,d\vec{r}_1\,d\vec{r}_k}{\iint \varphi_1 \varphi_k\,d\vec{r}_1\,d\vec{r}_k} \right\}. \tag{7.23}$$

In this eq. the neighbours of cell 1 are denoted by n whilst the others are denoted by k. The range of χ is restricted to small mutual distances hoping that the error can be kept small which is caused by neglecting the correlation between three or more particles. Then functional derivation can again be used to determine those functions φ and χ which make $F^{(1)}$ a minimum.

As fig. 7.1 shows, good results are obtained with this model at low and medium densities. But at high densities it does not work well for the following reason. Consider in fig. 7.2 b the correlated motion of particles 1 and 2 in the average potential produced by the others. Contrary to the uncorrelated model the φ-function allows now the particles to move in their cells within a volume which is too large, so that forbidden configurations, as the overlapping of particles 2 and 3 can occur rather frequently. An improvement of the self-consistent model necessitates a better approximation for the restrictive influence of other particles on the possible configurations of pairs.

7.3. Pair Distribution Function Theories

We have seen in chapter 4 and in section 5.3.4, that the pair distribution function $g(R)$ is essentially given by $e^{-\varepsilon/kT}$ plus an important correction which expresses the positioning of the second molecule due to the influence of other neighbours. It is this influence of the surroundings which is hard to take account of properly.

7.3.1. Yvon-Born-Green Theory [18–21]

The s-tuplet distribution functions were defined in eq. (4.9) by

$$g(1,2,\ldots,s) = \frac{1}{(N-s)!}\frac{1}{\rho^s}\frac{1}{Q_N}\int\ldots\int\exp\left\{-\frac{1}{kT}\sum \varepsilon_{ij}\right\}d\vec{r}_{s+1}\ldots d\vec{r}_N. \qquad (7.24)$$

By differentiating eq. (7.24) with respect to the position coordinate of particle 1, a hierarchy of integro-differential equations can be derived connecting always the s-tuplet distribution function with the corresponding interaction potential plus the $(s+1)$-tuplet distribution function. We show this in detail for $s = 2$

$$\nabla_1 g(1,2) = \frac{1}{(N-2)!}\frac{1}{\rho^2}\frac{1}{Q_N}\sum_{j=2}^{N}\int\ldots\int\left(-\nabla_1\frac{\varepsilon_{1j}}{kT}\right)\exp\left\{-\frac{1}{kT}\sum \varepsilon_{ij}\right\}d\vec{r}_3\ldots d\vec{r}_N. \qquad (7.25)$$

This sum contains two different types of integrals. For $j = 2$ the factor $\nabla_1 \varepsilon_{12}$ can be taken out of the integral. For $j = 3$ we keep at first particle 3 at a fixed position and write

$$\frac{\rho}{(N-2)}\int\left(-\nabla_1\frac{\varepsilon_{13}}{kT}\right)\left[\frac{1}{(N-3)!}\frac{1}{\rho^3}\frac{1}{Q_N}\int\ldots\int\exp\left\{-\frac{1}{kT}\sum \varepsilon_{ij}\right\}d\vec{r}_4\ldots d\vec{r}_N\right]d\vec{r}_3 =$$

$$= \frac{\rho}{(N-2)}\int\left(-\nabla_1\frac{\varepsilon_{13}}{kT}\right)g(1,2,3)d\vec{r}_3, \qquad (7.26)$$

as the expression in the square bracket is just $g(1,2,3)$. As the result does not depend on the special choice of particle 3, we get this term $(N-2)$ times. Multiplying eq. (7.25) by $kT/g(1,2)$ we get

$$-kT\nabla_1 \ln g(1,2) = \nabla_1 \varepsilon_{12} + \rho\int\frac{g(1,2,3)}{g(1,2)}\nabla_1 \varepsilon_{13}d\vec{r}_3, \qquad (7.27)$$

a relation which can also be derived in the grandcanonical formalism.
So far we have not gained very much, as for the calculation of the s-tuplet distribution function the knowledge of the $(s+1)$-tuplet distribution function is necessary. But if one assumes as a closure relation the superposition approximation

$$g(1,2,3) = g(1,2)g(2,3)g(3,1), \qquad (7.28)$$

and inserts this into eq. (7.27), one gets an equation containing only pair distribution functions, the so-called Yvon-Born-Green (YBG) equation.

7.3. Pair Distribution Function Theories

The physical meaning of the superposition approximation can perhaps be understood best with the concept of a potential of the mean force. If we take s particles out of the N at fixed positions and average over the positions of the other particles, the mean force acting on particle 1 is given by

$$-\nabla_1 \sum_{j=2}^{s} \varepsilon_{1j} - \rho \int \frac{g(1,2,\ldots,s+1)}{g(1,\ldots,s)} \nabla_1 \varepsilon_{1,s+1} \, d\vec{r}_{s+1}. \tag{7.29}$$

By comparison with eq. (7.27), or its generalization to an s-tuplet distribution function, it is seen that this mean force can be written as $kT\nabla_1 \ln g(1,\ldots,s)$. Thus we call

$$\psi(1,\ldots,s) = -kT \ln g(1,\ldots,s) \tag{7.30}$$

the mean force potential. On this basis the superposition approximation can be formulated as

$$\psi(1,2,3) = \psi(1,2) + \psi(2,3) + \psi(3,1), \tag{7.31}$$

which means additivity for the potentials of the mean force. In the low density case this will be a good assumption, whereas its meaning is not so clear at higher densities.
It will be shown in eq. (7.37) and (7.38), that for $g(1,2)$ an expansion in powers of density can be given

$$g(1,2) = e^{-\varepsilon_{12}/kT}[1 + g_1(1,2)\rho + g_2(1,2)\rho^2 + \cdots]. \tag{7.32}$$

By making a density expansion for $g(1,2)$ with undetermined coefficients and inserting this in the YBG equation, only g_1 is obtained in agreement with eq. (7.38).
The equations of state calculated from the YBG theory for a system of hard spheres is shown in fig. 7.5. The YBG theory gives worse results than the other two pair distribution function theories, which will be discussed later. But — and this point is still rather obscure — whereas the other theories indicate the melting transition at much too high densities or not at all, the YBG equation can be solved only to about those densities which are below the melting transition of hard spheres [7, 22].
The possible indication of the melting transition was the reason for trying improvements on the YBG theory. One attempt was to use a closure relation in the next equation out of the hierarchy of integro-differential equations, thus connecting $g(1,2,3)$ with $g(1,2,3,4)$. But this requires a lot of computational work and a complete result is not yet known [23, 24]. The second attempt concerns a correction of the superposition approximation in powers of density [25]. This led to an improvement of the equation of state, but the solution of the resulting equations was limited to relatively low densities.

7.3.2. Clusters [1–4]

The concept of clusters originated from the theory of real gases leading to eq. (7.1). For the pair distribution function a similar expansion can be done, and we will deal with the clusters occuring there. The definition of $g(1,2)$ [cf. eq. (4.9)] contains integration over factors $e^{-\varepsilon_{ij}/kT}$. For most pair potentials there exists only a relative short range where $e^{-\varepsilon_{ij}/kT}$ differs from 1 (cf. fig. 4.1a). Hence the function f_{ij} is introduced by

$$1 + f_{ij} = e^{-\varepsilon_{ij}/kT}. \tag{7.33}$$

Insertion of this into the integral occuring in the definition of $g(1,2)$ gives a sum of $2^{N(N-1)/2}$ integrals

$$\int \ldots \int \exp\left\{-\frac{1}{kT}\sum \varepsilon_{ij}\right\} d\vec{r}_3 \ldots d\vec{r}_N =$$

$$\int \ldots \int (1 + \sum f_{ij} + \sum f_{ij} f_{kl} + \sum f_{ij} f_{kl} f_{mn} + \cdots) d\vec{r}_3 \ldots d\vec{r}_N. \quad (7.34)$$

Each integral of this sum can be represented by a graph in the following way: the indexed particles are represented by points and the factors f_{ij} by lines connecting the particles i and j. A cluster contains all particles which are linked together directly or indirectly. Naturally many graphs consist of more than one cluster. It is well seen that any integral factorizes according to the clusters present in its graph, so that we have a one-to-one correspondence between integrals and clusters.

The particles 1 and 2 are singled out in eq. (7.34) by the fact that integration is not performed over $d\vec{r}_1 d\vec{r}_2$. Thus we find those clusters as a special class which contain both particles 1 and 2. They are called 1,2-clusters, the points 1 and 2 base points and all the other points field points. A typical 1,2-cluster is shown in fig. 7.3a. All single point attachments to the connection between

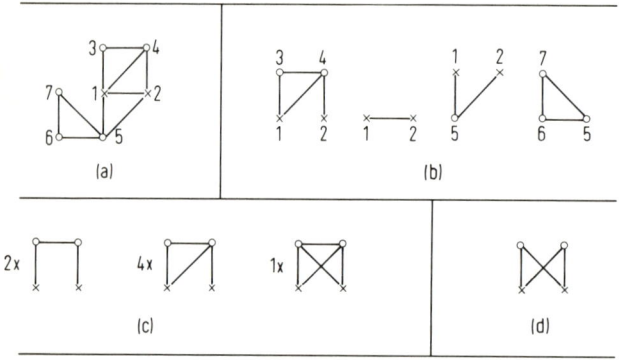

Fig. 7.3. a) An 1,2 cluster with five field points (○).
b) The cluster of fig. a) is snipped in order to give irreducible clusters.
c) All types of 1,2-irreducible clusters with two field points. The first and the second are node clusters, whilst the third is a bridge cluster. The different ways in which the node clusters can be realized are shown in fig. 7.4b.
d) The only parallel cluster with two field points.

the base points are so-called articulation points (point 5). A 1,2-cluster may be divided into simpler clusters by snipping a) simultaneously at both base points and b) at all articulation points, which means factorization of the corresponding integral. This is shown in fig. 7.3b. Those 1,2-clusters, which cannot be divided in this way are called 1,2-irreducible; the cluster f_{12} is not considered as irreducible *). The first and third cluster in fig. 7.3b are 1,2-irreducible. Moreover

---

* The definitions for the classification of clusters are not uniform in the literature [4]. Our definition of irreducibility follows ref. [2].

fig. 7.3c shows the different types of 1,2-irreducible clusters with two field points, indicating also the number of distinguishable but numerically equivalent clusters of the same type. We define now 1,2-irreducible cluster integrals $\gamma_k(1,2)$ by

$$\gamma_k(1,2) = \frac{1}{k!} \int \ldots \int d\vec{r}_3 \ldots d\vec{r}_{k+2} [\sum \prod f_{ij}], \qquad (7.35)$$

where the sum goes over all possible products of f-functions, which form an 1,2-irreducible cluster with k given field points. The first integrals are (cf. fig. 7.3c)

$$\gamma_1(1,2) = \int d\vec{r}_3 f_{13} f_{23} \qquad (7.36)$$

$$\gamma_2(1,2) = \frac{1}{2!} \iint d\vec{r}_3 d\vec{r}_4 f_{13} f_{34} f_{24} (2 + 4f_{14} + f_{14} f_{23}).$$

With these definitions it can be proved [2–4], that

$$g(1,2) = e^{-\varepsilon_{12}/kT} \exp\left\{ \sum_{k=1}^{\infty} \gamma_k \rho^k \right\}. \qquad (7.37)$$

This relation is closely connected to eq. (7.1) [26]. Here too, only the first γ functions are known, as the difficulties in performing the integration in eq. (7.35) increase rapidly with the number of field points. Thus $g(1,2)$ cannot be calculated from eq. (7.37) at the densities of a liquid. It may be noted that the coefficients of the density expansion in eq. (7.32) follow from eq. (7.37):

$$g_1(1,2) = \gamma_1(1,2) \qquad (7.38)$$

$$g_2(1,2) = \gamma_2(1,2) + \frac{1}{2} \gamma_1^2(1,2).$$

7.3.3. The Cluster Expansion Approach to Pair Distribution Function Theories

The expansion for the pair distribution function eq. (7.37) can also be considered as an expansion of the mean force potential in powers of the density

$$-\frac{1}{kT} [\psi(1,2) - \varepsilon(1,2)] = \sum_{k=1}^{\infty} \gamma_k \rho^k \equiv \text{all 1,2-irreducible clusters.} \qquad (7.39)$$

The equivalence sign means that this expression is obtained by summing the integrals which correspond to all possible clusters of the given class after having multiplied every integral with k field points by ρ^k. We are looking now for similar expansions of $h(1,2)$ and $c(1,2)$, the direct correlation function defined in eq. (4.30). For this reason we define as nodes field points through which all routes between the base points have necessarily to pass. Now we can distinguish two different classes of 1,2-irreducible clusters. Those which contain nodes are called node clusters; otherwise they are called bridge clusters. In fig. 7.3c the last cluster is the only bridge cluster with two field points. Every 1,2-irreducible cluster integral γ_k can then be divided into a part γ'_k containing the node clusters and a part γ''_k containing the bridge clusters. We define

$$n(1,2) = \sum_{k=1}^{\infty} \gamma'_k \rho^k \equiv \text{all node clusters} \qquad (7.40)$$

$$b(1,2) = \sum_{k=1}^{\infty} \gamma''_k \rho^k \equiv \text{all bridge clusters.} \qquad (7.41)$$

Inserting these expressions into eq. (7.37) we get

$$g(1,2) = (1 + f_{12}) \exp\{n(1,2) + b(1,2)\}, \qquad (7.42)$$

and expanding the exponential

$$g(1,2) = (1 + f_{12}) + (1 + f_{12})\left\{n + b + \frac{1}{2!}(n + b)^2 + \cdots\right\}. \qquad (7.43)$$

We can define now

$$p(1,2) = \frac{1}{2!}(n + b)^2 + \frac{1}{3!}(n + b)^3 + \cdots. \qquad (7.44)$$

The clusters corresponding to $p(1,2)$, called parallel clusters, are obtained as products of bridge and/or node clusters and are not 1,2-irreducible. The simplest of these is the cluster in fig. 7.3d. Thus we have

$$h(1,2) = f_{12} + (1 + f_{12})(n + b + p) \equiv \text{all 1,2-clusters without articulation points}. \qquad (7.45)$$

We will prove now that if we take the node clusters out of the expansion for $h(1,2)$, the remaining function is just the direct correlation function $c(1,2)$:

$$c(1,2) = h(1,2) - n(1,2) = f_{12}(1 + n + b + p) + b + p \equiv$$
$$\equiv \text{all 1,2-clusters without articulation points and nodes}. \qquad (7.46)$$

To give the proof we show the general representation of a node cluster, denoting the nodes by 3, 4, etc.:

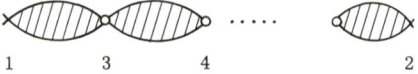

Splitting the diagram at point 3, i.e. making point 3 a base point in both parts, we get an 1,3-cluster and a 3,2-cluster. The 1,3-cluster must be a cluster out of the class corresponding to $c(1,3)$. On the other side the 2,3-cluster might have further nodes, hence it must be a cluster of the class corresponding to $h(2,3)$. So every node cluster can be thought of as a combination of a cluster belonging to c and a cluster belonging to h. As the node 3 can be any of the field points, we get the set of all node clusters by taking N times the combination of all clusters corresponding to $c(1,3)$ with all clusters corresponding to $h(2,3)$

$$n(1,2) = N\frac{1}{v}\int c(1,3)h(2,3)\,d\vec{r}_3. \qquad (7.47)$$

As this is the Ornstein-Zernike equation (4.30), the function $c(1,2)$ defined by eq. (7.46) is identical with the direct correlation function [26]. Fig. 7.4a shows all clusters for two field points (corresponding to the coefficient of ρ^2) for $c(1,2)$ whereas fig. 7.4b shows the additional node clusters needed for $h(1,2)$.

Eq. (7.42), (7.46), and (7.47) contain the four unknown functions h, c, n, and b. It is possible now to introduce well defined approximations to derive a complete set of equations. The Ornstein-Zernike equation is always retained.

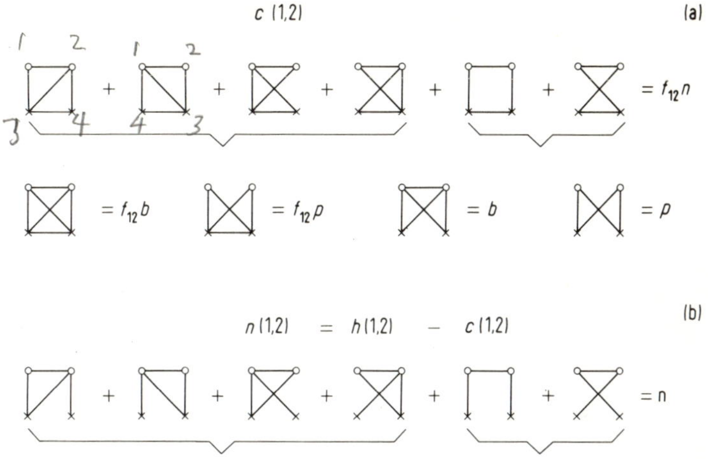

7.3. Pair Distribution Function Theories

Fig. 7.4. The clusters which determine the coefficient of ρ^2 in the density expansion (a) for the direct and (b) for the total correlation function and the corresponding contribution of n, b, and p-clusters. Those clusters which are bracketed give the same numerical value. Concerning this ρ^2-term, the clusters b and $f_{12}b$ are omitted in the HNC-theory, whereas the clusters b, $f_{12}b$, p, and $f_{12}p$ are omitted in the PY-theory.

The hypernetted chain (HNC) equation is obtained by the assumption $b(1,2) = 0$ [27–32]. Thus using eq. (7.42) and (7.46) we have

$$g(1,2) = (1 + f_{12})\exp\{h(1,2) - c(1,2)\}, \tag{7.48}$$

which gives together with the Ornstein-Zernike equation a selfcontained formalism for the determination of $g(1,2)$ and $c(1,2)$. In the HNC solution the bridge clusters are omitted, but also those node clusters, which contain bridge clusters between the nodes. This can be seen from the construction of the node clusters according to eq. (7.47), as neither $c(1,3)$ nor $h(2,3)$ can contain a bridge cluster. In the density expansion of $g(1,2)$ already the term in ρ^2 will be in error, as already two clusters of fig. 7.4a are excluded.

The Percus-Yevick (PY) equation is obtained, if in eq. (7.46) only the short ranging terms, namely those containing f_{12} are retained [33, 34]. Thus we get, considering first eq. (7.44) and then eq. (7.42),

$$c(1,2) = f_{12}e^{(n+b)} = \frac{f_{12}}{1 + f_{12}}g(1,2). \tag{7.49}$$

Algebraic rearrangements of eq. (7.49) lead to

$$g(1,2) = (1 + f_{12})(1 + h(1,2) - c(1,2)) \tag{7.50}$$

which gives the PY equation as a linearization of the exponential of the HNC equation. The clusters that are present in the solution of the PY equation can be found by inserting eq. (7.50) into the Ornstein-Zernike equation with an undetermined ansatz for $n(1,2) = h(1,2) - c(1,2)$ in powers of density

$$n(1,2) = \rho \int [f_{13}(1 + n(1,3))][(1 + f_{23})(1 + n(2,3)) - 1]\,d\vec{r}_3. \tag{7.51}$$

This way it can be shown that the solution for $g(1,2)$ is again wrong for the term in ρ^2 as four clusters of fig. 7.4a are missing. Though the PY solution contains less clusters than the HNC solution, it turns out strikingly, that it is superior to the HNC equation, as we will see in section 7.3.5.

7.3.4. The Functional Differentiation Approach [35–37]

The integral equations for the pair distribution function have been derived above with the cluster formalism. But there exists another approach using a quite different formalism which also offers some physical insight. Let us consider the case of an external potential, which takes the value $u(i) = u_i$ at $\vec{r}_i$. Then the s-particle distribution function will be given by

$$n(1,\ldots,s|u) = \frac{N!}{(N-s)!} \frac{\int \ldots \int e^{-\Sigma u_i/kT} e^{-\Sigma \varepsilon_{ij}/kT} \, d\vec{r}_{s+1} \ldots d\vec{r}_N}{\int \ldots \int e^{-\Sigma u_i/kT} e^{-\Sigma \varepsilon_{ij}/kT} \, d\vec{r}_1 \ldots d\vec{r}_N}, \qquad (7.52)$$

where the symbol $|u)$ denotes the presence of the external potential.

The external potential may be produced by an additional particle, which is identical with the other particles of the system and will be fixed at position 0. In this case the one-particle distribution function will be related to the two-particle distribution function of the same system without external potential by

$$n(1|u) = \frac{n(0,1)}{\rho}. \qquad (7.53)$$

This relation, which is intuitively obvious, follows from eq. (4.3) and (7.52).

By switching on an external potential u, various functions change which characterize the state of the assembly. If the change of any independent function φ is known to be $\Delta\varphi$ the change in a dependent functional*) F may be expressed by a Taylor series which uses functional differentiation [38].

$$F(1|u) = F(1|0) + \int \left[\frac{\delta F(1|u)}{\delta \varphi(2)}\right]_{\Delta\varphi=0} \Delta\varphi(2) \, d\vec{r}_2 +$$

$$+ \frac{1}{2!} \int\int \left[\frac{\delta^2 F(1|u)}{\delta \varphi(2) \delta \varphi(3)}\right]_{\Delta\varphi=0} \Delta\varphi(2) \Delta\varphi(3) \, d\vec{r}_2 \, d\vec{r}_3 + \cdots. \qquad (7.54)$$

Such an expansion is useful if the functional derivatives are easily accessible quantities, and if the series converges rapidly. For this a proper choice of the function φ is essential.

Thus one may ask, e.g., for the change of the one-particle distribution function, $n(1|u) = F(1|u)$, as this offers on the basis of eq. (7.53) a possibility to calculate the pair distribution function. Let us take at first for $\Delta\varphi$ the external potential u itself. A straightforward calculation, which will not be presented here, gives

$$\frac{\delta n(1|u)}{\delta u(2)} = -\frac{1}{kT}[n(1,2) - n(1)n(2) + n(1)\delta(1,2)]. \qquad (7.55)$$

The second derivative would then contain the three particle distribution function, and so on. But considering a weak external potential (small values of $\Delta\varphi$), the series can be broken off

---

* cf. footnote on page 46.

7.3. Pair Distribution Function Theories

after the first term. In the special case that the external potential is produced by an additional particle ($u(i) = \varepsilon(0, i)$), we obtain by combining eq. (7.53)–(7.55) and using $n(i|0) = n(i) = \rho$

$$\frac{n(0,1)}{\rho} = \rho - \int [n(1,2) - n(1)n(2) + n(1)\delta(1,2)] \frac{\varepsilon(0,2)}{kT} d\vec{r}_2, \tag{7.56}$$

or

$$\frac{1}{\rho^2}[n(0,1) - \rho^2] = h(0,1) = -\frac{\varepsilon(0,1)}{kT} + \rho \int \left(-\frac{\varepsilon(0,2)}{kT}\right) h(1,2) d\vec{r}_2. \tag{7.57}$$

This is just the Ornstein-Zernike equation, with the direct correlation function $c(1,2)$ given by $-\varepsilon_{12}/kT$. This is a very crude solution, for the interaction potential $\varepsilon(R)$ cannot be considered as a weak external potential, as it goes to infinity for $R \to 0$. Thus the cut-off of the Taylor expansion after the first term is not justified, $u(i) = \varepsilon(0, i)$ is not a good expansion parameter.

One can expect a better convergence by using the one particle distribution function as expansion parameter, as this function remains always finite, and its change will not be too drastic when switching on an external potential. We then ask for the change of the external potential necessary to produce a given change of the one particle distribution function, which is expressed by $\delta u/\delta n$, the inverse relation to eq. (7.55). It can be shown that this derivative is given by

$$\frac{\delta u(1)}{\delta n(2|u)} = -kT\left[\frac{1}{n(1)}\delta(1,2) - c(1,2)\right]. \tag{7.58}$$

The following relations are then easily obtained

$$\frac{\delta\left[\frac{1}{kT}u(1) + \ln n(1|u)\right]}{\delta n(2|u)} = c(1,2), \tag{7.59}$$

and

$$\frac{\delta[n(1|u)e^{u(1)/kT}]}{\delta n(2|u)} = [n(1|u)e^{u(1)/kT}c(1,2)]_{\substack{n(1|u)=\rho \\ u=0}} = \rho c(1,2). \tag{7.60}$$

These simple relations suggest an expansion of $F(1|u) = n(1|u)e^{u(1)/kT}$ or its logarithm in terms of $\Delta\varphi(i) = [n(i|u) - n(i|0)]$ and rapid convergence is to be expected. Again, if taking as external potential an additional particle and retaining only the first term in the Taylor expansion, a selfcontained formalism is found for an approximative determination of the pair distribution function. The use of eq. (7.60) leads to the PY equation and that of eq. (7.59) to the HNC equation.

The above examples show that there is some arbitrariness with respect to the functional which shall be expanded to give an equation for the pair distribution function. A sensible decision can only be made on physical grounds. For regions where $\varepsilon(R) = 0$ (e.g. for larger distances R and for the hard sphere potential), the PY choice for $F(1|\varepsilon) = n(1|\varepsilon) \exp\{\varepsilon/kT\}$ is a linear functional of n, and the cut-off of the expansion after the first term is exact. Thus one understands, e.g., that in the case of hard spheres the PY equation yields better results than the HNC equation.

The main advantage of the functional differentiation approach is that it shows a way for a second order correction if the first term has been constructed properly. The second term of eq. (7.54) has been constructed for the HNC and PY choice. It involves triplet distribution functions, which have to be related to pair distribution functions by some closure approximation. This is not so serious as in the YBG theory, as it enters only into a correction term. An improvement of the PY theory by evaluating the second order term will be mentioned in the next section under PY II [39, 40].

7.3.5. Numerical Results

The judgment on the pair distribution function theories depends to a certain extent on the way of testing them. A comparison between calculated pair distribution functions, and those obtained by machine experiments, is not very informative, as the thermodynamic properties are very sensitive to slight changes in $g(R)$ (cf. chapter 4).

The potential inversion method determines $g(R)$ and $c(R)$ by machine calculations, on the basis of a given pair potential. By putting these functions into the integral equations, the potential can be recalculated. The difference in pair potential between input and output is a rather good check for the theories, but not too much has been done in that field [41, 42], so that a thorough evaluation is not possible*[)].

The method which has been used most frequently is the calculation of the equation of state. This can be done either by the compressibility equation, eq. (4.24), or by the pressure equation, eq. (4.20). In the case of hard spheres there exists a third route to thermodynamics [43], but this will not be considered here. As the theories do not provide a correct $g(R)$, the results will differ, whether the pressure equation (p) or the compressibility equation (c) has been used. They can be compared in the low density region with the virial expansion, eq. (7.1), and at all densities with the results of machine calculations. If in the density expansion of $g(R)$ [eq. (7.32)] the terms are exact to g_i, then the virial coefficients (obtained either by the pressure or the compressibility equation) are exact to B_{i+2}. As in all three theories only g_1 is exact, they reproduce only the second and third virial coefficient correctly. For a further discussion we will deal with the results for the hard sphere and the 12/6 potential separately.

Table 7.1. The fourth and fifth virial coefficients (in reduced units) for the different theories in a hard sphere system.

	Exact	$(PY)_p$	$(PY)_c$	$(HNC)_p$	$(HNC)_c$	$(YBG)_p$	$(YBG)_c$
B_4/b_0^3	0.28695	0.25	0.2969	0.4453	0.2092	0.2252	0.3424
B_5/b_0^4	0.1104	0.0859	0.1211	0.1447	0.0493	0.0475	0.1335

For a system of hard spheres the first seven virial coefficients have been calculated [5] and, as was shown in chapter 3, the equation of state is known exactly by machine calculations. It turned out that by chance [44, 45] the equation of state of the fluid can be represented in an analytic form by a Padé approximant, which uses the first six virial coefficients:

$$\frac{Pv}{NkT} = 1 + \frac{4\eta + 1.01611\eta^2 + 1.10906\eta^3}{1 - 2.24597\eta + 1.30101\eta^2}, \tag{7.61}$$

where $\eta = \rho b_0/4$, $b_0 = \frac{2}{3}\pi\sigma^3$ being the second virial coefficient B_2 (per molecule). Melting occurs at $\eta = 0.494$ [7].

A remark is necessary about the use of the pressure equation in the case of hard spheres. Let us consider $y(R) = e^{\varepsilon/kT} g(R)$, which is given by [cf. eq. (7.32)]:

---

* The method is of special interest for determining effective potentials of metals, where $g(R)$ and $c(R)$ are taken from experimental scattering intensities.

$$y(R) = 1 + g_1(R) + g_2(R)\rho^2 + \cdots. \tag{7.62}$$

As the functions $g_i(R)$ are sums of products of 1,2-irreducible cluster integrals they are continuous at $R = \sigma$, and so is $y(R)$. Furthermore, $e^{-\varepsilon/kT}$ is the Heaviside step function and its derivative is the Dirac δ-function. So the integral occuring in the pressure equation can be rewritten as follows:

$$\int R \frac{\partial \varepsilon}{\partial R} g(R) d\vec{r} = -kT \int R y(R) \frac{\partial}{\partial R} [e^{-\varepsilon/kT}] d\vec{r} = -4\pi k T \sigma^3 y(\sigma). \tag{7.63}$$

As for $R \geq \sigma$ the functions $y(R)$ and $g(R)$ are identical, the pressure equation is given now by

$$\frac{Pv}{NkT} = 1 + \frac{2\pi}{3} \rho \sigma^3 g(\sigma). \tag{7.64}$$

Let us consider now the solutions of the integral equations. An analytic solution was found for the PY equation in the case of hard spheres [46–48]. With this solution the pressure and compressibility equations are given by*)

$$\left(\frac{Pv}{NkT}\right)_p = \frac{1 + 2\eta + 3\eta^2}{(1-\eta)^2} \tag{7.65}$$

$$\left(\frac{Pv}{NkT}\right)_c = \frac{1 + \eta + \eta^2}{(1-\eta)^3}, \tag{7.66}$$

and all virial coefficients can also be given explicitely. The HNC and YBG equation had to be solved numerically [22, 51, 52] and the first virial coefficients have also been determined [53, 54]. Fig. 7.5 shows the equations of state obtained by the three theories and table 7.1 compiles the virial coefficients. As can be seen, the equations of state obtained from the PY theory are the best and even at high densities rather close to the "experimental" curve, especially the compressibility solution. The HNC solutions deviate more, but still bracket the "experimental" curve, whereas the YBG solutions are rather poor at high densities. Considering now the virial coefficients one observes that the PY equation gives nearly exact fourth virial coefficients. Furthermore it is surprising that the coefficients of the HNC equation are worse than those of the YBG equation. Thus it can be concluded that a good theory should have necessarily good virial coefficients, whereas on the other hand good virial coefficients are not sufficient for a theory to give good high density results.

For the PY and the HNC solution no phase transition could be found. There is some indication [55], that the PY equation might have more solutions at high densities, but no conclusive investigation has been made.

Let us consider now a system with a 12/6 potential. As the hard sphere system can be considered as its high temperature limit, the high temperature results of the 12/6 system are similar to those discussed above. Considering now lower temperatures, the YBG equation cannot be solved at higher densities for temperatures below $kT/\varepsilon^* = 1.40$ [52]. The HNC theory becomes superior

---

* An empirical combination of both equations

$$\frac{Pv}{NkT} = \frac{1 + \eta + \eta^2 - \eta^3}{(1-\eta)^3} \tag{7.67}$$

is now regarded as the best analytic expression for the equation of state of a hard sphere fluid [49]. Eq. (7.66) has been derived before along the arguments of the scaled particle theory [50].

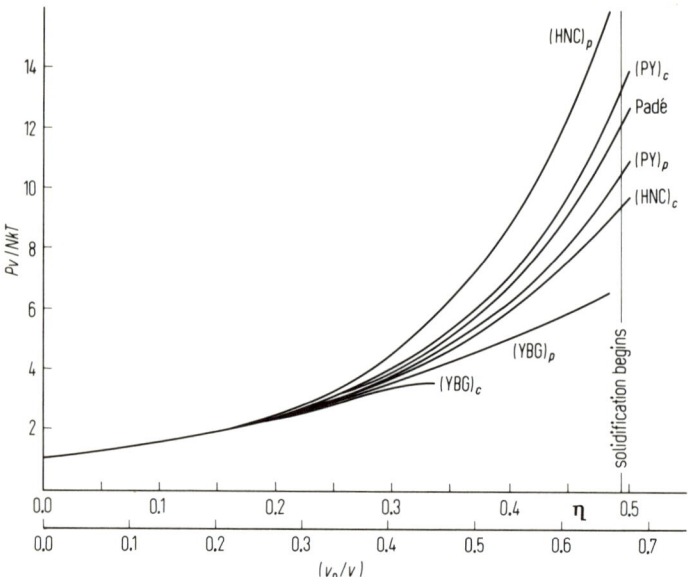

Fig. 7.5. The equations of state for a system of hard spheres obtained from pair distribution function theories using either the pressure or the compressibility equation. The Padé approximant is based on the first six virial coefficients and in the graphic representation identical with the molecular dynamics result.

to the PY theory for densities corresponding to the gaseous state. In fig. 7.6 and 7.7 we show the HNC and PY solutions [52, 56, 57] in comparison with the results of Monte Carlo calculations [58–60] at higher densities for $kT/\varepsilon^* = 1.35$ and $kT/\varepsilon^* = 1.15$. In the latter case the PY theory is again better, but the results of both theories are rather unsatisfactory. Only the pressure equation of PY II shows at $kT/\varepsilon^* = 1.35$ rather good agreement with the Monte Carlo results up to medium densities (fig. 7.6) [58]. Therefore the critical constants of a 12/6 system can be determined with its help and regarded as reliable. These are shown in table 7.2 together with the critical constants of the PY and the HNC theory.

Table 7.2. The critical data for a 12/6 potential according to some theories.

	kT_c/ε^*	$\left(\dfrac{Pv}{NkT}\right)_c$	$(v_0/v)_c$
$(PY\ II)_p$	1.36	0.31	0.36
$(PY\ II)_c$	1.33	0.34	0.33
$(PY)_p$	1.25	0.30	0.29
$(PY)_c$	1.32	0.36	0.28
$(HNC)_p$	1.25	0.38	0.26
$(HNC)_c$	1.39	0.30	0.28

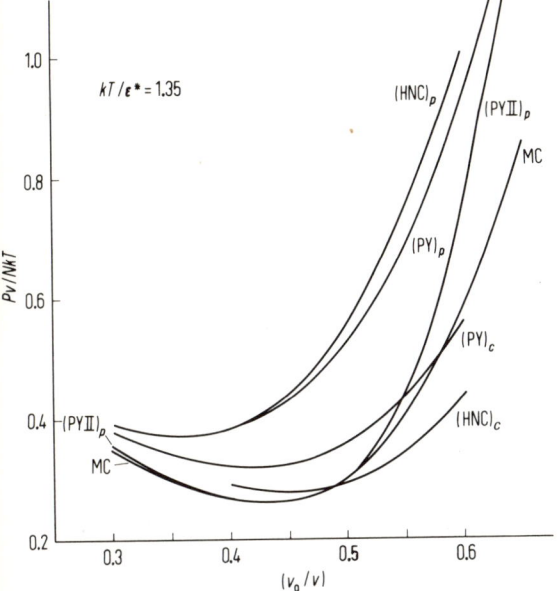
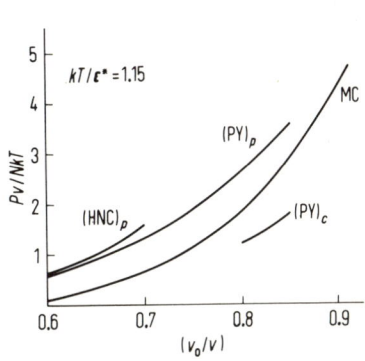

Fig. 7.6. The equations of state for a 12/6 system at a reduced temperature $kT/\varepsilon^* = 1.35$ and medium densities obtained from pair distribution function theories in comparison with Monte Carlo results (MC). The volume v_0 is obtained for f.c.c. packing with nearest neighbour distance R^*. The results of the second order theory (PY II)$_p$ show rather good agreement up to the density $(v_0/v) = 0.5$.

Fig. 7.7. The equation of state for a 12/6 system at a reduced temperature $kT/\varepsilon^* = 1.15$ and high densities obtained from pair distribution function theories in comparison with Monte Carlo results (MC). All theories yield rather poor results in this case.

7.4. Perturbation Theory

The equation of state for a hard sphere system is represented rather well by the PY solution as was shown in the last section. On the other hand, the results given by all theories for more realistic potentials, as e.g. the 12/6 potential, are poor in the truly liquid region. This suggests to take the hard sphere potential as a reference potential and to use perturbation techniques. Any realistic potential differs from the hard sphere potential by (a) that it has an attractive part and (b) that the repulsive potential is not infinitely steep. Some earlier attempts have considered solely the one or the other fact [61—65]. Recently a perturbation approach by Barker and Henderson (BH) [66—68] which takes account of both facts, proved to be successful. This theory will be described in more detail.

The basic idea of the perturbation theory is to assume that all relevant quantities of a reference system with pair interaction $\varepsilon^0(R)$ are known. Then the configurational partition function of the perturbed system with energy interaction $\varepsilon^0(R) + \varepsilon^1(R)$ is given by

$$Q = \frac{1}{N!} \int \ldots \int \exp\left\{-\frac{E_0}{kT}\right\} \exp\left\{-\frac{E_1}{kT}\right\} d\vec{r}_1 \ldots d\vec{r}_N, \tag{7.68}$$

where $E_0 = \sum \varepsilon_{ik}^0$ and $E_1 = \sum \varepsilon_{ik}^1$. (7.69)

By expanding the second exponential and using ensemble averaging with respect to the unperturbed ensemble, which is denoted by $\langle \rangle_0$, we obtain [cf. eq. (1.27)]

$$Q = Q_0 \left[1 - \frac{1}{kT} \langle E_1 \rangle_0 + \frac{1}{2(kT)^2} \langle E_1^2 \rangle_0 - \cdots \right], \quad (7.70)$$

where Q_0 is the partition function of the unperturbed system. With the relation for the free energy [cf. eq. (1.29)]

$$f - f_0 = -kT \ln Q/Q_0 \quad (7.71)$$

we get by expanding the logarithm of the bracket

$$f - f_0 = \langle E_1 \rangle_0 - \frac{1}{2kT} [\langle E_1^2 \rangle_0 - \langle E_1 \rangle_0^2] + O\left\{\frac{1}{(kT)^2}\right\}. \quad (7.72)$$

The average of the perturbing energy $\langle E_1 \rangle_0$ can be expressed by using the pair distribution function $g_0(R)$ of the unperturbed system

$$\langle E_1 \rangle_0 = \frac{\int \cdots \int (\sum \varepsilon_{ik}^1) \exp\left\{-\dfrac{E_0}{kT}\right\} d\vec{r}_1 \ldots d\vec{r}_N}{\int \cdots \int \exp\left\{-\dfrac{E_0}{kT}\right\} d\vec{r}_1 \ldots d\vec{r}_N} = \frac{N}{2} \rho \int \varepsilon^1(R) g_0(R) d\vec{r}. \quad (7.73)$$

The term of the order $\dfrac{1}{kT}$, which considers the energy fluctuations, can also be expressed by distribution functions. But this would already need the triplet and quadruplet distribution function of the unperturbed system, which are hardly accessible even with machine calculations. Nevertheless an approximate expression is provided by the BH theory. For convenience we assume for the moment, that only discrete sets of intermolecular distances R_i are possible. If N_i denotes the number of intermolecular distances R_i we have

$$\langle E_1^2 \rangle_0 = \sum_{i,j} \varepsilon^1(R_i)\varepsilon^1(R_j) \langle N_i N_j \rangle$$
$$\langle E_1 \rangle_0^2 = \sum_{i,j} \varepsilon^1(R_i)\varepsilon^1(R_j) \langle N_i \rangle \langle N_j \rangle. \quad (7.74)$$

As an approximation we neglect correlations between the occurence of different distances and use for the fluctuation of particles at equal distances the grandcanonical fluctuation expression, eq. (1.69):

$$\langle N_i N_j \rangle - \langle N_i \rangle \langle N_j \rangle = 0 \quad \text{for} \quad i \neq j$$
$$\langle N_i^2 \rangle - \langle N_i \rangle^2 = \langle N_i \rangle kT \left(\frac{\partial \rho}{\partial P}\right). \quad (7.75)$$

By taking the expression for the mean number of a certain intermolecular distance

$$\langle N_i \rangle = \frac{N}{2} \rho g_0(R_i) 4\pi R_i^2 (R_{i+1} - R_i) \quad (7.76)$$

and changing back to a continuous description we obtain

$$\langle E_1^2 \rangle_0 - \langle E_1 \rangle_0^2 = \frac{N}{2} \int kT \left(\frac{\partial \rho}{\partial P}\right) \rho g_0(R) [\varepsilon^1(R)]^2 d\vec{r}. \quad (7.77)$$

7.4. Perturbation Theory

The higher order terms in eq. (7.72) contain the higher moments of E_1 and thus are also determined by fluctuations of the occurrence of mutual distances. At higher densities these fluctuations will be small, so that a rapid convergence of the perturbation expansion is expected here.
Thus, using the exact first order term (7.73) and the approximate second order term (7.77), the perturbation theory yields

$$f = f_0 + \frac{N\rho}{2} \int \varepsilon^1(R) g_0(R) d\vec{r} - \frac{N\rho}{4} \int [\varepsilon^1(R)]^2 g_0(R) \left(\frac{\partial \rho}{\partial P}\right)_0 d\vec{r} . \tag{7.78}$$

The first order term gives the energy caused by $\varepsilon^1(R)$, if the molecules are kept at the unperturbed positions, the second order term takes account of the additional change of the molecular positions. Therefore, its proportionality to the compressibility is quite reasonable, but one could expect even better results by replacing $g_0(R)\left(\dfrac{\partial \rho}{\partial P}\right)_0$ by the local compressibility $\left(\dfrac{\partial \rho g(R)}{\partial P}\right)_0$.

In applications of the perturbation theory to any realistic potential, the question arises, what should be considered as reference potential and what as perturbing potential. It would be most easy to take the whole positive branch itself as reference potential, thus to make the division at the zero passing of the potential*⁾. If μ is the distance for which $\varepsilon(\mu) = 0$ the reference potential might be

$$\begin{aligned} \varepsilon^0(R) &= \varepsilon(R) \quad \text{for} \quad R \leq \mu \\ \varepsilon^0(R) &= 0 \quad \text{for} \quad R > \mu \end{aligned} \tag{7.79}$$

and $\varepsilon^1(R) = \varepsilon(r) - \varepsilon^0(R)$, i.e. the negative branch, is taken as perturbing potential (fig. 7.8.a–c). If the reference potential $\varepsilon^0(R)$ would be well investigated, the perturbation theory could be

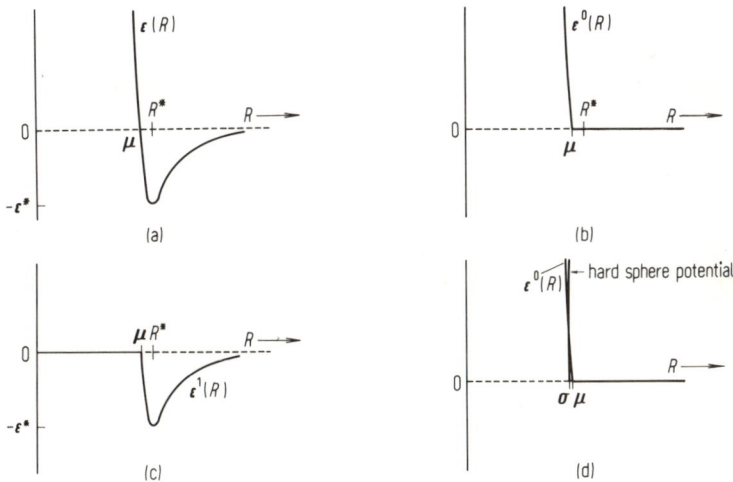

Fig. 7.8. The decomposition of a potential in the Barker-Henderson perturbation theory. The (a) original potential ε is divided at a breakpoint μ into (b) the reference potential ε^0 and (c) the perturbing potential ε^1. The reference potential is then replaced (d) by a hard sphere potential with diameter σ.

---

* It shall be noted that this prescription may be replaced by any other reasonable choice, as it is somewhat arbitrary [45].

applied at this stage. But usually this is not the case and we look for the best choice of a hard sphere system, by which the reference system could be replaced. It is intuitively clear, that for a very steep $\varepsilon^0(R)$ the best suited hard sphere diameter σ will be close to μ, but for a softer $\varepsilon^0(R)$ it will be smaller. By making an expansion in inverse powers of steepness, the BH theory finds as prescription for the hard sphere diameter the following temperature dependent expression

$$\sigma(T) = \int_0^\mu (1 - e^{-\varepsilon/kT}) dR. \tag{7.80}$$

Reviewing the procedure in the BH theory for any realistic potential, we can say: The potential is divided at a breakpoint μ into a perturbing potential and a reference potential. The latter is replaced by a hard sphere potential with a diameter according to eq. (7.80) (cf. fig. 7.8d). The thermodynamic properties of the system are then obtained from eq. (7.78), where the quantities with index 0 are those of the hard sphere system. Furthermore $\varepsilon^1(R)$ may be replaced by $\varepsilon(R)$, if μ is taken as lower limit of integration in eq. (7.78).

The approximations of the BH theory have been investigated very carefully for the 12/6 potential, by Monte Carlo calculations with the complete 12/6 potential as well as with the reference potential of eq. (7.79) [59]. It turned out that eq. (7.80), the prescription of replacing $\varepsilon^0(R)$ of eq. (7.79) by a hard sphere potential, is an excellent approximation. Moreover, it is sufficient to consider only the first two terms in the perturbation expansion; the contribution of the exact

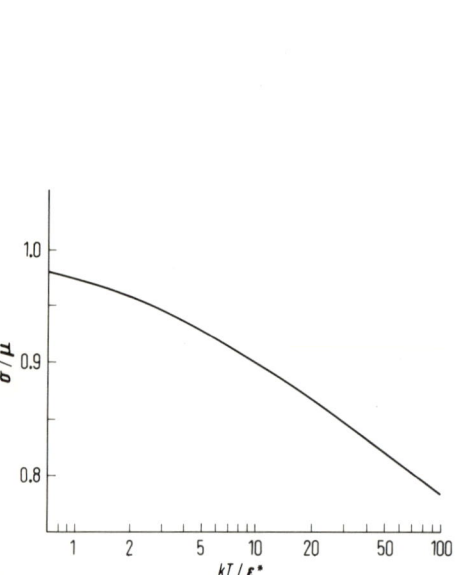

Fig. 7.9. The ratio of the hard sphere diameter σ to the zero of the 12/6 potential as a function of the reduced temperature according to eq. (7.80). After ref. 66.

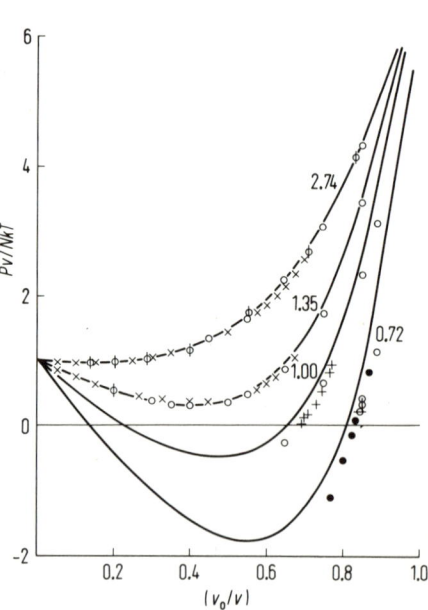

Fig. 7.10. Equations of state for the 12/6 potential. The full drawn curves are obtained from perturbation theory and are labeled with the reduced temperature kT/ε^*. The points denote either results of machine calculations (○, ⌽ and ●) or experimental results for argon (×, ⊖ and +). After ref. 66.

first term is much larger than that of the second term. The approximation to the second term, eq. (7.77), is qualitatively correct, but needs quantitative correction [59, 69].
For the 12/6 potential the hard sphere diameter as a function of the temperature is shown in fig. 7.9; some equations of state and the coexistence curve are shown in fig. 7.10 and fig. 7.11, respectively. One realizes good agreement with machine calculations, even at the lowest temperatures.

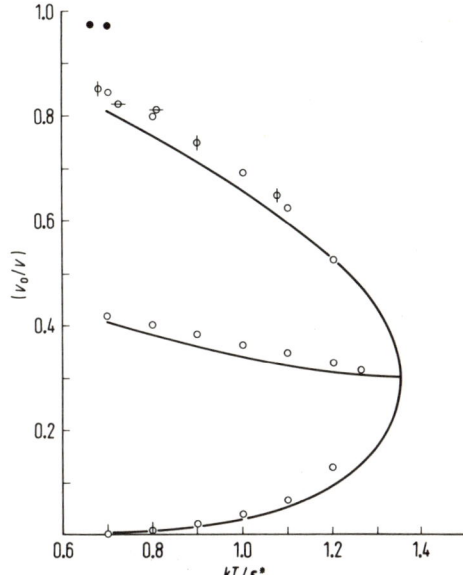

Fig. 7.11. Coexistence curve for the 12/6 potential. The full drawn curve is obtained from perturbation theory. The points denote either results of machine calculation (ϕ and $\ominus$) or experimental results for argon ($\bigcirc$ and $\bullet$). After ref. 66.

It might be of interest that agreement at lower densities is not as good. E.g., the third and fourth virial coefficient is only reproduced in a crude way [66].

7.5. References

1. J. E. Mayer and M. G. Mayer, Statistical Mechanics, Wiley, New York 1940.
2. E. E. Salpeter, Ann. Phys. (New York) 5, 183 (1958).
3. G. E. Uhlenbeck and G. W. Ford in: J. de Boer and G. E. Uhlenbeck (Eds.), Studies in Statistical Mechanics, Vol. 1, North-Holland, Amsterdam 1962.
4. G. Stell, in: H. L. Frisch and J. L. Lebowitz (Eds.), The Equilibrium Theory of Classical Fluids, Benjamin, New York 1964.
5. F. H. Ree and W. G. Hoover, J. Chem. Phys. 46, 4181 (1967).
6. J. A. Barker, P. J. Leonard, and A. Pompe, J. Chem. Phys. 44, 4206 (1966).
7. W. G. Hoover and F. H. Ree, J. Chem. Phys. 49, 3609 (1968).
8. J. P. Hansen and L. Verlet, Phys. Rev. 184, 151 (1969).
9. J. E. Lennard-Jones and A. F. Devonshire, Proc. Roy. Soc. (London) A 163, 53 (1937), A 165, 1 (1938).
10. M. Diaz Peña and M. Lombardero, An. Real. Soc. Espan. Fis. Quim. B 62, 945 (1966).
11. M. Weissmann, J. Chem. Phys. 40, 175 (1964).
12. D. Henderson and J. A. Barker, Mol. Phys. 14, 587 (1968).
13. J. A. Barker, Lattice Theories of the Liquid State, Pergamon Press, Oxford 1963.
14. J. G. Kirkwood, J. Chem. Phys. 18, 380 (1950).
15. W. W. Wood, J. Chem. Phys. 20, 1334 (1952).

16. J. S. Dahler and J. O. Hirschfelder, J. Chem. Phys. *32*, 330 (1960).
17. J. Fischer, F. Kohler, and F. Weissenböck, J. Chem. Phys. *53*, 2004 (1970).
18. J. Yvon, Actualités Scientifiques et Industrielles, Vol. 203, Hermann et Cie., Paris 1935.
19. J. G. Kirkwood, J. Chem. Phys. *3*, 300 (1935).
20. M. Born and M. S. Green, Proc. Roy. Soc. (London) *A 188*, 10 (1946).
21. N. Bogolyubov, J. Phys. (U.S.S.R.) *10*, 257 (1946).
22. J. G. Kirkwood, E. K. Maun, and B. J. Alder, J. Chem. Phys. *18*, 1040 (1950).
23. G. H. A. Cole, Adv. Phys. *8*, 205 (1959).
24. I. Z. Fisher, Statistical Theory of Liquids, University of Chicago Press, Chicago 1964.
25. S. A. Rice and L. Lekner, J. Chem. Phys. *42*, 3559 (1965).
26. G. S. Rushbrooke and H. I. Scoins, Proc. Roy. Soc. (London) *A 216*, 203 (1953).
27. J. M. van Loeuwen, J. J. Groenefeld, and J. de Boer, Physica *25*, 792 (1959).
28. M. S. Green, Hughes Aircraft Report (1959).
29. E. Meeron, J. Math. Phys. *1*, 192 (1960).
30. T. Morita and K. Hiroike, Progr. Theor. Phys. (Kyoto) *23*, 1003 (1960).
31. G. S. Rushbrooke, Physica *26*, 259 (1960).
32. L. Verlet, Nuovo Cimento *18*, 77 (1960).
33. J. K. Percus and G. J. Yevick, Phys. Rev. *110*, 1 (1958).
34. G. Stell, Physica *29*, 517 (1963).
35. J. Yvon, Nuovo Cimento *9*, Supplement 144 (1958).
36. J. K. Percus, Phys. Rev. Letters *8*, 462 (1962).
37. J. K. Percus, in: H. L. Frisch and J. L. Lebowitz (Eds.), ref. 4.
38. V. Volterra, Theory of Functionals, Dover, New York 1959.
39. L. Verlet, Physica *30*, 95 (1964).
40. L. Verlet, Physica *31*, 959 (1965).
41. L. Verlet, Phys. Rev. *165*, 201 (1968).
42. P. Hutchinson, Disc. Faraday Soc. *43*, 53 (1967).
43. J. S. Rowlinson, Mol. Phys. *10*, 533 (1966).
44. F. H. Ree and W. G. Hoover, J. Chem. Phys. *40*, 939 (1964).
45. W. G. Hoover, H. Ross, K. W. Johnson, D. Henderson, J. A. Barker, and B. C. Brown, J. Chem. Phys. *52*, 4931 (1970).
46. E. Thiele, J. Chem. Phys. *39*, 474 (1963).
47. M. S. Wertheim, Phys. Rev. Letters *10*, 321 (1963).
48. M. S. Wertheim, J. Math. Phys. *5*, 643 (1964).
49. N. F. Carnahan and K. E. Starling, J. Chem. Phys. *51*, 635 (1969).
50. E. Helfand, H. L. Frisch, and J. L. Lebowitz, J. Chem. Phys. *34*, 1037 (1961); H. Reiss, Adv. Chem. Phys. *9*, 1 (1965).
51. M. Klein, J. Chem. Phys. *39*, 1388 (1963).
52. D. Levesque, Physica *32*, 1985 (1966).
53. P. Hutchinson and G. S. Rushbrooke, Physica *9*, 675 (1963).
54. Gen. ref. 10.
55. H. N. V. Temperley, Proc. Phys. Soc. (London) *84*, 339 (1964).
56. M. Klein and M. S. Green, J. Chem. Phys. *39*, 1367 (1963).
57. F. Mandel, R. J. Bearman, and M. Y. Bearman, J. Chem. Phys. *52*, 3315 (1970).
58. L. Verlet and D. Levesque, Physica *36*, 254 (1967).
59. D. Levesque and L. Verlet, Phys. Rev. *182*, 307 (1969).
60. J. P. Hansen and L. Verlet, Phys. Rev. *184*, 159 (1969).
61. R. W. Zwanzig, J. Chem. Phys. *22*, 1420 (1954).
62. E. B. Smith and B. J. Alder, J. Chem. Phys. *30*, 1190 (1959).
63. H. L. Frisch, J. L. Katz, E. Praestgaard and J. L. Lebowitz, J. Phys. Chem. *70*, 2016 (1966).
64. D. A. McQuarrie and J. L. Katz, J. Chem. Phys. *44*, 2393 (1966).
65. J. S. Rowlinson, Mol. Phys. *7*, 349 (1964); *8*, 107 (1964).
66. J. A. Barker and D. Henderson, J. Chem. Phys. *47*, 2856; 4714 (1967).
67. J. A. Barker and D. Henderson, J. Chem. Educ. *45*, 2 (1968).
68. J. A. Barker and D. Henderson, Can. J. Phys. *45*, 3959 (1969).
69. J. A. Barker, D. Henderson, and W. R. Smith, Mol. Phys. *17*, 579 (1969).

8. Non-Equilibrium Properties: Transport Coefficients

8.1. Introduction

A treatment of non-equilibrium properties of liquids would fill a very elaborate monograph and is outside the scope of this book. Also, some very good monographs and reviews on theoretical aspects of this subject appeared recently [1–5]. The aim of this and the following chapter is, therefore, a sketchy and brief survey of the theory of transport coefficients and of liquid dynamics, with emphasis on the facts which can enhance our qualitative understanding.

There are essentially two different, though interconnected routes to transport properties of liquids. The first is based on kinetic equations which give the change of the momentum and positional distribution of the molecules with time. The basis of the kinetic theories is the (reversible) Liouville equation. In deriving kinetic equations from it irreversibility is introduced in a somewhat intuitive manner by means of additional assumptions, like chaos assumptions.

The second approach gives the connection between the fluctuations of certain quantities — as a response to small disturbances of equilibrium — and the corresponding dissipations, which are governed by transport processes. It is based on experimental determinations of the dynamical behaviour of liquids and will be dealt with in chapter 9.

In this chapter we will start with a formulation of the transport coefficients in terms of macroscopic and microscopic quantities (section 8.2). Then the kinetic equations (Boltzmann and Enskog equation) are discussed which are based on a balance of binary collisions (section 8.3). In section 8.4 the kinetic equation is dealt with which is derived from the theory of Brownian motion. In section 8.5 the results of the Rice-Allnatt theory are discussed, which is, in some respects, a combination of the Enskog theory and the theory for Brownian motion. Finally (section 8.6), it is shown how transport coefficients can be written in terms of the theorem of corresponding states, and a comparison is carried out for different liquids.

8.2. General Formulation of Transport Coefficients

8.2.1. Macroscopic Theory

The theory of continuous media yields a series of differential equations for the time dependence of density, momentum, and energy, which follow from the conservation laws. These hydrodynamic equations only describe the mechanical behaviour of the fluid and have nothing to do with irreversibility. In order to get the behaviour of the system for any deviations from equilibrium, additional empirical relations are needed which connect the quantities of the hydrodynamic equations to the macroscopic variables by which the non-equilibrium state of the fluid is defined (e.g. gradient of temperature or concentration). These empirical relations are: (1) Fourier's law of heat conduction

$$\vec{q} = -\lambda \nabla T, \tag{8.1}$$

which connects the vector of energy flux to the temperature gradient, (2) Fick's law of diffusion

$$\vec{j}_m = -D \nabla c, \tag{8.2}$$

which connects the vector of mass flux to the concentration gradient, and (3) Newton's expression of the stress tensor $\vec{\sigma}$

$$\vec{\sigma} = \left[-P + \left(\phi - \frac{2}{3}\eta\right)\nabla \cdot \vec{u} \right]\vec{1} + 2\eta\, \text{sym}(\nabla \vec{u}). \tag{8.3}$$

Eq. (8.1 – 3) are thought to be valid for a volume element, which is supposed to be large enough to contain many molecules (so that all quantities may be treated as continuous), but small enough to represent a point in comparison to the measuring instrument. The notation is as follows: P means the pressure, $\vec{u}$ the mean velocity of the volume element, $\vec{1}$ is the unity tensor, and sym$(\nabla \vec{u})$ is the symmetrical part of the tensor $(\nabla \vec{u})$,

$$[\text{sym}(\nabla \vec{u})]_{ij} = \frac{1}{2}\left(\frac{\partial u_i}{\partial x_j} + \frac{\partial u_j}{\partial x_i}\right).$$

These equations (8.1 – 3) define the phenomenological transport coefficients: λ is the coefficient of heat conduction, D the diffusion coefficient, and η and ϕ the coefficients of shear and bulk viscosity[*], resp.

Now we will shortly review the conservation laws [2–4] of mass, momentum, and energy, leading to the hydrodynamic equations mentioned earlier. We always restrict ourselves to fluids consisting of a single component. The total mass of the fluid within a volume v may then be written

$$M = \int_v \rho_m(\vec{r},t)\mathrm{d}\vec{r}, \tag{8.4}$$

where $\rho_m(\vec{r},t)$ is the mass density. The change of mass within the fixed volume v (not moving with the fluid) must be entirely due to the flow of mass over the boundary of v. This yields the equation of continuity

$$\frac{\partial \rho_m}{\partial t} + \nabla \cdot (\rho_m \vec{u}) = 0. \tag{8.5}$$

In a similar way the total momentum $\vec{p}$ may be written

$$\vec{p} = \int_v \rho_m \vec{u}\, \mathrm{d}\vec{r}, \tag{8.6}$$

and its change per second is given by the total force (volume and surface forces) acting on the mass in v minus the flow of momentum through the boundary of v. This consideration leads to the hydrodynamic equation of motion:

$$\frac{\partial}{\partial t}(\rho_m \vec{u}) + \nabla \cdot (\rho_m \vec{u}\, \vec{u}) = \vec{f} + \nabla \cdot \vec{\sigma}. \tag{8.7}$$

Here $\vec{f}$ is the density of any volume force (e.g. gravity), and $\vec{\sigma}$ the stress tensor, which is a symmetric tensor [cf. eq. (8.3)].

To obtain the equation of energy transport the change of the total energy contained in the volume v

$$\int_v (E_i + \rho_m \vec{u}^2/2)\mathrm{d}\vec{r} \tag{8.8}$$

---

[*] The existence of a non-zero value of the bulk viscosity implies that in compressions or expansions equilibrium is not installed instantaneously, but lags behind the volume change, thus giving rise to dissipation. Such a lag is well known in the case of more complicated molecules, where attainment of equilibrium between translational modes and vibrational modes needs time. But even in the case of monoatomic fluids the bulk viscosity has non-zero values.

must be considered, where E_i is the internal energy density. This change must equal the total work per unit time performed by the volume and surface forces minus the total energy current through the boundary of v, which consists of convection and conduction. If $\vec{q}$ is the vector of conduction of energy density (energy flux), the equation of energy transport is given by

$$\frac{\partial E_i}{\partial t} + \nabla \cdot (E_i \vec{u}) = \vec{\sigma} : \nabla \vec{u} - \nabla \cdot \vec{q} . \tag{8.9}$$

In the next section we will indicate how these hydrodynamic equations can be obtained in terms of statistical expressions of microscopic quantities.

8.2.2. Microscopic Theory

We will no longer be concerned with a continuous fluid as in the last section but rather with a system of N identical particles each having only 3 translational degrees of freedom. The instantaneous state of the system may be described by a point in the $6N$ dimensional phase space where the coordinates of that point are the position and momentum coordinates of all N particles of the system. According to Gibbs we construct an ensemble consisting of a large number of systems, each represented by a point in the $6N$ phase space, such that the distribution of these points is representative of the macroscopic state of the ensemble. This ensemble is characterized by the N-particle distribution function $f^{(N)}(\vec{r}^N, \vec{p}^N, t)$ which gives the relative density of representative points in phase space [6]. $\vec{r}^N$ is composed of $\vec{r}_1, \vec{r}_2, \ldots, \vec{r}_N$, and the same holds for $\vec{p}^N$. The distribution function $f^{(N)}$ is normalized to unity

$$\int f^{(N)}(\vec{r}^N, \vec{p}^N) \mathrm{d}\vec{r}^N \mathrm{d}\vec{p}^N = 1 . \tag{8.10}$$

We may define reduced distribution functions by

$$f^{(n)}(\vec{r}^n, \vec{p}^n) = \frac{N!}{(N-n)!} \int f^{(N)}(\vec{r}^n, \vec{r}^{N-n}; \vec{p}^n, \vec{p}^{N-n}) \mathrm{d}\vec{r}^{N-n} \mathrm{d}\vec{p}^{N-n} , \tag{8.11}$$

where the factor $N!/(N-n)!$ gives the number of different ways of choosing n particles out of N. Integration over the momentum coordinates yields the molecular distribution functions defined in eq. (4.3)

$$n^{(n)}(\vec{r}^n) = \int f^{(n)}(\vec{r}^n, \vec{p}^n) \mathrm{d}\vec{p}^n . \tag{8.12}$$

$f^{(N)}$ changes with time according to the Liouville equation:

$$\frac{\partial f^{(N)}}{\partial t} = \sum_{k=1}^{N} \left[-\frac{\vec{p}_k}{m} \cdot \nabla_{\vec{r}_k} f^{(N)} - \vec{F}_k \cdot \nabla_{\vec{p}_k} f^{(N)} \right] . \tag{8.13}$$

$\vec{F}_k$ is the total force acting on the kth molecule.

Let $\alpha(\vec{r}_1, \ldots, \vec{r}_N, \vec{p}_1 \ldots \vec{p}_N)$ be a dynamical property of the system which does not depend on time explicitely. Then the result of a macroscopic measurement gives no value for α itself but an averaged value which may be expressed by the ensemble average

$$\langle \alpha | f^{(N)} \rangle = \int \alpha(\vec{r}^N, \vec{p}^N) f^{(N)}(\vec{r}^N, \vec{p}^N, t) \mathrm{d}\vec{r}^N \mathrm{d}\vec{p}^N . \tag{8.14}$$

The bracket notation $\langle \ldots | f^{(N)} \rangle$ is due to Irving and Kirkwood [7]. The rate of change of this expectation value is given by the average over $\partial f^{(N)}/\partial t$

$$\frac{\partial}{\partial t} \langle \alpha | f^{(N)} \rangle = \left\langle \alpha \left| \frac{\partial f^{(N)}}{\partial t} \right. \right\rangle , \tag{8.15}$$

where $\partial f^{(N)}/\partial t$ is determined by the Liouville equation (8.13). Now equation (8.15) can be used to derive the equations of hydrodynamics for the average values, which correspond to densities of mass, momentum and energy.

Following Irving and Kirkwood [9] the probability density for the kth molecule to be at $\vec{r}_k$ at time t is given by the $(6N-3)$ fold integral

$$\int f^{(N)}(\vec{r}^N, \vec{p}^N, t)\,\mathrm{d}\vec{r}_1 \ldots \mathrm{d}\vec{r}_{k-1}\,\mathrm{d}\vec{r}_{k+1} \ldots \mathrm{d}\vec{r}_N \mathrm{d}\vec{p}_1 \ldots \mathrm{d}\vec{p}_N, \qquad (8.16)$$

where the integration is over all position coordinates except $\vec{r}_k$ and over all momentum coordinates. With the help of Dirac's δ-function this integration may be extended to include also $\vec{r}_k$, so that the probability density for the kth molecule to be at $\vec{r}$ at time t is

$$\underset{6N}{\int \ldots \int} \delta(\vec{r}_k - \vec{r}) f^{(N)}(\vec{r}^N, \vec{p}^N, t)\,\mathrm{d}\vec{r}^N \mathrm{d}\vec{p}^N = \langle \delta(\vec{r}_k - \vec{r}) | f^{(N)} \rangle \qquad (8.17)$$

The mass density at $\vec{r}$ due to all molecules is therefore given at time t by

$$\rho_m(\vec{r}, t) = m n^{(1)}(\vec{r}, t) = m \sum_{k=1}^{N} \langle \delta(\vec{r}_k - \vec{r}) | f^{(N)} \rangle \qquad (8.18)$$

Similar considerations [7] lead to the mean molecular velocity

$$\vec{u}(\vec{r}, t) = \frac{1}{\rho_m(\vec{r}, t)} \sum_{k=1}^{N} \langle \vec{p}_k \delta(\vec{r}_k - \vec{r}) | f^{(N)} \rangle \qquad (8.19)$$

and to the total energy density $E(\vec{r}, t)$:

$$E(\vec{r}, t) = E_K(\vec{r}, t) + E_\psi(\vec{r}, t) + E_V(\vec{r}, t) \qquad (8.20)$$

$$E_K(\vec{r}, t) = \sum_{k=1}^{N} \left\langle \frac{p_k^2}{2m} \delta(\vec{r}_k - \vec{r}) \middle| f^{(N)} \right\rangle \qquad (8.21)$$

$$E_\psi(\vec{r}, t) = \sum_{k=1}^{N} \langle \psi(\vec{r}_k) \delta(\vec{r}_k - \vec{r}) | f^{(N)} \rangle \qquad (8.22)$$

$$E_V(\vec{r}, t) = \frac{1}{2} \sum \sum_{j \neq k} \langle \varepsilon_{jk} \delta(\vec{r}_k - \vec{r}) | f^{(N)} \rangle. \qquad (8.23)$$

In eq. (8.20), the energy density is made up by contributions of the kinetic energy, the potential energy due to an external potential ψ, and due to intermolecular interactions. E_ψ is connected to the force density in eq. (8.7) which is

$$\vec{f} = - \sum_{k=1}^{N} \langle [\nabla_{\vec{r}_k} \psi(\vec{r}_k)] \delta(\vec{r}_k - \vec{r}) | f^{(N)} \rangle. \qquad (8.24)$$

The intermolecular interaction is taken as sum of pair interactions, and these are assumed to be localized half at each molecule.

After having found statistical expressions for the densities of mass, momentum, and energy, we will indicate the verification of the equations of hydrodynamics with these quantities. If we compare the definition of ρ_m in eq. (8.18) with the general expression of averages in eq. (8.14), we see that α must be taken as

$$\alpha_m = \sum_{k=1}^{N} m \delta(\vec{r}_k - \vec{r}) \qquad (8.25)$$

in order to obtain ρ_m. Inserting this α_m into eq. (8.15) and making use of the Liouville eq. (8.13), $\partial \rho_m/\partial t$ is given in terms of averages of α_m over space and momentum derivatives of $f^{(N)}$. Integration by parts gives then $\partial \rho_m/\partial t$ in terms of $\langle \nabla_{\vec{r}} \cdot \vec{p}_k \delta(\vec{r}_k - \vec{r}) | f^{(N)} \rangle$ which leads to the equations of continuity [cf. (8.5) and (8.19)]. In a similar way one can proceed with $\vec{u}(\vec{r},t)$ and $E(\vec{r},t)$ given in eq. (8.19) and (8.20) and obtain equations equivalent to the equation of motion (8.7) and the energy transport equation (8.9). For details we refer to the original work [7] and to monographs [2, 4]. The comparison of the resultant equations with the hydrodynamic equations furnishes statistical expressions for the stress tensor $\overleftrightarrow{\sigma}$ and the heat current density $\vec{q}$. These expressions can be separated into kinetic contributions and contributions due to particle interactions. They can be greatly simplified, as eq. (8.18) to (8.23) do not necessitate the simultaneous consideration of positions and momenta of all particles. Remembering the meaning of the bracket notation [eq. (8.14)], we can integrate over the phase space of $N - 1$ molecules in the kinetic contributions and over the subspace of $N - 2$ molecules in the contributions due to pair interactions, thus reducing $f^{(N)}$ to the single- and two-particle distribution functions defined in equation (8.11). This gives finally

$$\overleftrightarrow{\sigma} = \overleftrightarrow{\sigma}_K + \overleftrightarrow{\sigma}_V \tag{8.26}$$

$$\overleftrightarrow{\sigma}_K(\vec{r},t) = -m \int \left(\frac{\vec{p}_1}{m} - \vec{u}\right)\left(\frac{\vec{p}_1}{m} - \vec{u}\right) f^{(1)}(\vec{r},\vec{p}_1) \mathrm{d}\vec{p}_1 \tag{8.27}$$

$$\overleftrightarrow{\sigma}_V(\vec{r},t) = \frac{1}{2} \int \vec{R}_{12} \nabla_{\vec{R}_{12}} \varepsilon(\vec{R}_{12}) f^{(2)}(\vec{r},\vec{p}_1;\vec{r}+\vec{R}_{12},\vec{p}_2) \mathrm{d}\vec{R}_{12} \mathrm{d}\vec{p}_1 \mathrm{d}\vec{p}_2 \tag{8.28}$$

$$\vec{q} = \vec{q}_K + \vec{q}_V \tag{8.29}$$

$$\vec{q}_K(\vec{r},t) = \frac{m}{2} \int \left(\frac{\vec{p}_1}{m} - \vec{u}\right)^2 \left(\frac{\vec{p}_1}{m} - \vec{u}\right) f^{(1)}(\vec{r},\vec{p}_1) \mathrm{d}\vec{p}_1 \tag{8.30}$$

$$\vec{q}_V(\vec{r},t) = \frac{1}{2} \int (\varepsilon_{12} \overleftrightarrow{1} - \vec{R}_{12} \nabla_{\vec{R}_{12}} \varepsilon_{12}) \cdot \left(\frac{\vec{p}_1}{m} - \vec{u}\right) f^{(2)}(\vec{r},\vec{p}_1;\vec{r}+\vec{R}_{12},\vec{p}_2) \mathrm{d}\vec{p}_1 \mathrm{d}\vec{p}_2 \mathrm{d}\vec{R}_{12} \tag{8.31}$$

The right hand sides of these equations depend on time over the time variation of the local velocity $\vec{u}$ and of the distribution functions. $\vec{R}_{12}$ is an abbreviation for $\vec{r}_1 - \vec{r}_2$. The kinetic contributions $\overleftrightarrow{\sigma}_K$ and $\vec{q}_K$ are caused by the flux of momentum or kinetic energy due to the translational random movement of individual particles seen from a center of mass coordinate system moving with $\vec{u}$. The contributions of $\overleftrightarrow{\sigma}_K$ and $\vec{q}_K$ are dominating in the theory of gases.

On the other hand, the interaction terms $\overleftrightarrow{\sigma}_V$ and $\vec{q}_V$ are the dominating terms in dense fluids*). $\overleftrightarrow{\sigma}_V \cdot \mathrm{d}\vec{S}$ is the force acting across a boundary $\mathrm{d}\vec{S}$ due to the interaction of molecules on opposite sides of the boundary. $\vec{q}_V$ represents the current density of (1) potential energy (the term that involves ε_{12}) plus (2) the work done on molecules on one side of $\mathrm{d}\vec{S}$ by molecules on the other (the term that involves $\nabla_{\vec{R}_{12}} \varepsilon_{12}$) because a molecule moves in the force field of the others.

Nothing has appeared in the theory so far to account for the irreversible behaviour of the fluid. We have seen that also in the macroscopical theory the description of the irreversible behaviour needed the empirical relations (8.1) and (8.3) which connected stress tensor $\overleftrightarrow{\sigma}$ and heat flux $\vec{q}$ to gradients of thermodynamic variables.

We shall see that in order to derive the functions $f^{(1)}$ and $f^{(2)}$ in the microscopical theory additional assumptions are necessary (chaos assumptions) which bring irreversibility into the theory. Once $f^{(1)}$ and $f^{(2)}$ are known one can transform $\overleftrightarrow{\sigma}$ and $\vec{q}$ in eq. (8.26–31) into a form which may be directly compared to the form of eq. (8.1) and (8.3). This comparison gives explicit expressions

---

* In the course of the derivation of $\overleftrightarrow{\sigma}_V$ and $\vec{q}_V$, $f^{(2)}$ has been assumed to be a slowly varying function of position. This assumption is good for higher densities, where $\overleftrightarrow{\sigma}_V$ and $\vec{q}_V$ attain non-negligible values.

for the transport coefficients λ, η and ϕ. The diffusion coefficient D will be treated separately. Thus the problem has reduced to that of obtaining $f^{(1)}$ and $f^{(2)}$ with which we shall be concerned during the next sections. We note that the two-particle function $f^{(2)}$ is only necessary if the interaction terms $\vec{\sigma}_V$ and $\vec{q}_V$ are to be evaluated.

8.2.3. The Kinetic Equations

The theoretically most satisfying way of obtaining expressions for the distribution functions $f^{(1)}$ and $f^{(2)}$ starts with the Liouville equation (8.13), which is an equation for the N-particle distribution function $f^{(N)}$. As $f^{(N)}$ depends on the position and momenta of all particles and on time, an explicit solution of the Liouville equation for $f^{(N)}$ is impossible. But fortunately, we need only $f^{(1)}$ and $f^{(2)}$ for the evaluation of transport coefficients. Thus one can integrate the Liouville equation over the phase space of the other particles. The result is a hierarchy of differential equations where the equation for $f^{(1)}$ contains also $f^{(2)}$, that for $f^{(2)}$ contains $f^{(3)}$ and so on. One must cut off this hierarchy with the help of additional assumptions in order to obtain a single differential equation for $f^{(1)}$ and another equation for $f^{(2)}$. These equations are the so-called kinetic equations and describe the time and phase space dependence of the corresponding distribution function. They contain already irreversibility, which has been introduced by the assumptions needed to truncate the hierarchy [4, 8, 9]. With other words the loss of information in going over from $f^{(N)}$ to $f^{(1)}$ or $f^{(2)}$ makes additional conceptions necessary like that of equilibrium or non-equilibrium.

It is not possible to make this step from a reversible hierarchy of N equations to irreversible kinetic equations for $f^{(1)}$ and $f^{(2)}$ in full generality so that the theory is applicable to all fluid states and all possible intermolecular interactions. Every attempt to formulate these additional assumptions which are necessary to truncate the hierarchy is only the starting point of a model which is applicable for a rather restricted range of fluid states. In the next sections we will sketch briefly some of these models. But instead of following the general concept which has been indicated we will obtain $f^{(1)}$ and $f^{(2)}$ in a more intuitive way, and not by truncating the hierarchy of the reduced distribution functions which emerge from the Liouville equation. Of course, essentially the same additional assumptions are underlying the kinetic equations in both cases, but we hope that the somewhat more intuitive way gives the advantage of an easier understanding of the physical content otherwise hidden behind formal derivations.

8.3. The Kinetic Equations Derived from a Balance of Binary Collisions

8.3.1. The Boltzmann-Equation for Dilute Gases

For a detailed outline of this theory the interested reader is referred to several excellent treatments of this problem [10, 11]. As we shall see, it is sufficient for the simple case of a dilute gas to know $f^{(1)}$ alone. We note that $f^{(1)} d\vec{r} d\vec{p}$ is proportional to the number of particles located at time t in the 6 dimensional one particle phase space within the element $d\vec{r}$ around $\vec{r}$ and $d\vec{p}$ around $\vec{p}$.

To get the equation of motion for $f^{(1)}$ we first consider the case that no collisions occur at all. Then all molecules in a volume element $(d\vec{r} d\vec{p})$ at $(\vec{r}, \vec{p})$ at time t will be found in the volume

8.3. The Kinetic Equations Derived from a Balance of Binary Collisions

element $(\mathrm{d}\vec{r}'\mathrm{d}\vec{p}')$ located at $(\vec{r} + \frac{\vec{p}}{m}\mathrm{d}t, \vec{p} + \vec{K}\mathrm{d}t)$ an infinitesimal time element $\mathrm{d}t$ later, and therefore

$$f^{(1)}\left(\vec{r} + \frac{\vec{p}}{m}\mathrm{d}t, \vec{p} + \vec{K}\mathrm{d}t, t + \mathrm{d}t\right)\mathrm{d}\vec{r}'\mathrm{d}\vec{p}' = f^{(1)}(\vec{r},\vec{p},t)\mathrm{d}\vec{r}\,\mathrm{d}\vec{p}\,. \tag{8.32}$$

$\vec{K}$ is the external force acting on the particles.

If collisions occur during $\mathrm{d}t$, some molecules which are around $(\vec{r},\vec{p})$ at time t will not reach the volume element around $\left(\vec{r} + \frac{\vec{p}}{m}\mathrm{d}t, \vec{p} + \vec{K}\mathrm{d}t\right)$. Also other particles from outside, which have not been in $(\mathrm{d}\vec{r}\,\mathrm{d}\vec{p})$ around $(\vec{r},\vec{p})$ at time t, will enter $(\mathrm{d}\vec{r}'\mathrm{d}\vec{p}')$ during $\mathrm{d}t$. Denoting those collisional contributions per unit time as $J(f(1),f(2))$, equation (8.32) takes the form (since $\mathrm{d}\vec{r}\,\mathrm{d}\vec{p} = \mathrm{d}\vec{r}'\mathrm{d}\vec{p}'$ from Liouville's theorem)

$$f^{(1)}\left(\vec{r} + \frac{\vec{p}}{m}\mathrm{d}t, \vec{p} + \vec{K}\mathrm{d}t, t + \mathrm{d}t\right) - f^{(1)}(\vec{r},\vec{p},t) = J(f(1),f(2))\mathrm{d}t\,. \tag{8.33}$$

The l.h.s. of this equation is now expanded to order $\mathrm{d}t$ and one finds

$$\left(\frac{\partial}{\partial t} + \frac{\vec{p}}{m}\cdot\nabla_{\vec{r}} + \vec{K}\cdot\nabla_{\vec{p}}\right)f^{(1)}(\vec{r},\vec{p},t) = J(f(1), f(2))\,. \tag{8.34}$$

For an explicit evaluation of the collision term J some assumptions are necessary. The two most important of these are:

1. Only binary collisions are considered. By this assumption the theory is restricted to dilute gases.

2. There is no correlation between position and velocity of the colliding particles. Formally this condition may be written

$$f^{(2)}(\vec{r}_1,\vec{p}_1;\vec{r}_2,\vec{p}_2) = f^{(1)}(\vec{r}_1,\vec{p}_1)f^{(1)}(\vec{r}_2,\vec{p}_2)\,. \tag{8.35}$$

It is called the assumption of molecular chaos or "Stoßzahlansatz". As has been mentioned in section 8.2.3 an analogous assumption must be made also when the Boltzmann equation is derived from the (reversible) Liouville equation. Eq. (8.35) introduces irreversibility into the theory.

A detailed analysis of binary encounters is necessary to derive the expression for $J(f(1),f(2))$ (see [10–12]), which will be omitted here. We only note that it is linear in $f(1)$ and $f(2)$.

The equilibrium state is characterized by vanishing of the l.h.s. of eq. (8.34):

$$0 = J(f_0(1),f_0(2)) \tag{8.36}$$

where the equilibrium or unperturbed distribution function is denoted by f_0. If the statistical expressions of mass, density, velocity, and kinetic energy are taken into account [derived from (8.18)–(8.21) by integration over the subspace of $N-1$ molecules] [*]

$$\rho_m(\vec{r}) = m\int f_0(\vec{r},\vec{p}_1)\mathrm{d}\vec{p}_1 = m n^{(1)}(\vec{r}) \tag{8.37}$$

$$\vec{u}(\vec{r}) = \frac{1}{\rho_m}\int \vec{p}_1 f_0(\vec{r},\vec{p}_1)\mathrm{d}\vec{p}_1 \tag{8.38}$$

---

[*] In the general non-equilibrium case f_0 must be replaced by f in eq. (8.37) – (8.40).

$$E_K(\vec{r}) = \frac{1}{2m} \int p_1^2 f_0(\vec{r}, \vec{p}_1) \mathrm{d}\vec{p}_1 , \qquad (8.39)$$

and if the temperature is defined by its relation to the kinetic energy of random motion

$$T(\vec{r}) = \frac{1}{3k\rho_m} \int (\vec{p}_1 - m\vec{u})^2 f_0 \mathrm{d}\vec{p}_1 , \qquad (8.40)$$

eq. (8.36) yields the Maxwellian equilibrium function:

$$f_0 = \rho \left(\frac{1}{2\pi mkT} \right)^{3/2} \exp\left\{ -\frac{(\vec{p} - m\vec{u})^2}{2mkT} \right\} . \qquad (8.41)$$

Passing now from the equilibrium case to the non-equilibrium state, the l.h.s. of eq. (8.34) may be conceived as the perturbation which changes f_0 to a perturbed distribution function f. Using such a perturbation method Enskog and Chapman [10] have shown how the Boltzmann equation (8.34) may be solved for the special case, that the time dependence of $f^{(1)}$ is completely determined by that of the local variables, which are density, mean velocity, and temperature [*]:

$$\frac{\partial f}{\partial t} = \frac{\partial f}{\partial \rho_m} \frac{\partial \rho_m}{\partial t} + \nabla_{\vec{u}} f \cdot \frac{\partial \vec{u}}{\partial t} + \frac{\partial f}{\partial T} \frac{\partial T}{\partial t} . \qquad (8.42)$$

The method consists in an expansion of f,

$$f = f_0 + \lambda f_1 + \lambda^2 f_2 + \cdots , \qquad (8.43)$$

where λ is the perturbation parameter. In the following we consider only the first order correction f_1 (which is assumed $\ll f_0$). Eq. (8.43) is inserted into the Boltzmann equation and yields

$$\lambda \left[\frac{\partial}{\partial t} + \frac{\vec{p}_1}{m} \cdot \nabla_{\vec{r}} + \vec{K} \cdot \nabla_{\vec{p}} \right] f_0(1) = J(f_0(1) + \lambda f_1(1), f_0(2) + \lambda f_1(2)) . \qquad (8.44)$$

Equating all terms of equal power in λ gives:

$$(\lambda^0): \quad 0 = J(f_0(1), f_0(2)) \qquad (8.45\,\mathrm{a})$$

$$(\lambda^1): \quad \left(\frac{\partial}{\partial t} + \frac{\vec{p}_1}{m} \cdot \nabla_{\vec{r}} + \vec{K} \cdot \nabla_{\vec{p}} \right) f_0(1) = J(f_0(1), f_1(2)) + J(f_1(1), f_0(2)) . \qquad (8.45\,\mathrm{b})$$

Eq. (8.45b) allows to determine f_1. From eq. (8.45a) we confirm that f_0 is the unperturbed distribution function (8.41) which is already known. So the l.h.s. of eq. (8.45b) can be worked out. The quantity $\partial f_0/\partial t$ is given by eq. (8.42). The time derivatives of ρ_m, $\vec{u}$ and T must be taken from the hydrodynamic equations (8.5), (8.7), and (8.9). E.g., the energy transport equation (8.9) becomes after insertion of the temperature

$$\frac{3k\rho_m}{2m} \left(\frac{\partial}{\partial t} + \vec{u} \cdot \nabla \right) T = \bar{\sigma}_K : \nabla \vec{u} - \nabla \cdot \vec{q}_K . \qquad (8.46)$$

A difficulty arises, since the stress tensor $\bar{\sigma}_K$ [eq. (8.27)] and heat flux vector $\vec{q}_K$ [eq. (8.30)] [**] which appear in the hydrodynamic equations are defined with the help of the unknown total distribution function f. In the first approximation f in eq. (8.27) and (8.30) is replaced by the equilibrium function f_0 which gives the equilibrium values $(\bar{\sigma}_K)_0 = -\rho k T \vec{1}$ and $(\vec{q}_K)_0 = 0$.

---

[*] That means that we consider only time elements (hydrodynamic region) which are much longer than those of molecular fluctuations (kinetic region).
[**] The subscript K remembers that in dilute gases only the kinetic contributions to $\bar{\sigma}$ and $\vec{q}$ are important.

Carrying out all manipulations as indicated, eq. (8.45b) becomes

$$f_0\left\{\left(\frac{mC^2}{2kT} - \frac{5}{2}\right)\vec{C}\cdot\nabla\ln T + \frac{m}{kT}\left(\vec{C}\vec{C} - \frac{1}{3}C^2\vec{I}\right):\nabla\vec{u}\right\} =$$
$$= J(f_0(1), f_1(2)) + J(f_1(1), f_0(2)) \,. \tag{8.47}$$

The abbreviation

$$\vec{C} = \frac{\vec{p}}{m} - \vec{u} \tag{8.48}$$

has been used for the velocity relative to the mean velocity. Now the perturbation term f_1 can be expressed by

$$f_1 = f_0\left[A(C)\vec{C}\cdot\nabla\ln T + B(C)(\vec{C}\vec{C} - C^2\vec{I}/3):\nabla\vec{u}\right], \tag{8.49}$$

where the functions A and B are expanded in a set of orthogonal polynomials and the expansion coefficients evaluated from eq. (8.47) [10].

With the help of the perturbed distribution function $f = f_0 + f_1$ the stress tensor $\vec{\sigma}_K$ [eq. (8.27)] and the heat flux vector $\vec{q}_K$ (8.30) can be calculated. The contributions $\vec{\sigma}_V$ and $\vec{q}_V$ (which depend on pair interactions and would need the introduction of two-particle distribution functions) are neglected, which is justified for low densities.

Comparing the calculated quantities $\vec{\sigma}_K$ and $\vec{q}_K$ with the empirical eq. (8.1) and (8.3), the transport coefficients λ, η, and ϕ can be evaluated. We do not give the results, as we are not interested in dilute gases. We have only outlined the theory as a model of a kinetic equation for $f^{(1)}$, which is basic for the more realistic models of Enskog, and Rice-Allnatt. With the extension to more dense systems we will deal in the following sections.

One remark is necessary at this point. It has been shown by Kirkwood that the Boltzmann eq. (8.34), when it is derived from the Liouville equation as indicated above, is not satisfied by the exact distribution function $f^{(1)}$ but by a function $\bar{f}^{(1)}$, averaged over a time interval and defined by [8]

$$\bar{f}^{(1)}(\vec{r},\vec{p},t) = \frac{1}{\tau}\int_0^\tau f^{(1)}(\vec{r},\vec{p},t+s)\,ds\,. \tag{8.50}$$

This procedure is called coarse graining (in time) and means, that the information, necessary to extract transport coefficients, is contained in a distribution function smoothed over an interval τ. With the help of this method the duration of correlations for a pair of colliding molecules may be formally limited to time intervals τ so that successive time intervals are uncorrelated. This is strongly connected to the way how irreversibility is introduced into the theory. For a deeper discussion of these problems the reader is referred to ref. [4]. In the case of the Boltzmann equation, τ must be taken long compared to the collision time but short compared to times between successive collisions. In the case of hard spheres, τ may approach zero.

8.3.2. Dense Gases: The Enskog-Theory

The derivation of the Boltzmann equation has been restricted to binary collisions. This restriction is certainly not valid at higher densities because of the long ranging attractive part of the real pair potential. This difficulty does not arise in the case of a model fluid of hard spheres which was treated first by Enskog and Chapman [10, 13]. In dense systems the ratio of core diameter σ to mean free path between collisions becomes large, whereas in the dilute Boltzmann case the average distance travelled between collisions is large so that the finite size of the molecules could be neglected. The extension of the Boltzmann equation to the case of a hard sphere fluid at any density may be indicated as follows:

(1) The collision time between hard spheres is zero and so is the probability of three and more particle collisions. As in the Boltzmann case it is sufficient to treat binary collisions.

(2) The molecular chaos assumption [eq. (8.35)] has to allow for correlations of position at higher densities:

$$f^{(2)}(\vec{r}_1,\vec{p}_1;\vec{r}_2,\vec{p}_2) = f^{(1)}(\vec{r}_1,\vec{p}_1)f^{(1)}(\vec{r}_2,\vec{p}_2)g(R_{12}). \tag{8.51}$$

In eq. (8.51) positional correlation has been introduced by the pair distribution function $g(R_{12})$ [cf. eq. (4.12)]. Correlations of the momenta of the colliding molecules are still neglected.

(3) Because of the finite diameter σ of the hard spheres the distribution functions $f^{(1)}$ in the collisional integral eq. (8.34) must not be taken at the same space point $\vec{r}$ at the moment of collision, but at space points which are the distance σ apart. If $\vec{k}$ denotes the unit vector in the direction of the line of centres of colliding hard spheres the Enskog equation may be written

$$\left[\frac{\partial}{\partial t} + \frac{\vec{p}_1}{m}\cdot\nabla_{\vec{r}_1} + \vec{K}\cdot\nabla_{\vec{p}_1}\right]f^{(1)} = J(f(\vec{r}_1,\vec{p}_1,t), f(\vec{r}_1 \pm \sigma\vec{k},\vec{p}_2,t)). \tag{8.52}$$

This equation can be solved along similar lines as the Boltzmann equation.

(4) In dense gases not only the translational motion between collisions [described by $\vec{\sigma}_K$ and $\vec{q}_K$ in eq. (8.26)–(8.31)] will contribute to the total stress tensor $\vec{\sigma}$ and energy flux $\vec{q}$, but momentum and energy is transported also for a distance σ by an encounter. This "collisional transfer" may be best understood if one considers a plane in space and two molecules on each side but almost in contact with each other. If they collide, momentum and energy is transported across the plane even if both molecules remain on their side. This results in an appreciable contribution to $\vec{\sigma}_V$ and $\vec{q}_V$ [from eq. (8.26)–(8.31)]. The expressions for $\vec{\sigma}_V$ and $\vec{q}_V$ have to be modified to take account of the singularity of $\partial\varepsilon/\partial R$ which arises in the hard sphere potential. Accounting for all these complications the following expressions result for the transport properties [10]

$$\eta = \frac{\eta_0}{g(\sigma)}\left[1 + \frac{4}{15}\pi\rho\sigma^3 g(\sigma)\right]^2 + \frac{3}{5}\phi \tag{8.53}$$

$$\eta_0 = \frac{5}{16}\frac{(\pi mkT)^{1/2}}{\sigma^2\pi} \tag{8.54}$$

$$\phi = \frac{4}{9}(\pi mkT)^{1/2}\rho^2\sigma^4 g(\sigma) \tag{8.55}$$

$$\lambda = \frac{\lambda_0}{g(\sigma)}\left[1 + \frac{2}{5}\pi\rho\sigma^3 g(\sigma)\right]^2 + \frac{3k}{2m}\phi \tag{8.56}$$

$$\lambda_0 = \frac{25}{32}\frac{(\pi mkT)^{1/2}}{\sigma^2\pi}\frac{3k}{2m} \tag{8.57}$$

The quantities η_0 and λ_0 are taken in the limit of low densities. Eq. (8.53) and (8.56) give the corrections for higher densities. The quantity $3k/2m$ is the heat capacity per unit mass. The bulk viscosity ϕ is proportional to the square of the number density ρ and contributes to the higher order corrections of λ and η.

Nothing has been said yet about the coefficient of self-diffusion. For this quantity, the kinetic equations have to be formulated in terms of a two component mixture, the two components having the same properties. This more complicated formulation will not be given here and only the results will be reported. But it should be remarked that the diffusion flux contains, differently from momentum or energy flux, only a kinetic and no collisional part. The result is

$$D = \frac{3}{8} \frac{(\pi m k T)^{1/2}}{\sigma^2 \pi} \frac{1}{\rho_m} \frac{1}{g(\sigma)}. \tag{8.58}$$

As eq. (8.58) shows, the diffusion coefficient is inversely proportional to the density. The factor $1/g(\sigma)$ is the only correction which distinguishes the diffusion coefficient at higher densities from the low density limit, where $g(\sigma) = 1$.

The comparison of the calculated values for η and λ with experimental data for real gases depends on the values of σ and $g(\sigma)$. Values of these parameters have been chosen from several points of view [14]. We give here briefly the procedure adopted by Dymond and Alder [15]. Values of $g(\sigma)$ are taken from the hard sphere fluid (cf. chapter 3 and 7). For σ, adjusted values are calculated on the basis of a van der Waals treatment of equilibrium properties. Writing

$$\frac{PV}{N_A k T} = f\left(\frac{V}{N_A \sigma^3}\right) - \frac{a}{N_A k T V} \tag{8.59}$$

where $f(V/N_A \sigma^3)$ is the expression for a hard sphere fluid and $-a/V$ is the uniform background potential (cf. section 2.3 and 6.4), it is seen that a plot of the experimental values of $PV/N_A kT$ at constant volume versus $1/T$ should lead to a straight line whenever the van der Waals model is applicable. From the intercept $f(V/N_A \sigma^3)$ the parameter σ can be obtained. In practice the data for the rare gases lie on a slight curve, so that the value of the intercept depends on the temperature at which the tangent to the curve is drawn. By this way temperature-dependent values of σ can be deduced. The treatment gives nearly volume-independent values of σ only at densities above the critical densities. At lower densities (and temperatures not exceeding $3 T_c$) σ turns out to be volume dependent also, which indicates a breakdown of the van der Waals model. The resultant σ for higher densities are shown in fig. 8.1. With these values, a remarkable fit of the transport properties for higher densities and not too low temperatures has been obtained (fig. 8.2 to 8.4).

There are two limitations in the application of the Enskog theory to real dense fluids which concern the regions of high and of low temperatures. The first point is the chaos assumption made in eq. (8.51). At high densities the particles are practically reflected by the cage of next neighbours, so that the chaos assumption with respect to the momenta of the colliding particles becomes inadequate. This back scattering (dominating anticorrelation of the momenta of colliding particles) has been proved to occur preferentially for high densities by molecular dynamics calculations [15]. Its effect on the transport properties is a slowing down of transport. The second point concerns the neglect of "attractive" collisions. The application of the Enskog model to real dense fluids implies the validity of the van der Waals model, i.e. the absence of any inhomogeneity in the attractive background potential. The particles are allowed to travel in straight lines between the hard core collisions. The deviation of the particle trajectories from straight lines due to the inhomogeneous attractive fields of neighbouring particles will become the more pronounced, the lower the kinetic energy, i.e. the lower the temperature. The effect

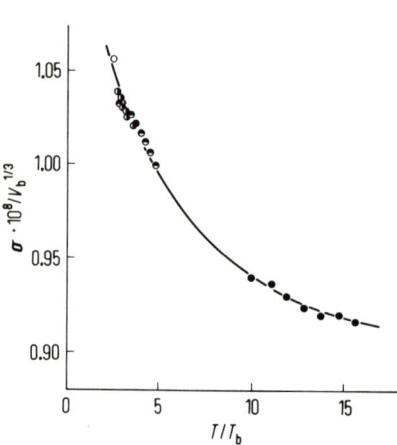

Fig. 8.1. The effective hard core diameter as a function of temperature, relative to volume and temperature at the boiling point, for the rare gases (● neon, ◐ argon, ◑ krypton, ○ xenon). After [15].

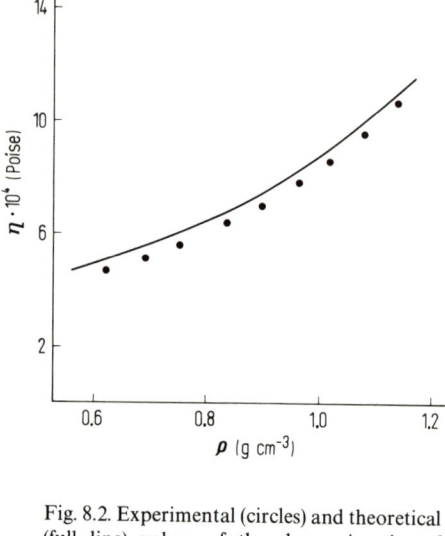

Fig. 8.2. Experimental (circles) and theoretical (full line) values of the shear viscosity of argon at 258 K as a function of density. After [15].

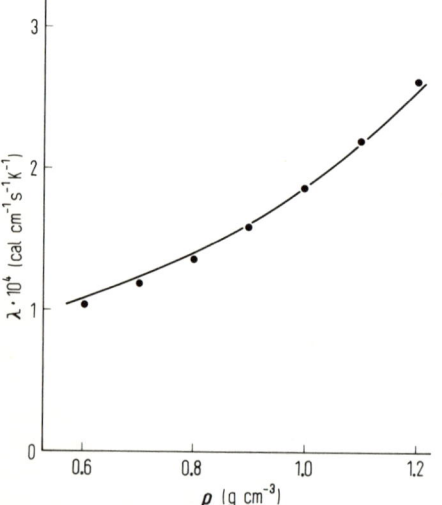

Fig. 8.3. Experimental (circles) and theoretical (full line) values of the thermal conductivity of argon at 258 K as a function of density. After [15].

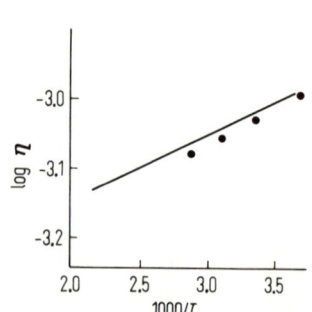

Fig. 8.4. Experimental (dots) and theoretical (full line) values of the logarithm of shear viscosity of argon at a constant pressure of 1373.5 atm as a function of reciprocal temperature. After [15].

of these "attractive" collisions is to break the correlation between successive hard core collisions, i.e. counteracting the effect of back scattering. So at high temperatures deviations from the Enskog theory occur on account of back scattering, whereas the influence of "attractive" collisions make the quantitative application of the Enskog theory impossible at low temperatures, i.e. in the truly liquid range. At temperatures around the critical and slightly above both corrections nearly cancel each other, and this explains the remarkable agreement between Enskog theory and experiment discussed above.

8.4. The Brownian Motion

A theory which avoids the explicit solution of collision integrals is the theory of Brownian motion for heavy particles immersed in a solvent and moving slowly under the influence of many successive collisions with solvent molecules. This theory is based on the concept of random processes [16]. Kirkwood has shown that this theory can be extended to treat also the molecular motion of an one-component fluid.

Let us consider three successive states of a particle, (y_1, t), $(y_2, t + \tau)$, and $(y_3, t + 2\tau)$, where y may stand for a position or momentum coordinate. Then the theory of Brownian motion is restricted to processes, where the probability of finding the particle at y_3 at time $t + 2\tau$ depends only on the population of states y_2 at time $t + \tau$ and the transition probability from y_2 to y_3 within the interval τ, but not on the state (y_1, t). Such processes are called Markov processes. With other words, the process at time $t + \tau$ has forgotten the value of y at time t, and the future of the process depends only on y_2 at $t + \tau$ and not on earlier stages of the process.

Such characteristics of a process are typical for the case that the time resolution τ of our observation is much coarser than the characteristic time τ_c during which changes in y occur. Thus we must have $\tau_c \ll \tau$. We have already dealt with such a process in the discussion of Monte Carlo calculations (section 3.2). Denoting again transition probabilities from one state y_i to another state y_j within a time interval τ by $p_{ij}(\tau)$, or more elaborately by $p(y_i|y_j;\tau)$, we obtain in analogy to eq. (3.7) the so-called Smoluchowsky equation

$$p(y_1|y_3; t + \tau) = \int p(y_1|y_2; t) p(y_2|y_3; \tau) \mathrm{d}y_2 . \tag{8.60}$$

If we assume that the changes $y_3 - y_2 = \Delta y$ during τ are small compared to $y_3 = y$ itself we may write (incorporating the sign into the integration limits)

$$p(y_1|y; t + \tau) = \int p(y_1|y - \Delta y; t) p(y - \Delta y|y; \tau) \mathrm{d}\Delta y . \tag{8.61}$$

Eq. (8.61) may be further transformed by multiplication with the probability $u(y_1)$ of the state y_1 and integration over y_1 [cf. eq. (3.5)] to give

$$u(y; t + \tau) = \int u(y - \Delta y; t) p(y - \Delta y|y; \tau) \mathrm{d}\Delta y . \tag{8.62}$$

Eq. (8.62) is another version of Smoluchowsky's equation. From (8.62) or (8.61) kinetic equations for the distribution functions u or p can be deduced, which are called Fokker-Planck equations.

8.4.1. Particle Motion in Configurational Space

As a first example of the application of Smoluchowsky's equation we treat the molecular motion in configurational space. The state variable $y(t)$ is identified with the position $\vec{r}(t)$ of a particle.

The random process $\vec{r}(\tau)$ will be Markovian if the time scale τ of our observation is much greater than the characteristic time τ_c during which changes of $\vec{r}(t)$ occur. In this case τ_c is the relaxation time of density fluctuations. The theory will fail for time intervals smaller or comparable to τ_c. We will change the notation for the transition probability p in eq. (8.61) and define

$$p(0|\vec{r};t) = G_s(\vec{r},t). \tag{8.63}$$

$G_s(\vec{r},t)d\vec{r}$ is the probability of finding the particle in a volume element $d\vec{r}$ around $\vec{r}$ at time t, if it has been at the origin at time 0. It follows that

$$G_s(\vec{r},0) = \delta(\vec{r}) \tag{8.64}$$
$$\int G_s(\vec{r},t)d\vec{r} = 1. \tag{8.65}$$

$G_s(\vec{r},t)$ is called selfcorrelation function. Now eq. (8.61) may be written

$$G_s(\vec{r},t+\tau) = \int G_s(\vec{r}-\Delta\vec{r},t) G_s(\Delta\vec{r},\tau) d\Delta\vec{r}, \tag{8.66}$$

where $\Delta\vec{r}$ is the displacement of the particle during τ. $G_s(\vec{r}, t+\tau)$ is now expanded in a Taylor series in powers of τ and $G_s(\vec{r}-\Delta\vec{r},t)$ in powers of $\Delta\vec{r}$:

$$G_s(\vec{r},t) + \frac{\partial G_s(\vec{r},t)}{\partial t}\tau + \cdots = \int d\Delta\vec{r}\, G_s(\Delta\vec{r},\tau)\left[G_s(\vec{r},t) - \nabla_{\vec{r}}\cdot(\Delta\vec{r}\, G_s(\vec{r},t)) + \right.$$

$$\left. + \frac{1}{2}\nabla_{\vec{r}}\nabla_{\vec{r}}:(\Delta\vec{r}\,\Delta\vec{r})\, G_s(\vec{r},t) + \cdots \right]. \tag{8.67}$$

Thus

$$\tau \frac{\partial G_s(\vec{r},t)}{\partial t} = -\nabla_{\vec{r}}\cdot(\langle\Delta\vec{r}\rangle G_s(\vec{r},t)) + \frac{1}{2}\nabla_{\vec{r}}\nabla_{\vec{r}}:(\langle\Delta\vec{r}\,\Delta\vec{r}\rangle G_s(\vec{r},t)), \tag{8.68}$$

where $\langle\,\rangle$ denotes the corresponding averages over $G_s(\Delta\vec{r},\tau)$. Eq. (8.68) is the Fokker-Planck equation for $G_s(\vec{r},t)$. It can be simplified by noticing that $\langle\Delta\vec{r}\rangle$ is zero for equilibrium conditions. Furthermore diffusion is isotropic, so that

$$\langle\Delta\vec{r}\,\Delta\vec{r}\rangle = \frac{1}{3}\langle(\Delta\vec{r})^2\rangle \overleftrightarrow{1}. \tag{8.69}$$

With these simplifications one obtains from eq. (8.68)

$$\frac{\partial G_s(\vec{r},t)}{\partial t} = \frac{\langle(\Delta\vec{r})^2\rangle}{6\tau}\cdot\nabla_{\vec{r}}^2 G_s(\vec{r},t). \tag{8.70}$$

If the quantity

$$D = \frac{\langle(\Delta\vec{r})^2\rangle}{6\tau} \tag{8.71}$$

is considered to be independent of τ, the diffusion equation

$$\frac{\partial G_s(\vec{r},t)}{\partial t} = D\nabla_{\vec{r}}^2 G_s(\vec{r},t). \tag{8.72}$$

is obtained.

Eq. (8.71) gives an expression for the self diffusion coefficient which is correct at long times τ such that the displacements of the same particle during two successive time intervals of length τ are uncorrelated. The shortest τ for which this assumption holds may be called a characteristic self diffusion time τ_0, and may be estimated on the basis of computer calculations (cf. section 9.3.2).

8.4.2. Kinetic Equations in Phase Space

We have seen that consecutive positions $\vec{r}(t)$ of a particle may be treated as a Markov process. The same holds for the phase point $(\vec{r}(t), \vec{p}(t))$ in the single particle phase space if we consider only time intervals $\tau \gg \tau_c$ where τ_c is the characteristic time of molecular fluctuations. On the other hand we assume that τ is short so that the relative momentum change of the particle during τ is small; then we may write (to order τ^2)

$$\Delta \vec{r} = \frac{\vec{p}}{m} \tau . \tag{8.73}$$

This is an important restriction which we shall have to comment on later. With this assumption the Smoluchowsky equation (8.62) may be rewritten for the single particle distribution function $f^{(1)}$:

$$f^{(1)}(\vec{r}, \vec{p}, t + \tau) = \int f^{(1)}\left(\vec{r} - \frac{\vec{p}}{m}\tau, \vec{p} - \Delta \vec{p}; t\right) \cdot p\left(\vec{r} - \frac{\vec{p}}{m}\tau, \vec{p} - \Delta \vec{p} \mid \vec{r}, \vec{p}; \tau\right) d\Delta p . \tag{8.74}$$

Similar to eq. (8.67), we expand $f^{(1)}$ and the transition probability p in Taylor series. We shall not give the details of this calculation [4] but only quote the result. If only first order terms in τ are retained, the Fokker-Planck equation for $f^{(1)}$ becomes:

$$\tau \left\{ \frac{\partial f^{(1)}}{\partial t} + \frac{\vec{p}}{m} \cdot \nabla_{\vec{r}} f^{(1)} \right\} = -\nabla_{\vec{p}} \cdot (\langle \Delta \vec{p} \rangle f^{(1)}) + \frac{1}{2} \nabla_{\vec{p}} \nabla_{\vec{p}} : (\langle \Delta \vec{p} \Delta \vec{p} \rangle f^{(1)}) . \tag{8.75}$$

Eq. (8.75) could have been obtained directly from the Liouville equation, or more precisely from the hierarchy of the reduced distribution functions $f^{(n)}$ [8, 17]. The introduction of the statistical concept of a Markov process by eq. (8.74), which is responsible for the irreversible nature of the Fokker-Planck equation (8.75), is again connected with time smoothing of the distribution functions involved [see eq. (8.50)], so the Fokker-Planck equation (8.75) is fulfilled for the smoothed function $\bar{f}^{(1)}$. The smoothing interval τ must be chosen much larger than τ_c.

In a similar way a second differential equation of Fokker-Planck type for the binary distribution function $f^{(2)}$ can be obtained [4]; $f^{(1)}$ and $f^{(2)}$ are necessary for the evaluation of transport coefficients.

In order to solve eq. (8.75) further assumptions concerning $\langle \Delta \vec{p} \rangle$ and $\langle \Delta \vec{p} \Delta \vec{p} \rangle$ have to be introduced. This parallels eq. (8.71), where $\langle (\Delta \vec{r})^2 \rangle / \tau$ has been set independent of τ. Now we assume that $\langle \Delta \vec{p} \Delta \vec{p} \rangle / \tau$ is a constant multiple of the unit tensor. It can then be shown [4] that $\langle \Delta \vec{p} \rangle / \tau$ must be linear in $\vec{p}$,

$$\frac{\langle \Delta \vec{p} \rangle}{\tau} = -\frac{\zeta}{m} \vec{p} + \vec{K}(\vec{r}) , \tag{8.76}$$

where $\vec{K}(\vec{r})$ represents an external conservative force field, if existing. The constant ζ is called friction constant for reasons which will become clear in section 8.4.3.

Then eq. (8.75) can be solved explicitly as well as the corresponding Fokker-Planck equation for $f^{(2)}$ [4]. For simplicity the friction constant appearing in the equation for the doublet distribution function is set equal to the singlet friction constant. We list only the results for the transport coefficients which have been first obtained by Kirkwood and collaborators [18, 19, 20]:

$$\eta = \eta_K + \eta_V \tag{8.77}$$

$$\eta_K = \rho \frac{mkT}{2\zeta} \tag{8.78}$$

$$\eta_V = \frac{\pi \zeta}{15kT} \rho^2 \int_0^\infty R^3 \frac{d\varepsilon}{dR} g_0(R) \psi_2(R) \, dR \tag{8.79}$$

$$\phi = \phi_V = \frac{\pi \zeta}{9kT} \rho^2 \int_0^\infty R^3 \frac{d\varepsilon}{dR} g_0(R) \psi_0(R) \, dR \tag{8.80}$$

$$\lambda = \lambda_K + \lambda_V \tag{8.81}$$

$$\lambda_K = \frac{k^2 T}{2\zeta} \rho - \frac{k^2 T^2}{6\zeta} \left(\frac{\partial \rho}{\partial T} \right)_P \tag{8.82}$$

$$\lambda_V = \frac{\rho^2 \pi kT}{\zeta} \left\{ \frac{1}{3} \int_0^\infty R^3 \left(R \frac{d\varepsilon}{dR} - \varepsilon \right) g_0^{(2)}(R) \frac{d}{dR} \left(\frac{\partial \ln g_0}{\partial T} \right)_P dR \right. $$
$$\left. + \int_0^\infty R^2 \left(\varepsilon - \frac{1}{3} R \frac{d\varepsilon}{dR} \right) \left(\frac{\partial g_0}{\partial T} \right)_P dR \right\} \tag{8.83}$$

$$D = \frac{kT}{\zeta}. \tag{8.84}$$

The indices K and V denote the kinetic contribution (originating from $\vec{\sigma}_K$ and $\vec{q}_K$), and the intermolecular force contribution [originating from $\vec{\sigma}_V$ and $\vec{q}_V$, cf. eq. (8.26)–(8.31)] resp. The equilibrium value of the pair distribution function is $g_0(R)$; $\psi_0(R)$ and $\psi_2(R)$ describe the perturbation of g during the transport processes and can be derived from differential equations given by Kirkwood et al. [18]. The evaluation of approximate values of ζ is discussed in section 8.4.4.

Table 8.1 shows the theoretical values for liquid argon near the boiling point in comparison to experimental determinations. The estimation of the friction constant used is rather inaccurate. A better value of ζ [bringing D_{calc} to the experimental value according to eq. (8.84)] would improve the agreement of all other transport properties though still considerable discrepancies would remain.

Table 8.1. Comparison of calculated (theory of Brownian motion) and experimental values for argon at $T = 89$ K.

	calc.	exp.
η (mP)	0.73 [19]	2.40 [21]
ϕ (mP)	0.36 [19]	2.01 [4][a) b)]
$\lambda \cdot 10^4$ (cal/g·s·K)	4.1 [20]	2.9 [20]
$D \cdot 10^5$ (cm^2/s)	4.3 [19]	2.43 [22][a)]
$\zeta \cdot 10^{10}$ (g/s)	2.85 [19]	5.11 [22][a)]

a) These values are obtained for 90 K and 1.3 atm.
b) The exp. value for the bulk viscosity is calculated from η/ϕ data from ultrasonic measurements and experimental shear viscosities.

8.4.3. The Langevin Equation

Another approach to the theory of Brownian motion starts with the equation of motion of one particle and proceeds to the kinetic equations by some additional assumptions [16]. We will not show the equivalence of both descriptions in detail (see, e.g., [4]) but only indicate some essential characteristics of the second approach.

If no external forces $\vec{K}(\vec{r})$ are acting on the particle, the equation of motion — the Langevin equation — has the form

$$m\frac{d\vec{v}}{dt} = \vec{F}(t) = -\zeta\vec{v} + \vec{X}(t) \tag{8.85}$$

which assumes that it is possible to separate the total force $\vec{F}(t)$ acting on the particle in a frictional force $\zeta\vec{v}$ and a stochastic driving force $\vec{X}(t)$. $\vec{v}$ is the velocity of the molecule. The stochastic force comes from fast changes of the potential energy due to fluctuations of the surroundings which are assumed to occur over much smaller time intervals than the damping time m/ζ of the velocity. In order that eq. (8.76) with vanishing external forces $\vec{K}(\vec{r})$ is equivalent to eq. (8.85), the ensemble average of $\vec{X}(t)$ must vanish:

$$\langle \vec{X}(t) \rangle = 0. \tag{8.86}$$

Furthermore the force correlation function is assumed to be a stationary function*), which is, as assumed above, damped out for intervals τ_1 still small compared to the damping time m/ζ (which is characteristic for the momentum change of the particle):

$$\begin{aligned} \langle \vec{X}(t) \cdot \vec{X}(t+\tau) \rangle &\neq 0 \quad \text{for} \quad \tau < \tau_1 \ll m/\zeta \\ &= 0 \quad \text{for} \quad \tau > \tau_1 \end{aligned} \tag{8.87}$$

The assumption $\tau_1 \ll m/\zeta$ means that the molecular fluctuations have a time scale which is widely separated from the time scale for momentum change. A similar assumption led to the Smoluchowsky equation (8.61) and consequently to the Fokker-Planck equation in our first approach to this problem. This assumption that $\vec{p}(\tau_1) - \vec{p}(0)$ will be small may be easily verified for heavy particles immersed in a fluid. But in the case of molecular motion it is not trivial. Ross [17] has shown, that it is fulfilled for weakly interacting particles for which the kinetic energy is much greater than the potential energy at all times. The assumption is certainly not generally valid in liquids and not true for strongly repulsive encounters.

If the Langevin eq. (8.85) is multiplied by $e^{\zeta t/m}$ and integrated by parts, the relation is obtained

$$\vec{v}(\tau) = \vec{v}(0) e^{-\zeta\tau/m} + \frac{e^{-\zeta\tau/m}}{m} \int_0^\tau e^{\zeta t/m} \vec{X}(t) \, dt . \tag{8.88}$$

Multiplying by $\vec{v}(0)$ and averaging this product both over all particles with the same $\vec{v}(0)$ and over all values of $\vec{v}(0)$ gives the velocity correlation function $\langle \vec{v}(\tau) \cdot \vec{v}(0) \rangle$

$$\langle \vec{v}(\tau) \cdot \vec{v}(0) \rangle = \langle \vec{v}^2(0) \rangle e^{-\zeta\tau/m} . \tag{8.89}$$

The second term on the r.h.s. of eq. (8.88) vanishes for sufficiently large values of τ [$\tau > \tau_1$, according to eq. (8.87); but τ may still be much shorter than m/ζ]. Integration of the velocity correlation function, and insertion of $\langle v^2(0) \rangle = 3kT/m$, yields

$$\int_0^\infty \langle \vec{v}(\tau) \cdot \vec{v}(0) \rangle \, d\tau = 3 \frac{m}{\zeta} \frac{kT}{m} . \tag{8.90}$$

---

* By stationarity it is implied that a correlation function $\langle A(t+\tau) B(t) \rangle$ is independent of t.

In the next chapter, we will see that the integral over the velocity correlation function is equal to three times the diffusion coefficient:

$$\int_0^\infty \langle \vec{v}(\tau) \cdot \vec{v}(0) \rangle \, d\tau = 3D. \tag{8.91}$$

Anticipating this result, it follows again that $D = kT/\zeta$.

8.4.4. Estimation of the Friction Constant

For stationary conditions[*], the velocity correlation function is connected to the force correlation function:

$$\int_0^\tau \langle \vec{F}(t) \cdot \vec{F}(0) \rangle \, dt = -m^2 \frac{d}{d\tau} \langle \vec{v}(\tau) \cdot \vec{v}(0) \rangle, \tag{8.92}$$

which may be verified with the help of eq. (8.85). It follows [from eq. (8.89) and $\langle v^2(0) \rangle = 3kT/m$] that

$$\int_0^\tau \langle \vec{F}(0) \cdot \vec{F}(t) \rangle \, dt = 3kT\zeta e^{-\zeta\tau/m}, \tag{8.93}$$

which is equal to a plateau value $3kT\zeta$ for $\tau_1 < \tau \ll m/\zeta$ and zero for times τ much larger than m/ζ [8]. We will see in the next chapter, that in real liquids neither the exponential decay of the velocity correlation function [eq. (8.89)] is observed, nor is a plateau value of the integral (8.93) found for time τ much shorter than m/ζ. This is another indication that momentum changes occur too fast to provide different time scales τ_1 and m/ζ. If we proceed nevertheless by inserting the decay of the velocity correlation function [eq. (8.89)], a similar decay is obtained for the force correlation function

$$\langle \vec{F}(0) \cdot \vec{F}(t) \rangle = \langle \vec{F}(0) \cdot \vec{F}(0) \rangle e^{-\zeta t/m}. \tag{8.94}$$

Differentiating eq. (8.93) and inserting eq. (8.94) gives

$$\zeta^2 = \frac{m}{3kT} \langle \vec{F}(0) \cdot \vec{F}(0) \rangle = \frac{m}{3kT} \langle (\nabla E)^2 \rangle = \frac{m}{3} \langle \nabla^2 E \rangle, \tag{8.95}$$

where we have expressed the total force acting on the molecule by the gradient of the total potential energy[**]. If E is split into the sum of pair potentials $\varepsilon(R)$ one can show that

$$\zeta^2 = \frac{4\pi\rho_m}{3} \int_0^\infty R^2 g(R) \frac{d^2 \varepsilon(R)}{dR^2} \, dR. \tag{8.96}$$

There are many other estimates of ζ reported in the literature [23, 4], but no really satisfying treatment has been given yet. Eq. (8.95) has the advantage that $\langle \nabla^2 E \rangle$ is directly accessible experimentally. So a possible error due to $g(R)$ or $\varepsilon(R)$ in eq. (8.96) can be eliminated. The experiment giving the mean square force is the measurement of isotopic separation in liquid-vapour equilibrium. The theoretical calculation of the isotopic separation factor α, which is defined by the ratio of the isotopes in a phase A to their ratio in the phase B

$$\alpha = \frac{(N_1/N_2)_A}{(N_1/N_2)_B}, \tag{8.97}$$

---

[*] See footnote on p. 149.
[**] For the transformation $\langle (\nabla E)^2 \rangle = kT \langle \nabla^2 E \rangle$ see [25].

is based on first order quantum corrections of the partition function and will not be given here (see [24]). The result is

$$\ln \alpha = \frac{\hbar^2}{24(kT)^2}\left(\frac{1}{m_1} - \frac{1}{m_2}\right)(\langle \nabla^2 E \rangle_B - \langle \nabla^2 E \rangle_A) \ . \tag{8.98}$$

If phase A can be approximated by an ideal gas the expression $\langle \nabla^2 E \rangle_A$ vanishes and $\langle \nabla^2 E \rangle_B$ in the liquid phase may be determined. Of course, the other assumptions underlying eq. (8.95) cannot be eliminated by this method.

8.5. Rice-Allnatt Theory of Transport Properties

It has been mentioned that the Brownian theory is not applicable to strongly repulsive encounters. The Enskog theory deals exclusively with this part of molecular interaction but neglects the zig-zag of the trajectories of the particles between repulsive encounters, which is caused by the changing attractive fields of the surroundings. Rice and Allnatt [26, 4] have combined both concepts to a theory which at the moment is the most successful theory of transport processes in simple liquids.

In this theory the true intermolecular potential (see fig. 8.5) is separated into a strongly repulsive part ε^H and a soft negative part ε^S. The repulsive part is short-ranging and is responsible for the large momentum and energy transfer. The duration of such a collision is very short so that

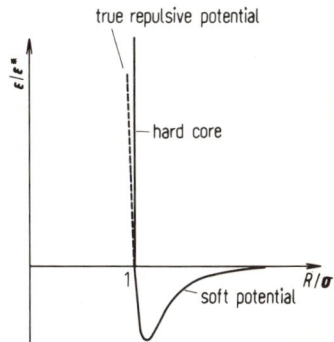

Fig. 8.5. The hard core and the soft potential used in the theory of Rice-Allnatt [4] in comparison to a realistic pair potential.

consideration of two-body collisions suffices as in the Enskog case. Without introducing appreciable error, ε^H can be approximated by a hard core potential. The negative soft potential is long-ranging and all contributions of the surrounding molecules produce a rapidly fluctuating soft force field. Thus a Brownian motion of the central molecule is generated which is characterized by frequent and small momentum and energy changes.

The main assumption of the theory is now that this random zig-zag between hard core encounters is effective enough to destroy the momentum correlation between them. Indeed, a detailed analysis [4] has shown that the equilibrium momentum distribution is reached much faster due to the random motion in the fluctuating soft force field than due to hard core encounters alone. The corresponding relaxation time of the first process is 20 times shorter than that of the latter. Thus a combined dynamical event, which consists of a hard core collision followed by a Brownian motion is independent of the preceding events, and therefore forms a Markov process.

Along these lines kinetic equations for $\bar{f}^{(1)}$ and $\bar{f}^{(2)}$ are derived. The collisional contribution is a sum of the collisional integral appearing on the r.h.s. of the Enskog equation (8.52), and a Fokker-Planck term given essentially by terms of the r.h.s. of eq. (8.75). The same holds for the corresponding equation for $\bar{f}^{(2)}$.

The resulting equations of the transport properties may be presented as combinations of the Enskog eq. (8.53–8.57) and the Brownian motion eq. (8.77–8.83):

$$D = \frac{kT}{\zeta^H + \zeta^S} \tag{8.99}$$

$$\eta = \rho \frac{mkT}{6\zeta^H/5 + 2\zeta^S}\left[1 + \frac{4}{15}\pi\rho\sigma^3 g(\sigma)\right]^2 + \frac{3}{5}\phi^H - \frac{37}{70}\frac{2\pi\rho^2\sigma^3 g(\sigma)}{15}\zeta^S\psi_2(\sigma) + \eta_V^S \tag{8.100}$$

$$\phi = \frac{2\pi\rho^2\sigma^3 g(\sigma)}{15}\zeta^S\frac{4\psi_2(\sigma) - 35\psi_0(\sigma)}{42} + \phi^H + \phi^S \tag{8.101}$$

$$\lambda = \rho\frac{k^2 T}{8\zeta^H/25 + 6\zeta^S/5}\left[1 + \frac{2}{5}\pi\rho\sigma^3 g(\sigma)\right]^2 + \frac{3k}{2m}\phi^H + \lambda^S \tag{8.102}$$

$$\zeta^H = \frac{8}{3}\rho(\pi mkT)^{1/2}\sigma^2 g(\sigma) . \tag{8.103}$$

In eq. (8.99)–(8.103) all coefficients with superscript H are due to the hard core contributions of the intermolecular potential and are given by Enskog's results (8.53)–(8.57). All coefficients with superscript S arise from the soft potential and are given by the Brownian motion results (8.77)–(8.84), if in these equations ζ is replaced by ζ^S.

The comparison with experiments in argon will be made in two steps. As the estimation of ζ^S, which may be based in principle on eq. (8.96), is the most critical step in applying the Rice-Allnatt theory to experimental data, we will use experimental values of ζ^S derived from measurements of diffusion coefficients taking account of the calculated hard sphere contributions ζ^H [eq. (8.103)]. The values of ζ^H and ζ^S for four different states of fluid argon are shown in table 8.2. The table

Table 8.2. Experimental values for ζ (from measurements of self-diffusion coefficients [22]) and calculated values for ζ^H (from [4]).

T	ρ_m	P	$\zeta_{exp} \cdot 10^{10}$	$\zeta^H \cdot 10^{10}$	$\zeta^S \cdot 10^{10}$
K	g/cm³	atm	g/s	g/s	g/s
90	1.38	1.3	5.11	0.64	4.47
128	1.12	50	2.94	0.94	2.00
133.5	1.12	100	3.13	1.00	2.13
185.5	1.12	500	3.20	1.52	1.68

also shows the small change of the diffusion coefficient with T at constant density. The evaluation of the transport properties with the values of ζ^S from table 8.2 is done on the basis of a 12/6 potential ($\varepsilon^*/k = 124$ K, $\sigma = 3.418$ Å). The pair distribution function is calculated from the pair potential by means of the superposition approximation and slightly shifted ("scaled") in order to get the correct equilibrium pressure [4]. This somewhat arbitrary correction effects η_V^S, ϕ^S and λ_V^S, mostly at 90 K. The same manipulation was used in the comparison to the theory of Brownian motion. The comparison of the transport properties with experiment is shown in table 8.3 for the states characterized in table 8.2. The experimental values of the bulk viscosity are

Table 8.3. Comparison of computed and measured transport coefficients in liquid argon at 4 different states (from [4]). The computed values are scaled (see text of this page).

T	K	90	128	133.5	185.5
P	atm	1.3	50	100	500
η_{calc}	(mP)	1.74	0.727	0.730	0.874
η_{exp}	(mP)	2.39	0.835	0.843	0.869
ϕ_{calc}	(mP)	6.863	0.933	0.970	1.123
ϕ_{exp}	(mP)	2.01	–	–	–
$\lambda_{calc} \cdot 10^4$	(cal/cm·s·K)	1.645	1.692	1.589	1.696
$\lambda_{exp} \cdot 10^4$	(cal/cm·s·K)	2.96	1.89	1.86	1.87

from measurements of the excess ultrasonic sound absorption yielding ϕ/η and experimental values for η. The following remarks seem appropriate: (1) The "scaling" procedure of the pair distribution functions improves (mostly at 90 K) the agreement with experiment for η and λ but overdoes apparently the correction for ϕ (which is at 90 K very sensitive to "scaling"). (2) The important contributions to the transport coefficients arise from the hard core terms involving ϕ_H, and the Brownian motion terms η_V^S, ϕ^S and λ_V^S. Table 8.4 shows the contributions of these terms in per cents for three temperatures. Kinetic contributions are almost negligible.

Table 8.4. Contributions of the most important terms to the total transport coefficients according to the Rice-Allnatt theory [4].

	Term	$T = 90$ K $P = 1.3$ atm	128 K 50 atm	184.5 K 500 atm
η	$\frac{3}{5}\phi^H$	16%	45%	61%
	η_V^S	72%	31%	14%
ϕ	ϕ^H	7%	59%	79%
	ϕ^S	72%	32%	14%
λ	$\frac{3k}{2m}\phi^H$	12%	14%	22%
	λ_V^S	86%	77%	63%

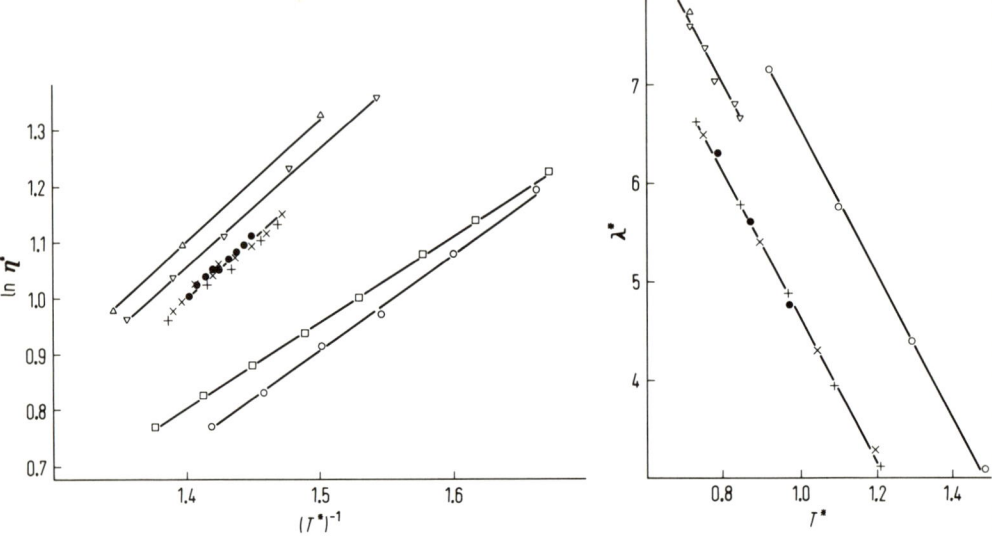

Fig. 8.6. The reduced viscosity as a function of the reciprocal reduced temperature: (+) Ar, (×) Kr, (●) Xe, (△) CO, (▽) N_2, (□) O_2, and (○) CH_4. After Rice, Boon and Davis in gen. ref. [4].

Fig. 8.7. The reduced thermal conductivity as a function of the reduced temperature. Key same as that for fig. 8.6. After Rice, Boon and Davis, gen. ref. [4].

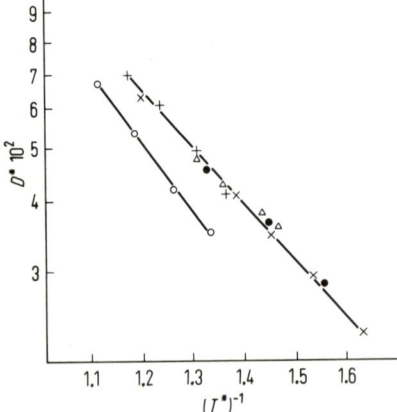

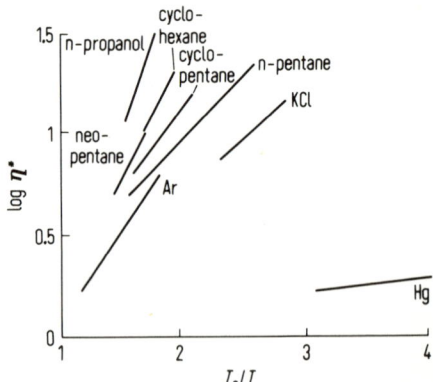

Fig. 8.8. The reduced self-diffusion coefficient as a function of the reciprocal reduced temperature for Ar, Kr, Xe, CO, and CH_4 (key as in fig. 8.6). After Rice, Boon and Davis in gen. ref. [4].

Fig. 8.9. The logarithm of the reduced shear viscosity as a function of the reciprocal reduced temperature for argon [28], mercury [29], potassium chloride [30], n-pentane, cyclopentane and neo-pentane [31], cyclohexane and n-propanol [32].

8.6. Reduced Transport Coefficients

As an exact theory of transport properties is still lacking it is important to see how far the principle of corresponding states is applicable to non-equilibrium properties and how the transport properties can be brought into a reduced form for such a comparison. Again we assume interaction in pairs (pair potential as two parameter equation of an universal functional form: $\varepsilon = \varepsilon^*(R/\sigma)$) and neglect quantum effects. In eq. (8.92), which will be derived in chapter 9, the diffusion coefficient is given in terms of the velocity correlation function. Writing $\vec{v}^* = \vec{v}\,(m/\varepsilon^*)^{1/2}$, we have $\vec{v}^*(0) \cdot \vec{v}^*(t) = (m/\varepsilon^*)\,\vec{v}(0) \cdot \vec{v}(t)$. Furthermore we have to introduce a reduced time $t^* = t\varepsilon^{*1/2}/\sigma m^{1/2}$, which gives

$$D = \frac{1}{3} \frac{\varepsilon^*}{m} \frac{m^{1/2}}{\varepsilon^{*1/2}} \int_0^\infty dt^* \langle \vec{v}^*(0) \cdot \vec{v}^*(t^*) \rangle = \frac{\sigma \varepsilon^{*1/2}}{m^{1/2}} D^* \tag{8.105}$$

so that

$$D^* = D m^{1/2}/\sigma \varepsilon^{*1/2}. \tag{8.106}$$

For the other transport coefficients correlation function expressions can also be obtained. Alternatively one could look at eq. (8.100)–(8.103) introducing D instead of the friction constant, and taking account of $\psi_2(R) = \sigma^2 \psi_2^*(r/\sigma)$, and similarly $\psi_0(R)$, and observing that $g_0^*(R/\sigma)$, $\psi_0^*(R/\sigma)$ and $\psi_2^*(R/\sigma)$ are also functions of the *reduced* temperature and volume. The results are [27]

$$\eta^* = \eta \sigma^2/(m\varepsilon^*)^{1/2} \tag{8.107a}$$

$$\phi^* = \phi \sigma^2/(m\varepsilon^*)^{1/2} \tag{8.107b}$$

$$\lambda^* = \lambda \sigma^2 m^{1/2}/k\varepsilon^{*1/2}. \tag{8.107c}$$

Some experimental results for simple liquids in reduced plots are shown in figs. 8.6 to 8.8. It is seen, that greater deviations from the principle of corresponding states occur more readily than in the case of equilibrium properties.

In fig. 8.9 a comparison for different classes of liquids is shown. In calculating the reduced quantities, ε^*/k has been replaced by T_c, and σ by $(v_c/N)^{1/3}$. Therefore, fig. 8.6 and fig. 8.9 are not strictly comparable.

8.7. References

1. J. G. Kirkwood, Selected Topics in Statistical Mechanics, ed. by R. W. Zwanzig, Gordon and Breach, New York 1967.
2. R. M. Mazo, Statistical Mechanical Theories of Transport Processes, Pergamon Press, Oxford 1967.
3. H. J. M. Hanley (Ed.), Transport Phenomena in Fluids, Dekker, New York 1969.
4. S. A. Rice and P. Gray, gen. ref. 10; S. A. Rice, J. P. Boon, and H. T. Davis in gen. ref. 4; P. Gray in gen. ref. 12.
5. I. Prigogine, Non-Equilibrium Statistical Mechanics, Interscience, New York 1962.
6. T. L. Hill, gen. ref. 5.
7. J. H. Irving and J. G. Kirkwood, J. Chem. Phys. *18*, 817 (1950) and ref. 1, p. 52.
8. J. G. Kirkwood, J. Chem. Phys. *14*, 180 (1946) and ref. 1, p. 1.
9. J. G. Kirkwood, J. Chem. Phys. *15*, 72 (1947) and ref. 1, p. 31.
10. S. Chapman and T. G. Cowling, The Mathematical Theory of Non-Uniform Gases, Cambridge University Press, 1953.

11. Gen. ref. 6.
12. Kerson Huang, Statistical Mechanics, Wiley, New York 1963.
13. S. A. Rice, J. G. Kirkwood, J. Ross, and R. W. Zwanzig, J. Chem. Phys. *31*, 575 (1959), and ref. 1, p. 172.
14. J. V. Sengers, Int. J. Heat and Mass Transfer *8*, 1103 (1965).
15. J. H. Dymond and B. J. Alder, J. Chem. Phys. *45*, 2061 (1966); B. J. Alder, Ber. Bunsenges. *70*, 968 (1966).
16. S. Chandrasekhar, Rev. Mod. Phys. *15*, 1 (1943); M. C. Wang and G. S. Uhlenbeck, Rev. Mod. Phys. *17*, 323 (1945). Both articles are also contained in N. Wax (Ed.), Selected Papers on Noise and Stochastic Processes, Dover, New York 1954. See also ref. 4.
17. J. Ross, J. Chem. Phys. *24*, 375 (1956), and ref. 1, p. 134.
18. J. G. Kirkwood, F. P Buff, and M. S. Green, J. Chem. Phys. *17*, 988 (1949), and ref. 1, p. 38.
19. R. W. Zwanzig, J. G. Kirkwood, K. F. Stripp, and I. Oppenheim, J. Chem. Phys. *21*, 2050 (1953), and ref. 1, p. 91.
20. R. W. Zwanzig, J. G. Kirkwood, I. Oppenheim, and B. J. Alder, J. Chem. Phys. *22*, 783 (1954), and ref. 1, p. 101.
21. G. A. Cook Ed., Argon, Helium and the Rare Gases, Vol. 1, Interscience, New York 1961.
22. J. Naghizadeh and S. A. Rice, J. Chem. Phys. *36*, 2710 (1962).
23. E. Helfand, Phys. Fluids *4*, 1 (1961).
24. G. Boato, G. Casanova, and A. Levi, J. Chem. Phys. *40*, 2419 (1964); W. Friedmann and W. A. Steele, J. Chem. Phys. *40*, 2669 (1964); G. Casanova and A. Levi, in gen. ref. 12.
25. L. D. Landau and E. M. Lifshitz, Statistical Physics, Pergamon Press, London 1958, § 33.
26. S. A. Rice and A. R. Allnatt, J. Chem. Phys. *34*, 2144 (1961); A. R. Allnatt and S. A. Rice, J. Chem. Phys. *34*, 2156 (1961).
27. H. Mori, Phys. Rev. *112*, 1829 (1958).
28. A. van Itterbeek, H. Zink, and O. van Paemel, Cryogenics *2*, 210 (1962); data of B. A. Lowry, S. A. Rice, and P. Gray, J. Chem. Phys. *40*, 3673 (1964), extrapolated to the saturation curve.
29. Gmelins Handbuch der anorgan. Chemie, 8. Aufl., *34 A 1*, Verlag Chemie, Weinheim 1960.
30. R. Lorenz, Z. Phys. Chem. *79*, 63 (1912).
31. Selected Values of Physical and Thermodynamic Properties of Hydrocarbons and Related Compounds (API Res. Proj. 44), Carnegie Press, Pittsburgh 1953.
32. J. Timmermans, Physicochemical Constants of Pure Organic Compounds, Elsevier, Amsterdam 1950; Vol. II, 1965.

9. Non-Equilibrium Properties: Liquid Dynamics

9.1. Introduction

The approach to equilibrium from a non-equilibrium state of the system has been treated so far in chapter 8 with the help of kinetic equations derived from plausible assumptions related to liquid structure. So the long time behaviour of the system, described by the macroscopic transport coefficients, could be evaluated. But nothing definite has been said about the short time behaviour of the molecules, which describes the dynamical state of the liquid, with fluctuations growing and decaying even without external non-equilibrium conditions.

In the following, we will concentrate on the short time behaviour of the molecules. There exists a very powerful though more formal approach, which brings the measurable response to disturbances in connection to the time correlation of certain quantities, and which relates this time correlation to energy dissipation and transport coefficients. So the way is open for the experimental determination of time correlation functions. Another way for their evaluation is given by computer calculations [1, 2]. Theoretical calculations of time correlation functions have been tried on the basis of kinetic equations or related models [3], but we will not deal with this problem.

The formal approach is based on the assumption that the response of the system (deviation of a measurable quantity from the equilibrium value) is linear in the external or internal (thermal) force, as is justified for small deviations from equilibrium. Therefore, it is called linear response theory. It will be sketched in section 9.2. In analogy to chapter 8 we will present first the macroscopic phenomenological laws and then the same formulations on the basis of statistical mechanics (microscopic theory). Again, the fundamental relations follow from a comparison of macroscopic and microscopic formulations. In section 9.2.3 to 9.2.5 we will give important examples of the time correlation formalism. It will become obvious that many interesting consequences of the theory are outside the scope of this book. Its application to single particle motions of monoatomic liquids will be given in section 9.3, whereas section 9.4 will deal with collective particle motions. The orientational motions of polyatomic molecules will be treated in chapter 11, after the description of the peculiarities of their equilibrium properties.

9.2. Linear Response Theory

9.2.1. Phenomenological Relations

According to eq. (1.1), the total differential of the entropy can be written

$$ds = \frac{1}{T}du + \frac{P}{T}dv - \frac{\mu}{T}dn. \tag{9.1}$$

In the presence of external fields, this equation may be generalized to [4]

$$ds = -\sum_i \frac{F_i}{T}d\alpha_i + \sum_l X_l d\alpha_l, \tag{9.2}$$

where the first term gives the change of the energy with $F_i = -\partial u/\partial \alpha_i$ being the external force conjugated to the extensive variable α_i, and where $TX_l = T(\partial s/\partial \alpha_l)_u$ may be termed the internal (or thermal) force conjugated to the thermodynamic variable α_l. Application of an external force leads to the possibility of storage of energy, which is not the case for an internal force. But application of both forces causes changes in the total entropy, i.e. dissipation of energy. As far as the dissipation of energy is concerned, the linear response theory has many features in common with the thermodynamics of irreversible processes [4]. But the latter deals mainly with macroscopic fluxes which might be understood as successive regressions of fluctuations, whereas the linear response theory considers time dependent forces connected with the generation and decay of fluctuations.

We will concentrate first on homogeneous external forces, generally varying with time, for which the instantaneous perturbing energy (the interaction Hamiltonian due to the perturbing force) can be written

$$H'(t) = -A v F(t) . \tag{9.3}$$

The quantity A, to which the force is coupling, contains the properties of the ensemble and depends on time only over the positions and momenta of all molecules. It is usually expressed as

$$A = \frac{1}{v} \sum_k a_k(\vec{r}_k) , \tag{9.4}$$

where a_k is the corresponding molecular property of the k-th molecule at $\vec{r}_k$. The factor $1/v$ serves to make $\langle A \rangle$ a property density so that the quantities derived from eq. (9.3) have the usual dimensions. The expectation value $\langle A \rangle v$ is defined as the deviation of the variable α_i from the case without perturbing force F_i, such that $\langle A \rangle = 0$ for the absence of external forces. E.g., if F is a time varying homogeneous electric field, A would be the component of the total electric dipole moment per unit volume, parallel to the applied field.

Let us measure now the response of the system due to an external perturbation in terms of a dynamical quantity B [5–10]; again $\langle B \rangle v$ gives the deviation of a variable α from the unperturbed case, for which $\langle B \rangle = 0$. For sufficiently small driving forces, the response will be linear in the forces, i.e.

$$\langle B(t) \rangle = \int_{-\infty}^{t} \phi_{BA}(t - t') F(t') dt' . \tag{9.5}$$

The function ϕ_{BA} is called "after effect function" since it relates $\langle B \rangle$ at time t to the forces at all previous times $t' < t$. Eq. (9.5) may be best understood if the force is applied as superposition of short pulses (δ-functions in time). The response to a pulse $F(t') = F_0 \delta(t' - t_0)$ applied at time t_0 is then $\langle B(t) \rangle = \phi_{BA}(t - t_0) F_0$ for $t > t_0$ and zero for $t < t_0$. Thus eq. (9.5) is an expression of causality: the response is related to all previous force pulses. Since $\langle B \rangle$ and $F(t)$ are real quantities, if follows from the definition that $\phi_{BA}(\tau)$ must be a real function*⁾.

Since eq. (9.5) is linear in the force it is convenient to take Fourier transforms. Each complicated time behaviour of the disturbance is then given by a superposition of monochromatic forces. If the Fourier transforms of $B(t)$ and $F(t)$ are defined in the following form

$$B(t) = \int_{-\infty}^{+\infty} \tilde{B}(\omega) e^{+i\omega t} d\omega \quad \tilde{B}(\omega) = \frac{1}{2\pi} \int_{-\infty}^{\infty} B(t) e^{-i\omega t} dt , \tag{9.6}$$

---

* If the external force is switched on in the infinite past, care must be taken that the integral (9.5) converges. Therefore, the forces considered are sometimes expressed by $\lim_{\varepsilon \to 0} F(t) e^{\varepsilon t}$ with $\varepsilon > 0$. We refer to the literature [5, 6, 9, 10] for this more rigorous language.

eq. (9.5) can be rewritten (with $\tau = t - t'$)

$$\langle B(t) \rangle = \int_0^\infty \phi_{BA}(\tau) F(t-\tau) \, dt = \int_{-\infty}^{+\infty} d\omega \tilde{F}(\omega) e^{i\omega t} \int_0^\infty d\tau \phi_{BA}(\tau) e^{-i\omega \tau} \, . \tag{9.7}$$

This gives with the definition

$$\chi_{BA}(\omega) = \int_0^\infty \phi_{BA}(t) e^{-i\omega t} dt \tag{9.8}$$

and the help of eq. (9.6) the result

$$\langle \tilde{B}(\omega) \rangle = \tilde{F}(\omega) \chi_{BA}(\omega) \, . \tag{9.9}$$

The function $\chi_{BA}(\omega)$ is an one-sided Fourier transform of the after effect function and is called a generalized susceptibility. It is, in general, a complex function

$$\chi_{BA}(\omega) = \chi'_{BA}(\omega) + i\chi''_{BA}(\omega) \, . \tag{9.10}$$

Without going into details, it should be remarked that $\chi'_{BA}(\omega)$ and $\chi''_{BA}(\omega)$ are connected by the so-called Kramers-Kronig (or dispersion) relations which are a consequence of the causal nature of the process [5–11]. A knowledge of $\chi'_{BA}(\omega)$ for all ω enables the calculation of $\chi''_{BA}(\omega)$ and vice versa.

We will show now that the complex part of χ_{AA} is connected to the dissipation of energy. The dissipated energy per unit time is given by the time derivative of the internal energy (or, which is the same, by the time derivative of its perturbation $\Delta u = \langle H'(t) \rangle$). When forming this derivative, time averaging is necessary over one period $\tau = 2\pi/\omega$ of the disturbing force, which is assumed to be monochromatic. The total time derivative of the averaged energy of the system is equal to the averaged partial time derivative of the energy [11]. Since only $F(t)$ in eq. (9.3) depends explicitely on time, the dissipated energy can be written

$$\frac{d\Delta u}{dt} = \left\langle \frac{dH'(t)}{dt} \right\rangle = -\frac{\omega}{2\pi} \int_0^{2\pi/\omega} \langle A \rangle v \frac{\partial F}{\partial t} \, dt \, . \tag{9.11}$$

Since the force must be real we have

$$F(t) = \operatorname{Re} \tilde{F}(\omega) e^{i\omega t} = \frac{1}{2} [\tilde{F}(\omega) e^{i\omega t} + \tilde{F}^*(\omega) e^{-i\omega t}] \, , \tag{9.12}$$

where $\tilde{F}^*$ is the complex conjugate of $\tilde{F}$. Inserting this into eq. (9.7), and using eq. (9.8), $\langle A(t) \rangle$ is given by

$$\langle A(t) \rangle = \int_0^\infty F(t-\tau) \phi_{AA}(\tau) \, d\tau = \frac{1}{2} [\tilde{F}(\omega) e^{i\omega t} \chi_{AA}(\omega) + \tilde{F}^*(\omega) e^{-i\omega t} \chi_{AA}(-\omega)] \, . \tag{9.13}$$

Since $\phi_{AB}(t)$ is a real quantity, it follows from the definition (9.8) that *)

$$\chi_{AA}(-\omega) = \chi^*_{AA}(\omega) \, . \tag{9.14}$$

Inserting now eq. (9.12) to (9.14) into eq. (9.11), and observing that the time average eliminates all terms which contain t only in $e^{\pm 2i\omega t}$, the result is

$$\frac{d\Delta u}{dt} = -\frac{\omega}{2} v \chi''_{AA}(\omega) \tilde{F}(\omega) \tilde{F}^*(\omega) = -\frac{\omega}{2} v \chi''_{AA}(\omega) |F(\omega)|^2 \, . \tag{9.15}$$

---

* There exist many other symmetry conditions for χ_{BA} [5–10].

A measurement of the energy dissipation under the influence of a monochromatic force $F(\omega)$ is equivalent to the determination of the spectral density of this process at the particular frequency ω. It is essentially given by $\chi''_{AA}(\omega)$. For a complete analysis the spectral density must be known for all ω. The part $\chi''_{AA}(\omega)$ of the susceptibility is called the dissipative part, whereas $\chi'_{AA}(\omega)$, connected to the storage of energy, is called the reactive part.

9.2.2. Microscopic Relations: Molecular Correlation Functions

In the microscopic description, the macroscopic quantities have to be related to the molecular properties. In general, microscopic description goes along the lines given in section 8.2.2. The ensemble averages are formed by means of the N-particle distribution function $f^{(N)}(\vec{r}^N, \vec{p}^N)$, which changes with time according to the Liouville equation (8.13). As the time (and space) dependence of A and B are solely caused by the molecular distributions readjusting under the influence of the perturbing force, we have to consider the time dependence of $f^{(N)}$ in some detail [5-7]*[)].

Let us split the distribution function $f^{(N)}$ into an equilibrium or unperturbed part $f_0^{(N)}$, which does not change with time, and a part $f_1^{(N)}$, which carries the deviation from $f_0^{(N)}$ under the influence of the perturbation. Similarly, the total energy is $H = H_0 + H'$, where H_0 is the energy given by the unperturbed or equilibrium distribution, and H' is the perturbation [eq. (9.3)]. Furthermore, the velocity $\vec{p}_k/m$ and the force $\vec{F}_k$ may be replaced in eq. (8.13) by $\nabla_{\vec{p}_k} H$ and $-\nabla_{\vec{r}_k} H$, respectively. This way the Liouville equation is written

$$\frac{\partial f_1^{(N)}}{\partial t} = -\sum_{k=1}^{N} \{\nabla_{\vec{p}_k}(H_0 + H') \cdot \nabla_{\vec{r}_k}(f_0^{(N)} + f_1^{(N)}) - \nabla_{\vec{r}_k}(H_0 + H') \cdot \nabla_{\vec{p}_k}(f_0^{(N)} + f_1^{(N)})\}. \tag{9.16}$$

In eq. (9.16), only the cross-terms of the products are of interest. The products between the gradients of H_0 and $f_0^{(N)}$ belong to the equilibrium distribution, and furnish $\partial f_0^{(N)}/\partial t$, which is zero. On the other hand, the perturbation should be small, so that the products between the gradients of H' and $f_1^{(N)}$ can be neglected. Furthermore, we have for the equilibrium distribution

$$f_0^{(N)} = \text{const.} \, e^{-H_0/kT}, \quad \text{or} \quad \nabla f_0^{(N)} = -\frac{f_0^{(N)}}{kT} \nabla H_0. \tag{9.17}$$

Thus eq. (9.16) can be rewritten

$$\frac{\partial f_1^{(N)}}{\partial t} = -\sum_{k=1}^{N} [\nabla_{\vec{p}_k} H_0 \cdot \nabla_{\vec{r}_k} f_1^{(N)} - \nabla_{\vec{r}_k} H_0 \cdot \nabla_{\vec{p}_k} f_1^{(N)}]$$
$$- \frac{f_0^{(N)}}{kT} \sum_{k=1}^{N} [\nabla_{\vec{p}_k} H_0 \cdot \nabla_{\vec{r}_k} H' - \nabla_{\vec{r}_k} H_0 \cdot \nabla_{\vec{p}_k} H']. \tag{9.18}$$

The only factor of H', which depends on the positions and momenta, is the factor A of eq. (9.3). As A depends on time only over $\vec{r}^N$ and $\vec{p}^N$, we have

$$\dot{A} = \sum_{k=1}^{N} \left[\frac{\partial A}{\partial \vec{r}_k} \cdot \dot{\vec{r}}_k + \frac{\partial A}{\partial \vec{p}_k} \cdot \dot{\vec{p}}_k \right] = \sum_{k=1}^{N} [\nabla_{\vec{p}_k} H \cdot \nabla_{\vec{r}_k} A - \nabla_{\vec{r}_k} H \cdot \nabla_{\vec{p}_k} A]. \tag{9.19}$$

---

* For complete quantum mechanical treatments we refer to the literature [5, 6, 8-10].

Eq. (9.18) becomes this way

$$\frac{\partial f_1^{(N)}}{\partial t} = - \sum_{k=1}^{N} [\nabla_{\vec{p}_k} H_0 \cdot \nabla_{\vec{r}_k} f_1^{(N)} - \nabla_{\vec{r}_k} H_0 \cdot \nabla_{\vec{p}_k} f_1^{(N)}] + \frac{v f_0^{(N)}}{kT} F(t) \dot{A}_0 .\qquad(9.20)$$

The notation $\dot{A}_0$ should indicate that only H_0 has been inserted in eq. (9.19), i.e. $\dot{A}_0$ refers to a reference system which has not been perturbed. As is shown in Appendix B, eq. (9.20) can be integrated to give

$$f_1^{(N)}(t) = \frac{v f_0^{(N)}}{kT} \int_{-\infty}^{t} F(t') e^{-i(t'-t)L_0} \dot{A}_0 \, dt' ,\qquad(9.21)$$

where $e^{-i(t'-t)L_0}$ is an operator which brings about a time shift for the interval $t'-t$ in the variables of $\dot{A}_0$ such that one starts with some value of $\dot{A}$ in the equilibrium ensemble and follows its evolution in time for that interval. The response $\langle B(t) \rangle$ can now be written as

$$\langle B(t) \rangle = \int B f_1^{(N)}(t) \, d\vec{r}^N d\vec{p}^N = \frac{v}{kT} \int_{-\infty}^{t} dt' F(t') \int f_0^{(N)} B e^{-i(t'-t)L_0} \dot{A}_0 \, d\vec{r}^N d\vec{p}^N .\qquad(9.22)$$

Again the operator $e^{-i(t'-t)L_0}$ indicates that the values of the variables of $\dot{A}$ in the equilibrium ensemble refer to a time instant which is for the interval $t'-t$ removed from the moment at which the variables of B have to be taken. Comparison with eq. (9.5), denoting $t-t'$ by τ, shows that

$$\phi_{BA}(\tau) = \frac{v}{kT} \langle B(0) \dot{A}(-\tau) \rangle = \frac{v}{kT} \langle B(\tau) \dot{A}(0) \rangle ,\qquad(9.23)$$

because the origin of the time scale is immaterial, as the average of the correlation function (9.23) is over the unperturbed ensemble, i.e. over the natural motion of the molecules, as has been expressed in eq. (9.22) by $f_0^{(N)}$ and $\dot{A}_0$.

Eq. (9.23) gives for the susceptibility [cf. eq. (9.8)]

$$\chi_{BA}(\omega) = \frac{v}{kT} \int_0^\infty e^{-i\omega t} \langle B(t) \dot{A}(0) \rangle \, dt .\qquad(9.24)$$

For studying the dissipation we need $\chi''_{AA}(\omega) = [\chi_{AA}(\omega) - \chi^*_{AA}(\omega)]/2i$. Now the complex conjugate is given by

$$\chi^*_{AA}(\omega) = \frac{v}{kT} \int_0^\infty e^{i\omega t} \langle \dot{A}(t) \dot{A}(0) \rangle \, dt = -\frac{v}{kT} \int_{-\infty}^{0} e^{-i\omega t} \langle \dot{A}(t) \dot{A}(0) \rangle \, dt .\qquad(9.25)$$

The last relation is obtained by considering that

$$\langle \dot{A}(0) \dot{A}(-t) \rangle = \frac{d}{dt} \langle A(0) A(-t) \rangle = -\frac{d}{dt} \langle A(0) A(t) \rangle = -\langle A(0) \dot{A}(t) \rangle .\qquad(9.26)$$

Thus it follows for $\chi''_{AA}(\omega)$ that

$$\chi''_{AA}(\omega) = \frac{v}{2ikT} \int_{-\infty}^{+\infty} e^{-i\omega t} \langle \dot{A}(t) \dot{A}(0) \rangle \, dt = \frac{v\omega}{2kT} \int_{-\infty}^{+\infty} e^{-i\omega t} \langle A(t) A(0) \rangle \, dt ,\qquad(9.27)$$

where the last relation is obtained by partial integration. Eq. (9.27) connects the dissipation [eq. (9.15)] to the time correlation function of the dynamical variable A and is called fluctuation-dissipation theorem[*].

---

[*] For its quantum mechanical formulation we refer to ref. [5, 6, 8–10].

One further remark concerns the case that the response is given by the flux of the dynamical variable A, i.e. B = Ȧ. Partial integration of eq. (9.24) gives then

$$\chi_{\dot{A}A}(\omega) = \frac{1}{i\omega}\left[\frac{\langle \dot{A}(0)^2\rangle v}{kT} - \chi_{\dot{A}\dot{A}}(\omega)\right], \tag{9.28}$$

where

$$\chi_{\dot{A}\dot{A}}(\omega) = \frac{v}{kT}\int_0^\infty e^{-i\omega t}\langle \dot{A}(t)\ddot{A}(0)\rangle\,dt, \tag{9.29}$$

in accord with the notation of eq. (9.24). This shows, that the real part $\chi'_{\dot{A}A}(\omega)$ is essentially given by the imaginary part $\chi''_{\dot{A}\dot{A}}(\omega)$. On the other hand, it can be shown again by partial integration of eq. (9.24), that the real part $\chi'_{\dot{A}A}(\omega)$ is determined by the imaginary part $\chi''_{AA}(\omega)$. Now $\chi'_{\dot{A}A}(\omega)$ is characteristic for a transport coefficient. Dissipation is thus given essentially by the real part of the transport coefficient or by the imaginary part of the susceptibilities χ_{AA} or $\chi_{\dot{A}\dot{A}}$.

Up to this point, we have presented a simplified treatment considering only time dependent, but not space dependent fluctuations. For space dependent fluctuations, Fourier transformation in space and separate consideration of the components with different Q provides the same convenience as the splitting of the fluctuations according to their frequencies [8, 9]. Here only the fundamental equations will be given in terms of Fourier transforms in space and time:
The perturbing energy [cf. eq. (9.3)] is written as

$$H'(t) = -\int_v F(\vec{r},t)\,A(\vec{r},t)\,d\vec{r}, \tag{9.30}$$

with

$$A(\vec{r},t) = \sum_{k=1}^{N} a_k(\vec{r}_k)\delta(\vec{r} - \vec{r}_k). \tag{9.31}$$

The response, eq. (9.5), takes then the form

$$\langle B(\vec{r},t)\rangle = \int_{-\infty}^{t}\int_v \phi_{BA}(\vec{r} - \vec{r}',t - t')\,F(\vec{r}',t')\,d\vec{r}'\,dt', \tag{9.32}$$

and the definition of the susceptibility (along the same lines leading to eq. (9.8)) becomes:

$$\chi_{BA}(\vec{Q},\omega) = \int_0^\infty\int_v \phi_{BA}(\vec{r},t)\,e^{i(\vec{Q}\cdot\vec{r} - \omega t)}\,d\vec{r}\,dt. \tag{9.33}$$

Considering the Q-component of a monochromatic external force, i.e.

$$F(\vec{r},t) = \mathrm{Re}\,F(\vec{Q},\omega)\,e^{-i(\vec{Q}\cdot\vec{r} - \omega t)}, \tag{9.34}$$

the dissipated energy [cf. eq. (9.15)] becomes

$$\frac{d\Delta u}{dt} = -v\frac{\omega}{2}\chi''_{AA}(\vec{Q},\omega)|F(\vec{Q},\omega)|^2. \tag{9.35}$$

Denoting the Fourier components in space of the dynamical variables by

$$B_Q(t) = \int_v B(\vec{r},t)\,e^{i\vec{Q}\cdot\vec{r}}\,d\vec{r} \quad\text{and}\quad A_{-Q}(t) = \int_v A(\vec{r},t)\,e^{-i\vec{Q}\cdot\vec{r}}\,d\vec{r}, \tag{9.36}$$

the Q-component of the after effect function becomes

$$\phi_{BA}(\vec{Q},t) = \int_v \phi_{BA}(\vec{r},t)\,e^{i\vec{Q}\cdot\vec{r}}\,d\vec{r} = \frac{1}{vkT}\langle B_Q(t)\,\dot{A}_{-Q}(0)\rangle, \tag{9.37}$$

and the fluctuation-dissipation theorem (9.27) is generalized to

$$\chi''_{AA}(\vec{Q},\omega) = \frac{\omega}{2vkT} \int_{-\infty}^{+\infty} e^{-i\omega t} \langle A_Q(t) A_{-Q}(0) \rangle \, dt . \tag{9.38}$$

In the limit $Q \to 0$ the quantity A_Q corresponds to our previous quantity Av, whereas $\chi(\vec{Q},\omega)$ and $\phi(\vec{Q},t)$ approach our previous quantities $\chi(\omega)$ and $\phi(t)$.

9.2.3. Electric Conduction

This section should serve as an example for the application of the time correlation formalism if the perturbation is exerted by an external force [5, 6, 10]. The force is given as an uniform electric field of strength E in the direction z. As there is no space dependence of the field we can apply the formalism in the limit $Q \to 0$. The electric field couples to the total electric moment in the direction of the field,

$$A = \frac{1}{v} M_z = \frac{1}{v} \sum_{k=1}^{N} \mu_{kz}(\vec{r}_k), \tag{9.39}$$

where $\mu_{kz}(\vec{r}_k)$ is the z-component of the moment of the k-th molecule. The current response in the direction z is then given by [cf. eq. (9.5)]

$$\langle J_z(t) \rangle = \int_{-\infty}^{t} \phi_{JM}(t-t') E(t') \, dt', \tag{9.40}$$

where the current is given in terms of molecular quantities by

$$J_z = \frac{1}{v} \sum_k \dot{\mu}_{kz}(\vec{r}_k) = \frac{1}{v} \dot{M}_z . \tag{9.41}$$

Eq. (9.9) is then Ohm's law, written as a function of frequency

$$\langle J_z(\omega) \rangle = \chi_{JM}(\omega) E_\omega, \tag{9.42}$$

so that $\chi_{JM}(\omega)$ can be identified with the conductance $\sigma(\omega)$. Inserting eq. (9.24), we have

$$\sigma(\omega) = \chi_{JM}(\omega) = \frac{v}{kT} \int_0^\infty e^{-i\omega t} \langle J_z(0) J_z(t) \rangle \, dt . \tag{9.43}$$

According to eq. (9.28) and the discussion of the case $B = \mathring{A}$, it is clear that the real part of $\sigma(\omega)$ is responsible for the power dissipation. Eq. (9.43) contains two theorems, if the ensemble average can be replaced by a time average. The first is the Wiener-Khintchine theorem, which defines $S(\omega)$, the spectral density of the current, proportional to $|\tilde{J}_z(\omega)|^2$ and shows that the time average of the correlation function is connected to $S(\omega)$ by [4, 11]

$$\overline{J_z(t) J_z(t+\tau)} = \int_{-\infty}^{+\infty} S(\omega) e^{i\omega\tau} d\omega \quad \text{or} \quad S(\omega) = \frac{1}{2\pi} \int_{-\infty}^{\infty} \overline{J_z(t) J_z(t+\tau)} \, e^{-i\omega\tau} d\tau . \tag{9.44}$$

The second theorem relates the spectral density, which can be shown to be a real quantity, to the real part of the conductance, which is responsible for dissipation:

$$\sigma'(\omega) = \frac{\pi S(\omega)}{kT} v . \tag{9.45}$$

This is the Nyquist theorem, one of the first examples of a fluctuation-dissipation theorem, formulated for electric noise [12, 10]. It shows that only the ohmic part of the resistor is responsible for the noise and that the current fluctuations increase with temperature.

9.2.4. Diffusion

This is an example for a perturbation exerted by an internal force. As the variable is the number of molecules of a certain component s, eq. (9.1) and (9.2) indicate that the proper force is given by the negative chemical potential $(-\mu_s)$ (under isothermal conditions) [6]. This shows already the difficulties with internal forces: μ_s cannot be applied at will like an electric field, and it makes sense only for an assembly in local equilibrium. In order to circumvent the difficulties, we will deal with a volume element where the particles are in local equilibrium, and where the chemical potential has the mean value μ_s in the time interval between 0 and t^*, but has different values in the surroundings. Furthermore, we are interested only in self-diffusion. The diffusing component s consists thus of tagged molecules, which are replaced by molecules of the same kind, but without tag. The quantity A is then given by $\rho_s = N_s/v$, particle number per volume, the response is the particle flux $\langle \vec{j}_s \rangle$ being given by

$$\vec{j}_s = \frac{1}{v} \sum_{k=1}^{N_s} \frac{\vec{p}_k}{m} . \tag{9.46}$$

Application of eq. (9.5) with ϕ_{BA} given by eq. (9.23) leads to

$$\langle \vec{j}_s \rangle = - \frac{v}{kT} \int_0^\infty \langle \vec{j}_s(\tau) \dot{\rho}_s(0) \rangle \mu_s \, d\tau . \tag{9.47}$$

The continuity eq. (8.5) gives $\dot{\rho}_s = -\nabla \cdot \vec{j}_s$, so that eq. (9.47) becomes

$$\langle \vec{j}_s \rangle = \frac{v}{kT} \int_0^\infty \langle \vec{j}_s(\tau)(-\nabla \cdot \vec{j}_s(0)) \rangle \mu_s \, d\tau . \tag{9.48}$$

Now the fact may be used that $\mu_s \vec{j}_s$ is a conserved quantity, so that $\mu_s \nabla \cdot \vec{j}_s = -\vec{j}_s \cdot \nabla \mu_s$. As our model assumes constancy of μ_s over the relevant time interval and as μ_s is already a mean value for the ensemble considered, we may take μ_s into the ensemble average and $\nabla \mu_s$ out of the integral to give

$$\langle \vec{j}_s \rangle = - \frac{v}{kT} \nabla \mu_s \cdot \int_0^\infty \langle \vec{j}_s(\tau) \vec{j}_s(0) \rangle \, d\tau . \tag{9.49}$$

Comparing eq. (9.49) with the definition of the diffusion coefficient,

$$\langle \vec{j}_s \rangle = -D \nabla \rho_s , \tag{9.50}$$

and using $\mu_s = kT \ln \rho_s + \text{const.}$, we have

$$D = \frac{v}{\rho_s} \int_0^\infty \langle \vec{j}_s(0) \vec{j}_s(t) \rangle \, dt = \frac{v}{3 \rho_s} \int_0^\infty \langle \vec{j}_s(0) \cdot \vec{j}_s(t) \rangle \, dt . \tag{9.51}$$

In eq. (9.51), the symmetry properties of a second-rank tensor correlation function in an isotropic fluid have been used [7].

---

* That means for the time interval in which the correlation function has non-zero values.

Further simplification can be obtained by inserting eq. (9.46) and neglecting correlations between the velocities of different molecules at different times ($\langle \vec{v}_l(0) \cdot \vec{v}_k(t)\rangle = 0$ for $l \neq k$)

$$D = \frac{1}{3}\int_0^\infty \langle \vec{v}(0) \cdot \vec{v}(t)\rangle \, dt, \tag{9.52}$$

where $\vec{v} = \vec{p}/m$ is the velocity of a certain molecule. Similar correlation function expressions can be found for all other transport coefficients [7, 13].

9.2.5. Experimental Determination of Correlation Functions and Neutron Scattering

We have seen that correlation functions are very useful in the $Q \to 0$ and $\omega \to 0$ limit, where the local variables change very slowly with space and time, and where the correlation functions are related to transport coefficients. Now the importance of correlation functions for the investigation of the short time behaviour in fluids should be emphasized. Table 9.1 summarizes several experiments which can be evaluated for the corresponding correlation functions.

Table 9.1. Experimental determination of time correlation functions. The functions $F_1(t)$ and $F_2(t)$ are discussed in section 11.2, whereas the intermediate scattering functions $F(Q,t)$ and $F_s(Q,t)$ are given in section 9.2.5. M_x is the x-component of the magnetization of the system.

Experiment	Time correlation function involved	
Self-diffusion	$z(t) = \langle \vec{v}(0) \cdot \vec{v}(t)\rangle$	Velocity autocorrelation function
Dielectric relaxation	$F_1(t) \propto \langle \cos \vartheta(t)\rangle$	Angle correlation function
Infrared absorption	$F_1(t)$	
Raman scattering	$F_2(t) \propto \langle 3\cos^2 \vartheta(t) - 1\rangle$	
	Angle correlation function	
N.m.r.-lineshape	$\langle M_x(0)M_x(t)\rangle$	
	The essential intramolecular contribution is $F_2(t)$	
Rayleigh and Brillouin light scattering (isotropic molecules)	$F(Q,t) = \langle \tilde{n}_{-Q}(0)\tilde{n}_Q(t)\rangle/N - 8\pi^3 \rho \delta(Q)$	
Depolarized Rayleigh scattering	$F_2(t)$	
Neutron scattering incoherent	$F_s(Q,t) = \frac{1}{N}\sum_i \langle e^{-i\vec{Q}\cdot\vec{r}_i(0)} e^{i\vec{Q}\cdot\vec{r}_i(t)}\rangle$	
coherent	$F(Q,t)$	
Moessbauer absorption	$F_s(Q,t)$	
Moessbauer scattering	$F(Q,t)$	

Considering now the dynamics of spherical particles, neutron scattering proves to be the most suitable experimental method [14–16]. As has been pointed out in section 4.4.2, X-ray scattering gives only information on the molecular structure but not on the energy levels or — what is equivalent — on the time dependence of molecular motion. Light scattering on the other hand provides only information on the time behaviour of modes of very long wave lengths (Q very small) [17].
Eq. (4.87) and (4.91) show that coherent neutron scattering gives the scattering law $S(\vec{Q},\omega)$, whereas incoherent neutron scattering gives the self-scattering law $S_s(\vec{Q},\omega)$. $S(\vec{Q},\omega)$ and $S_s(\vec{Q},\omega)$

contain all the information of the structural and dynamical properties of the scatterer which can be obtained by a scattering experiment. In the light of the preceding sections, it is clear that the scattering laws are connected to correlation functions. However, the construction of these correlation functions presents certain difficulties. First, the perturbation given by the interaction neutron-particle cannot be written down easily, second, the classical calculation is insufficient. Therefore, we quote only the results for the classical limit*) [18, 14–16]

$$S(\vec{Q},\omega) = \frac{1}{2\pi} \iint e^{i(\vec{Q}\cdot\vec{r}-\omega t)} [G(\vec{r},t) - \rho] \, d\vec{r} \, dt \tag{9.53}$$

$$S_s(\vec{Q},\omega) = \frac{1}{2\pi} \iint e^{i(\vec{Q}\cdot\vec{r}-\omega t)} G_s(\vec{r},t) \, d\vec{r} \, dt \tag{9.54}$$

$$G(\vec{r},t) = \frac{1}{N} \sum_{i,j} \int d\vec{r}' \langle \delta(\vec{r} + \vec{r}_i(0) - \vec{r}') \delta(\vec{r}' - \vec{r}_j(t)) \rangle = \frac{1}{N} \sum_{i,j} \langle \delta(\vec{r} + \vec{r}_i(0) - \vec{r}_j(t)) \rangle \tag{9.55}$$

$$G_s(\vec{r},t) = \frac{1}{N} \sum_i \int d\vec{r}' \langle \delta(\vec{r} + \vec{r}_i(0) - \vec{r}') \delta(\vec{r}' - \vec{r}_i(t)) \rangle = \frac{1}{N} \sum_i \langle \delta(\vec{r} + \vec{r}_i(0) - \vec{r}_i(t)) \rangle. \tag{9.56}$$

The first relation in eq. (9.55) and (9.56) follows from the quantum mechanical calculation. In the classical limit, the δ-functions commute and the second relation can be obtained. We have given both versions, because the first version is sometimes more convenient to use for classical calculations also.

Introducing the one-particle distribution function $n(\vec{r},t)$ [cf. eq. (4.5)] and its Fourier transform $\tilde{n}_Q(t)$ by

$$n = \sum_i \delta(\vec{r} - \vec{r}_i) \quad \text{and} \quad \tilde{n}_Q = \sum_i e^{i\vec{Q}\cdot\vec{r}_i}, \tag{9.57}$$

eq. (9.55) may be written as

$$G(\vec{r},t) = \frac{1}{N} \int d\vec{r}' \langle n(\vec{r}' - \vec{r},0) n(\vec{r}',t) \rangle = \frac{\langle n(0,0) n(\vec{r},t) \rangle}{\rho}, \tag{9.58}$$

which shows $G(\vec{r},t)$ as density-density correlation function. The limiting value for $G(\vec{r},t)$ for large $\vec{r}$ or t, where the local densities at space points 0 and $\vec{r}$ are not correlated, are thus found to be $G(\infty,t) = G(\vec{r},\infty) = \rho$. As already indicated in section 4.4.2, the limit $t \to 0$ gives $G(\vec{r},0) = \delta(\vec{r}) + \rho g(\vec{r})$, where the first term is the $t \to 0$ limit of the "self" part G_s and the second term of the "distinct" part $G_d = G - G_s$.

Fourier transformation of $G(\vec{r},t)$ in space gives the so-called intermediate scattering function, which is the relevant time correlation function for scattering processes:

$$F(\vec{Q},t) = \int [G(\vec{r},t) - \rho] e^{i\vec{Q}\cdot\vec{r}} d\vec{r} = \frac{1}{N} \sum_{ij} \langle e^{-i\vec{Q}\cdot\vec{r}_i(0)} e^{i\vec{Q}\cdot\vec{r}_j(t)} \rangle - 8\pi^3 \rho \delta(\vec{Q})$$

$$= \frac{1}{N} \langle \tilde{n}_{-Q}(0) \tilde{n}_Q(t) \rangle - 8\pi^3 \rho \delta(\vec{Q}) \tag{9.59}$$

$$F_s(\vec{Q},t) = \int G_s(\vec{r},t) e^{i\vec{Q}\cdot\vec{r}} d\vec{r} = \frac{1}{N} \sum_i \langle e^{-i\vec{Q}\cdot\vec{r}_i(0)} e^{i\vec{Q}\cdot\vec{r}_i(t)} \rangle. \tag{9.60}$$

---

* The equations differ from eq. (4.88) and (4.90) in the way the Fourier transform is taken. This is a matter of definition. Unfortunately, the common use is different when treating equilibrium problems or when dealing with time-dependent problems (cf. appendix). The term ρ in the integrand of eq. (9.53) excludes from $S(\vec{Q},\omega)$ the infinity for forward scattering, which is not always done in the literature (cf. ref. [16]).

Comparison with eq. (9.53) and (9.54) shows

$$S(\vec{Q},\omega) = \frac{1}{2\pi} \int_{-\infty}^{+\infty} e^{-i\omega t} F(\vec{Q},t) \, dt \qquad (9.61)$$

$$S_s(\vec{Q},\omega) = \frac{1}{2\pi} \int_{-\infty}^{+\infty} e^{-i\omega t} F_s(\vec{Q},t) \, dt . \qquad (9.62)$$

Therefore in terms of the fluctuation-dissipation theorem [eq. (9.27)], the scattering law is related to the imaginary part of the density-density susceptibility:

$$S(\vec{Q},\omega) = \frac{kT}{\rho \pi \omega} \chi''_{nn}(\vec{Q},\omega) - 8\pi^3 \rho \delta(\vec{Q}) \delta(\omega) . \qquad (9.63)$$

Some properties of $S(\vec{Q},\omega)$ can be seen from the inverse Fourier transformation of eq. (9.61):

$$F(\vec{Q},t) = \int_{-\infty}^{+\infty} e^{i\omega t} S(\vec{Q},\omega) \, d\omega . \qquad (9.64)$$

First, this gives, as already quoted in eq. (4.89),

$$\int_{-\infty}^{+\infty} S(\vec{Q},\omega) \, d\omega = F(\vec{Q},0) = \frac{1}{N} \sum_{i,j} \langle e^{-i\vec{Q}\cdot\vec{r}_i(0)} e^{i\vec{Q}\cdot\vec{r}_j(0)} \rangle - 8\pi^3 \rho \delta(\vec{Q}) = S(\vec{Q}). \qquad (9.65)$$

Similarly, the second moment can be derived

$$\int_{-\infty}^{+\infty} \omega^2 S(\vec{Q},\omega) \, d\omega = \lim_{t \to 0} -\frac{d^2}{dt^2} F(\vec{Q},t) = Q^2 \frac{kT}{m} . \qquad (9.66)$$

In the same way, eq. (4.92) can be obtained from $F_s(\vec{Q},t)$.

9.3. Single Particle Motion

The basic relations connect the mean square displacement of the particle, diffusion coefficient, velocity (auto-)correlation function, and the frequency spectrum. We will first review these relations, then try to give a feeling for the significance of velocity correlation functions and frequency spectra, then give some results of computer calculations, and finally deal with the experimental determination, applying the formalism of the last section. We start with the Einstein formula for the diffusion coefficient, eq. (8.71), written only in the x-component of the position vector $\vec{r}$:

$$\langle (r_x(\tau) - r_x(0))^2 \rangle = 2D\tau, \quad \tau \text{ large.} \qquad (9.67)$$

In general, the mean square displacement starts in some irregular fashion at very small time intervals and approaches then the linear relation, for which the diffusion coefficient is defined. On the other hand, differentiating $\langle (\Delta r(\tau))^2 \rangle$ twice with respect to τ gives the velocity correlation function. To show this, we observe first that the ensemble average of $r_x^2(\tau)$ does not depend on τ, but that the only term on the l.h.s. of eq. (9.67) varying with τ is the cross product $-\langle 2r_x(0)r_x(\tau)\rangle$. Second, we apply the stationarity condition, which expresses the fact that the choice of the time origin is immaterial, so that [cf. eq. (9.26)]

$$\langle \dot{r}_x(0) \dot{r}_x(\tau) \rangle = -\langle r_x(0) \ddot{r}_x(\tau) \rangle . \qquad (9.68)$$

The l.h.s. is one third of the total velocity correlation function, the r.h.s. can be obtained by differentiating $\langle(\Delta r_x(\tau))^2\rangle/2$ twice with respect to τ, so that

$$d^2 \langle(\Delta r_x(\tau))^2\rangle/d\tau^2 = \frac{2}{3} \langle \vec{v}(0) \cdot \vec{v}(\tau)\rangle . \tag{9.69}$$

Comparison with eq. (9.67) shows that the diffusion coefficient is given by the integral over the velocity correlation function:

$$D = \frac{1}{3} \int_0^\tau \langle \vec{v}(0) \cdot \vec{v}(\tau)\rangle \, d\tau, \tag{9.70}$$

where the upper limit is for large times, for which $\langle(\Delta \vec{r})^2\rangle$ is proportional to τ, and for which the velocity correlation function is zero, so that τ can be replaced by infinity, and eq. (9.70) becomes identical to eq. (9.52).

The frequency spectrum characterizing the particle motion is then given, according to the Wiener-Khintchine theorem, by the Fourier transform of the velocity correlation function:

$$\tilde{z}(\omega) = \frac{1}{2\pi} \int_{-\infty}^{+\infty} \langle \vec{v}(0) \cdot \vec{v}(\tau)\rangle e^{-i\omega\tau} d\tau \tag{9.71}$$

The limit $\tilde{z}(0)$ is, according to eq. (9.70),

$$\tilde{z}(0) = 3D/\pi . \tag{9.72}$$

The integral over the frequency spectrum gives

$$\int_{-\infty}^{+\infty} \tilde{z}(\omega) \, d\omega = \int_{-\infty}^{+\infty} \langle \vec{v}(0) \cdot \vec{v}(\tau)\rangle \int_{-\infty}^{+\infty} \frac{1}{2\pi} e^{-i\omega\tau} d\omega \, d\tau$$

$$= \int_{-\infty}^{+\infty} \langle \vec{v}(0) \cdot \vec{v}(\tau)\rangle \delta(\tau) \, d\tau = \langle \vec{v}^2(0)\rangle = 3kT/m . \tag{9.73}$$

This transformation makes use of the Fourier transform of the δ-function.

9.3.1. Some Illustrative Examples (Fig. 9.1)

Let us start with the perfect (non-colliding) gas, where $v_x(\tau)$ is the same as $v_x(0)$ and the velocity correlation function is equal to $3kT/m$ and constant with τ. The frequency spectrum is then $\tilde{z}(\omega) = (3kT/m)\delta(\omega)$, proportional to a δ-function at the origin, with an infinite diffusion coefficient. The mean square displacement is given by $\langle(\Delta r_x(\tau))^2\rangle = \tau^2(kT/m)$, so that the differential quotient is still proportional to τ and goes to infinity for larger τ.

How about the colliding gas? The mean square displacement will start as in the case of the non-colliding gas, but after the particles have travelled the mean free path, their displacement is slowed down by the collisions, so that $\langle(\Delta r(\tau))^2\rangle$ becomes ultimately linear with τ. Accordingly, the velocity correlation starts with $3kT/m$ and drops to zero. The frequency spectrum involves then lower frequencies too. At higher densities of the gas, the effect of the collisions will be appreciable at smaller time intervals and the diffusion coefficient will become smaller. Accordingly, the frequency spectrum starts at a lower value of $\tilde{z}(0)$ and is more broadened (as the area is given by $3kT/m$).

9.3. Single Particle Motion

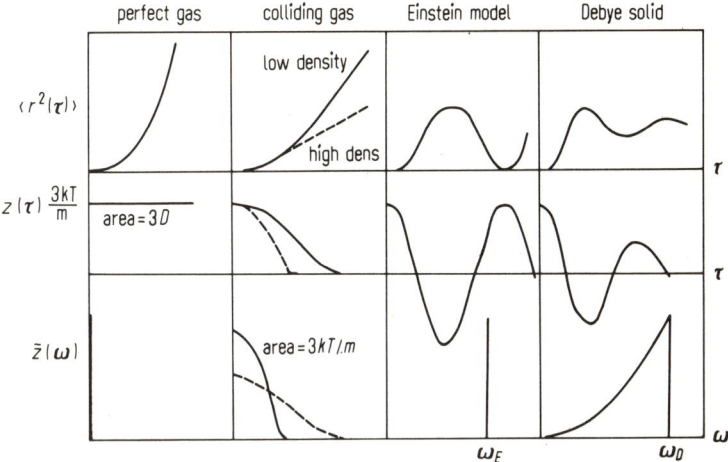

Fig. 9.1. Schematic comparison of the mean square displacement $\langle r^2(\tau)\rangle$, the velocity autocorrelation function $z(\tau)$ and the frequency spectrum $\tilde{z}(\omega)$ for an ideal gas, a real colliding gas, an Einstein-solid and a Debye-solid. The solid line is for a gas at low densities, the dotted line for a high density gas.

Let us now consider the other extreme, the Einstein model of a solid. Here $\langle(\Delta r(\tau))^2\rangle$ and the velocity correlation function are periodic, the frequency spectrum is a δ-function around the Einstein frequency ω_E: $\tilde{z}(\omega) = (3kT/m)\delta(\omega - \omega_E)$. The situation changes drastically when a more realistic model of a solid is employed. As a consequence of the frequency distribution, the different normal modes are superposed, so that the oscillations of the velocity correlation are strongly damped [19]. The same applies to the oscillations of the mean square displacement. As the diffusion coefficient is zero (or practically zero in a real solid), $\langle(\Delta r(\tau))^2\rangle$ becomes constant and the area under the velocity correlation function has to vanish.

9.3.2. Computer Calculation for Argon

Molecular dynamics calculations have been done for argon on the basis of a 12/6 potential [1]. From the mean square displacement the velocity correlation and the frequency spectrum has been obtained. The results are shown in fig. 9.2.

This figure was a surprise to most scientists, because one is too much accustomed to treat the liquid either as a dense gas or as an Einstein solid. But the gross features can be interpreted rather readily on the basis of the disorder model dealt with in section 2.6.1, where each molecule is considered to be surrounded by a fluctuating number of nearest neighbours. One might expect the given molecule to oscillate in the potential well caused by the neighbours until a fluctuation causes a defect in the nearest neighbour shell, thus enabling the central molecule to move translationally until it is stopped by the next-nearest neighbours and trapped again. Computer calculations on hard spheres show, however, that the distribution of mean free paths decreases monotonously with the length of the mean free path [20], i.e. there is no possibility to distinguish

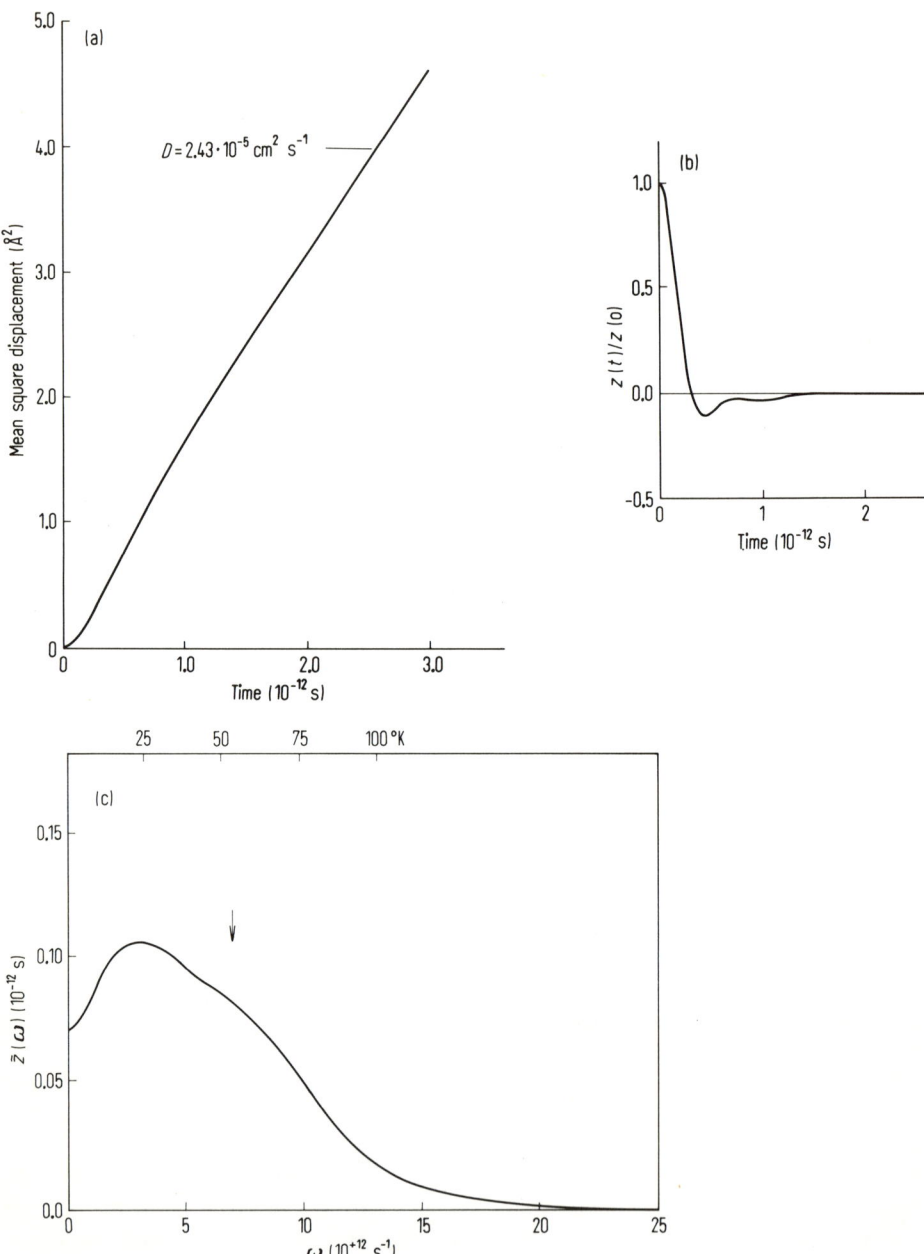

Fig. 9.2. Computer calculations of a system of 864 argon atoms:
a) Mean square displacement (after Rahman 1964 [1])
b) The velocity autocorrelation function normalized to unity at $t = 0$ (after Rahman 1966 [1])
c) The frequency spectrum. The area is normalized to unity (after Rahman 1966 [1]).
The temperature and density of the system were 94.4 K and 1.374 g cm^{-3} for a) and 85.5 K and 1.407 g cm^{-3} for b) and c).

between small movements and large jumps. Obviously the inhomogeneities are smoothed out in the liquid to a large extent. Such movements may correspond to a combination of a damped oscillation and a short translation. The ensemble average $\langle v_x(0) v_x(\tau) \rangle$ would be much more damped. The Fourier transform of a strongly damped oscillation is a strongly broadened spectral line, which may cut the ordinate at a non-zero value. Damping and diffusion are, therefore, interconnected.

Next we will consider the magnitude of the frequency at the maximum of the spectrum. The maximum frequency is about $\omega = 3 \cdot 10^{12}$ s^{-1}, which corresponds to a characteristic temperature of $\theta = 25$ K, a very small value. The calculated frequency spectrum seems to contain also a spurious peak (arrow in fig. 9.2c) at $\omega = 7 \cdot 10^{12}$ s^{-1} corresponding to $\theta = 54$ K. This would agree better with a simplified cell model calculation, where neighbours are assumed to be located in the centers of their cells [*].

One further comment is concerned with the calculated velocity correlation function of argon, which is shown for various densities in fig. 9.3. It is seen that the oscillation vanishes with decreasing

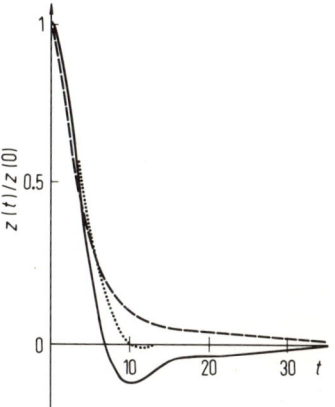

Fig. 9.3. The velocity correlation function of a Lennard-Jones fluid, normalized to unity for $t \to 0$, as obtained from molecular dynamics calculations for various fluid states:
solid line: $\rho = 0.85$, $T = 0.76$
dotted line: $\rho = 0.75$, $T = 1.12$
dashed line: $\rho = 0.65$, $T = 1.43$
Temperature and density are given in normalized units kT/ε^* and ρ/ρ^*, where ρ^* corresponds to a dense packing with neighbour distance R^*. Ten time units correspond to $4 \cdot 10^{-13}$ s in the case of argon (from Levesque and Verlet [1]).

density and increasing temperature. However, the function is then far from Gaussian, having a small half width but a very broad foot. An explanation for the long time behaviour of the velocity autocorrelation function at intermediate fluid densities has been proposed by Alder and Wainwright [21] based on molecular dynamics calculations for hard spheres. The molecule, moving with $\vec{v}$, causes a vortex flow of its surrounding molecules by which the room behind the moving molecule is filled. This results in a persistence of velocities and a decay of the correlation

---

[*] The order of magnitude of the diffusion coefficient can be estimated from the disorder model dealt with in section 2.6.1. Writing the diffusion coefficient in the simple form
$$D = \bar{v}\bar{\lambda}/3$$
$\bar{v}$ being the mean thermal velocity, $\bar{\lambda}$ the mean free path, and approximating $\bar{\lambda}$ by
$$\bar{\lambda} = xa(\sqrt{2} - 1)$$
(x fraction of broken nearest neighbour contacts, a distance to nearest neighbour shell), one obtains for argon $D = 1.1$ (in 10^5 cm^2 s^{-1}) at the triple point, and $D = 8.8$ at the critical point (the experimental figures are 1.6 and 9.7, resp.; but this agreement should not be taken too seriously).

function with $t^{-3/2}$ at long times. Hydrodynamic models confirm this explanation [22]. It is not yet clear if the same considerations hold for Lennard-Jones fluids at intermediate densities (Levesque and Verlet [1]).

9.3.3. Experimental Determinations of Single Particle Motions

The experimental determination depends on the evaluation of the self-scattering law $S_s(\vec{Q},\omega)$, i.e. on incoherent neutron scattering.

A large incoherent contribution to the total neutron scattering is present only in argon, sodium, and hydrogen-containing molecules (of the readily accessible liquids). Therefore, the experiments involving $S_s(\vec{Q},\omega)$ are to date restricted to these liquids [14]. Results on methane and hydrogen will be mentioned in chapter 11.

The division of the neutron scattering cross-section into a coherent and an incoherent contribution is a matter of knowing the appropriate scattering lengths. Only two remarks should be made here. First, an experimental division is possible for spin-dependent scattering lengths by using polarized neutrons, and recording those neutrons which are depolarized upon scattering. The depolarized cross-section is 2/3 of the incoherent contribution. For intensity reasons such experiments require an extremely high neutron flux and are very difficult to perform. The second remark concerns the possibility to distinguish for small τ between $G_s(\vec{r},\tau)$ and $G_d(\vec{r},\tau) = G(\vec{r},\tau) - G_s(\vec{r},\tau)$ which describe the time correlations of the same atom and of different atoms, respectively (d standing for distinct). The double Fourier transform of the total coherent *and* incoherent cross section [23] gives a weighted average of $G_s(\vec{r},\tau)$ and $G_d(\vec{r},\tau)$, namely $\langle b^2 \rangle G_s(\vec{r},\tau) + \langle b \rangle^2 G_d(\vec{r},\tau)$. But for small time intervals $G_s(\vec{r},\tau)$ is different from zero only in the neighbourhood of $\vec{r} = 0$, where $G_d(\vec{r},\tau)$ is certainly zero. On this basis, the two functions can be disentangled. For larger τ-values some additional assumptions, like Gaussian behaviour of $G_s(\vec{r},\tau)$ would be necessary.

In principle, the connection of $S_s(\vec{Q},\omega)$ to the quantities characteristic for the single particle motion is obtained from

$$\langle (\Delta \vec{r}(\tau))^2 \rangle = \int \vec{r}^2 G_s(\vec{r},\tau) d\vec{r} . \tag{9.74}$$

But the evaluation can be carried out without the double Fourier transformation from $S_s(\vec{Q},\omega)$ to $G_s(\vec{r},\tau)$. To show this we start with the equality

$$\frac{1}{2\pi} \int_{-\infty}^{+\infty} dt\, e^{-i\omega t} \frac{d^2}{dt^2} F_s(\vec{Q},t) = -\frac{\omega^2}{2\pi} \int_{-\infty}^{+\infty} dt\, e^{-i\omega t} F_s(\vec{Q},t), \tag{9.75}$$

which is obtained by integrating the left hand side partially twice and considering the fact, that the appearing time correlation functions vanish in the limit $t \to \infty$. Eq. (9.60) gives

$$\frac{d^2 F_s(\vec{Q},t)}{dt^2} = \frac{d^2}{dt^2} \langle e^{-i\vec{Q}\cdot\vec{r}(0)} e^{i\vec{Q}\cdot\vec{r}(t)} \rangle = \frac{d}{dt} \langle e^{-i\vec{Q}\cdot\vec{r}(0)} i\vec{Q}\cdot\vec{v}(t) e^{i\vec{Q}\cdot\vec{r}(t)} \rangle$$

$$= \frac{d}{dt} \langle e^{-i\vec{Q}\cdot\vec{r}(-t)} i\vec{Q}\cdot\vec{v}(0) e^{i\vec{Q}\cdot\vec{r}(0)} \rangle = - \langle \vec{Q}\cdot\vec{v}(0) e^{-i\vec{Q}\cdot\vec{r}(0)} \vec{Q}\cdot\vec{v}(t) e^{i\vec{Q}\cdot\vec{r}(t)} \rangle . \tag{9.76}$$

We have used the fact that the ensemble average is independent of the time origin and that all particles give the same contribution so that the sum cancels the factor $1/N$; $\vec{r}$ and $\vec{v} = \dot{\vec{r}}$ denote the position and velocity of an arbitrary molecule.

From eq. (9.62), (9.75), and (9.76) one finally gets

$$\omega^2 S_s(\vec{Q},\omega) = \frac{1}{2\pi} \int_{-\infty}^{+\infty} dt \, e^{-i\omega t} \langle \vec{Q} \cdot \vec{v}(0) \, e^{-i\vec{Q}\cdot\vec{r}(0)} \vec{Q} \cdot \vec{v}(t) \, e^{-i\vec{Q}\cdot\vec{r}(t)} \rangle . \tag{9.77}$$

Taking Q^2 out of the integral and then going with Q against zero, the result is

$$\lim_{Q \to 0} \frac{\omega^2 S_s(\vec{Q},\omega)}{Q^2} = \frac{1}{2\pi} \int_{-\infty}^{+\infty} dt \, \langle v_Q(0) \, v_Q(t) \rangle \, e^{-i\omega t} = \frac{1}{3} \tilde{z}(\omega) , \tag{9.78}$$

where v_Q denotes the component of $\vec{v}$ in the direction of $\vec{Q}$. Eq. (9.78) may serve to determine the frequency spectrum from the measured incoherent cross section. Methods have been worked out which eliminate the necessity of going to the limit $Q \to 0$ [24].

Such a determination has been carried out on liquid sodium [24]. The frequency spectrum is shown in fig. 9.4 together with a computer calculation. This is based on a potential with a

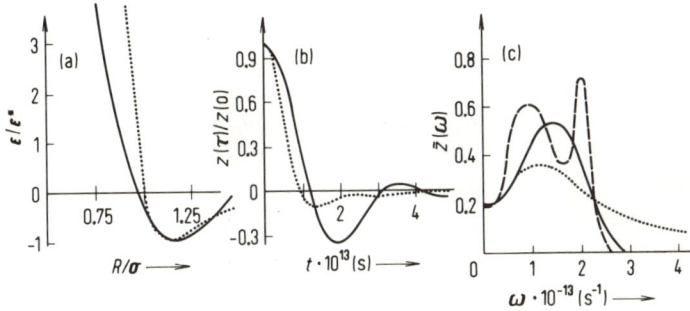

Fig. 9.4. Comparison of the experimental frequency spectrum as obtained from neutron scattering (dashed line in (c)) (Cocking [24]) with molecular dynamics calculations (Schiff [24]). For the computer calculations either a soft core potential (full curves) or a Lennard-Jones potential (dotted line) have been used (a). The computed autocorrelation function are shown in (b) for both cases. After Cocking [24].

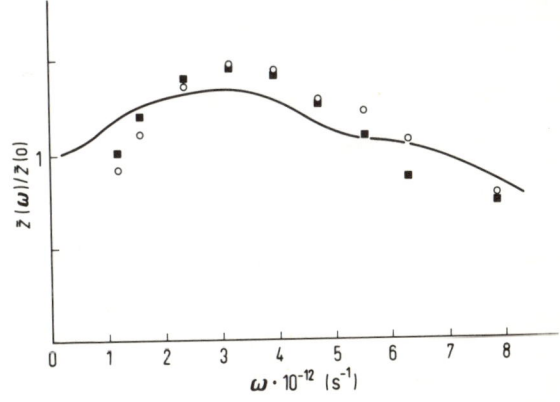

Fig. 9.5. The frequency spectrum for liquid argon according to neutron scattering experiments of Sköld and Larsson [25] evaluated by a quasi-crystalline model of Singwi (○) or Egelstaff (■), resp. [30]. Computer calculations by Rahman (1964) [1] are given for comparison (full line). After Larsson, Dahlborg, and Sköld [25].

soft repulsion, i.e. a wider bowl than the pair potential of argon. It is seen that the soft core produces a more solid-like behaviour of the frequency spectrum. However, the very pronounced appearance of two maximum frequencies (at about 7 and $19 \cdot 10^{12}$ s^{-1}, i.e. in about the ratio 1 : 3) in the experimental determination is difficult to explain [24].

Another possibility for deriving $\tilde{z}(\omega)$ is to fit a model to the whole $S(Q,\omega)$ surface (distinct as well as self terms), and to derive $\tilde{z}(\omega)$ from the model. Such an attempt has been reported by Sköld and Larsson for argon [25]. Their results are shown in fig. 9.5.

9.4. Collective Particle Motions

9.4.1. Description of the Behaviour of the Scattering Law $S(Q,\omega)$

The study of collective particle motions rests on the observation of coherent scattering at definite energy transfer $\hbar\omega$. This can be accomplished, for smallest momentum transfer $\hbar Q$, by light scattering, for which a high resolution is possible since the advent of laser beams. As has been mentioned already (chapter 1.5 and 6.5), the density fluctuations responsible for light scattering can be divided into a propagating part (propagating with the velocity of sound) and a non-propagating part (so-called entropy fluctuations where the density difference is compensated by a difference in molecular kinetic energy, such that the pressure remains constant). The light scattered on the non-propagating density fluctuation appears at the ingoing frequency $v_0 = \omega_0/2\pi$ (Rayleigh line), whereas the light scattered on the propagating density fluctuations appears at circular frequencies [17]

$$\omega = \omega_0 \pm cQ, \tag{9.78}$$

c being the (hypersonic) sound velocity, and $Q = \dfrac{2\pi}{\lambda} \sin \dfrac{\theta}{2}$ determining the momentum transfer which is related to the light wave length λ and the scattering angle θ. Eq. (9.78) might be considered

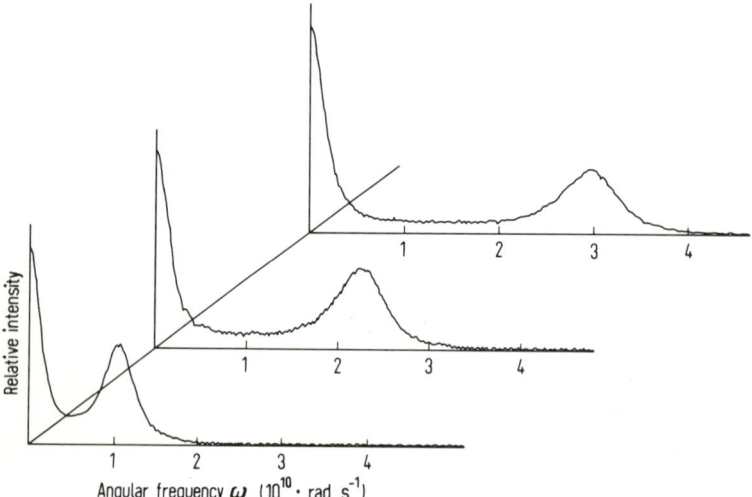

Fig. 9.6. Relative intensity of light scattered on liquid CCl$_4$ for scattering angles of 44°, 97°, and 155°, resp. From Gornall et al. [26].

as a Doppler effect. The shifted lines in the scattered spectrum are called Brillouin lines. As λ of laser light is of order $7 \cdot 10^{-5}$ cm, Q is of order 10^5 cm^{-1}, and as the sound velocity is of order 10^5 cm s^{-1}, the frequency shift amounts to 10^9 s^{-1}. Thus the sound velocity measured by the frequency difference between Rayleigh and Brillouin line corresponds to a frequency in the GHz range. An example [26] of such a scattered spectrum, which is essentially a plot of $S(Q,\omega)$ at small Q, is shown in fig. 9.6.

Let us now jump for about three orders of magnitude in momentum and energy transfer and consider $S(Q,\omega)$ at smallest Q-values accessible for neutron scattering [27]. An example is given in fig. 9.7. Though the resolution is much poorer than in the case of light scattering, it is well

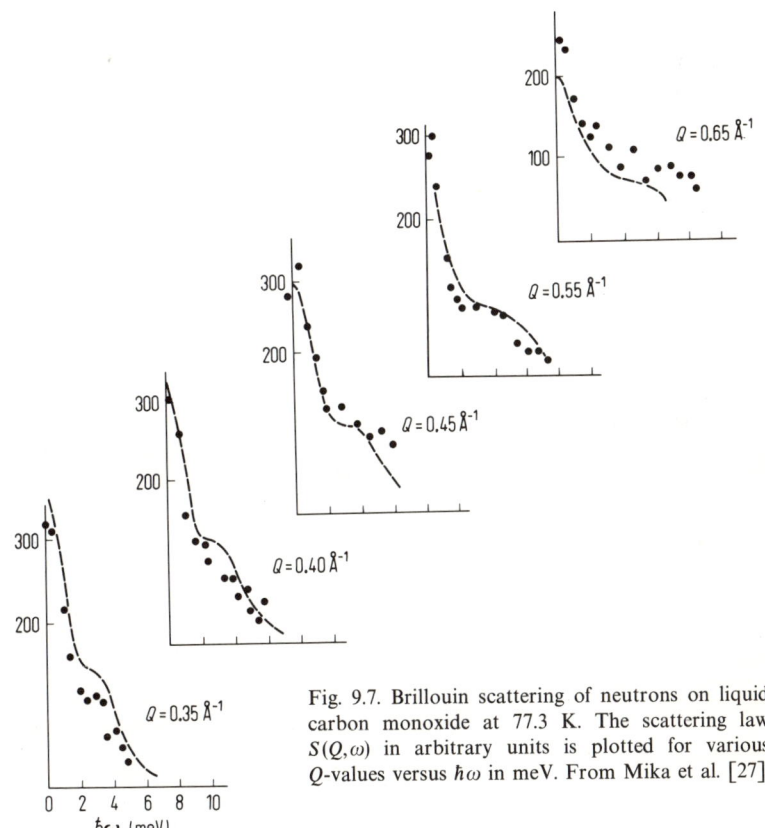

Fig. 9.7. Brillouin scattering of neutrons on liquid carbon monoxide at 77.3 K. The scattering law $S(Q,\omega)$ in arbitrary units is plotted for various Q-values versus $\hbar\omega$ in meV. From Mika et al. [27].

seen how the Brillouin wings appear more and more shifted with increasing Q. The dotted lines in fig. 9.7 are obtained from a hydrodynamic model (cf. section 9.4.3). It will be seen, that we are here just at the limit of the so-called hydrodynamic region where the continuum terminology with sound velocity and transport coefficients ceases to be applicable.

176   9. Non-Equilibrium Properties: Liquid Dynamics

Before looking at fig. 9.8, which gives the neutron scattering intensity for argon [28] for a larger range of Q, it is necessary to remember that in the case of argon an appreciable fraction of the neutrons is scattered incoherently. Therefore, fig. 9.8 displays a superposition of $S(Q,\omega)$ and $S_s(Q,\omega)$.

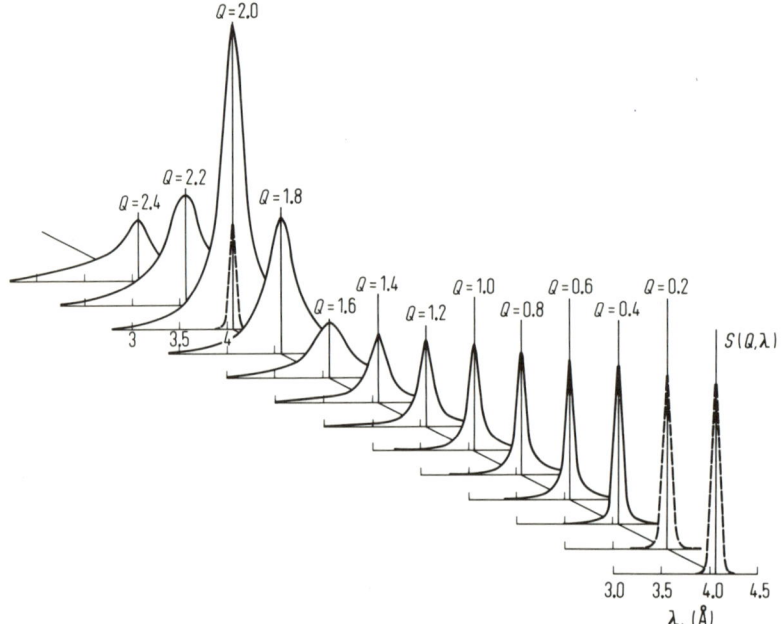

Fig. 9.8. The scattering surface (double differential cross section) for liquid argon (84.5 K, 550 mm Hg) for various Q as a function of the wavelength λ of scattered neutrons (the shift of λ at scattering is a measure of ω). After ref. 28.

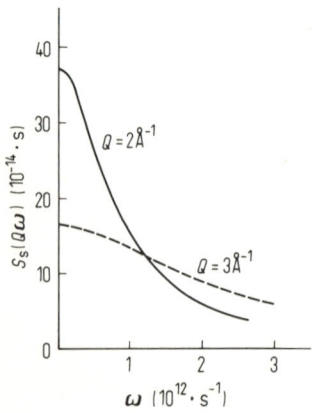

Fig. 9.9. The incoherent scattering law $S_s(Q,\omega)$ as a function of ω for different Q, obtained from molecular dynamics calculations for liquid argon ($T = 85.5$ K, $\rho = 1.407$ g cm^{-3}). For the evaluation the Gaussian approximation for $F_s(Q,t)$ has been used. (Data are taken from Nijboer and Rahman [1]).

In order to show the shape of $S_s(Q,\omega)$, fig. 9.9 presents examples for two different Q-values, calculated from the frequency spectrum according to the molecular dynamics calculations. For small Q-values, $S_s(Q,\omega)$ is a sharp peak with small wings at the bottom corresponding to the high frequencies in $\tilde{z}(\omega)$; at larger Q-values, $S_s(Q,\omega)$ flattens out (for small Q the half width is proportional to DQ^2, D being the coefficient of self diffusion). As the integral over ω has to be unity [eq. (4.92)], the flattening of $S_s(Q,\omega)$ reduces the peak height.

Now, part of the behaviour of the scattering surface of argon (fig. 9.8) is due to $S_s(Q,\omega)$, but nevertheless it can be seen that the wings of $S(Q,\omega)$ — the Brillouin lines in the hydrodynamic region — reduce from a certain Q-value on, which is about half of Q_0, where $S(Q)$ has its maximum. At Q_0, $S(Q_0,\omega)$ is quite narrow; for larger Q-values, $S(Q,\omega)$ becomes broader again. Due to the sharpening of $S(Q,\omega)$ in the neighbourhood of Q_0, the maximum of $S(Q)$ and $(S(Q,\omega=0))$ does not completely coincide. Because $S(Q)$ is the integral of $S(Q,\omega)$ over all ω [eq. (9.65)], its maximum is shifted from the maximum of $S(Q,0)$ to slightly higher Q-values, because then the $S(Q,\omega)$-function thickens again. Also, on account of the sharpening of the $S(Q,\omega)$ function versus ω around Q_0, $S(Q,\omega)$ has for a constant large ω-value a minimum versus Q at about Q_0, near the place where $S(Q,0)$ has a very sharp maximum. The maximum of $S(Q,0)$, and the minimum of $S(Q,\omega$ large) occurs near Q_0; both are slightly shifted, but for different amounts. The difference between the extremes of $S(Q)$, $S(Q,0)$ and $S(Q,\omega$ large) becomes more pronounced at the higher maxima.

9.4.2. Computer Calculations of the Current-Current Fluctuation

The general features of the behaviour of $S(Q,\omega)$ receive some qualitative explanation by computer results on the current-current fluctuation, obtained by molecular dynamics on an argon-like liquid (parameters of a 12/6-potential suitable for argon, $T = 76$ K [29]). The current $\vec{J}$ is defined as

$$\vec{J}(\vec{r},t) = \sum_{j=1}^{N} \vec{v}_j(t)\,\delta(\vec{r} - \vec{r}_j(t)), \tag{9.79}$$

where $\vec{v}_j$ is the velocity of the j-th particle at time t. The Fourier components for a wave vector $\vec{Q}$ become

$$\vec{J}(\vec{Q},t) = \sum_{j=1}^{N} \vec{v}_j(t)\, e^{i\vec{Q}\cdot\vec{r}_j(t)}. \tag{9.80}$$

The scalar product $\vec{Q}\cdot\vec{J}(\vec{Q},t)$ will be denoted by $J_\|(\vec{Q},t)$ and called the longitudinal component of $\vec{J}(\vec{Q},t)$. Such a definition is meaningful also at large values of Q (small values of $\lambda = 2\pi/Q$), where it is not possible any more to speak of a wave or a phonon. The "wave" length λ gives then the periodicity of the particle velocities being in phase, and this makes sense also for λ-values smaller than next particle distances. The correlation of the current fluctuation will be denoted by

$$C_\|(\vec{Q},\tau) = \langle J_\|(\vec{Q},0)\, J_\|^*(\vec{Q},\tau)\rangle/N. \tag{9.81}$$

It has the following properties. For time zero, we have

$$C_\|(\vec{Q},0) = Q^2 kT/m. \tag{9.82}$$

This result follows because positions and velocities are statistically independent. The factor Q^2 in comparison to the velocity autocorrelation function [eq. (9.73)] follows from the definition of $J_\|(\vec{Q},t)$. Then — similarly as the velocity correlation function is given as the second time derivative of the mean square displacement [eq. (9.69) and (9.76)] — the current correlation function is given as the negative second time derivative of the intermediate scattering function [cf. eq. (9.59)]

$$C_{\|}(\vec{Q},\tau) = -\frac{d^2}{d\tau^2} F(\vec{Q},\tau). \tag{9.83}$$

Upon Fourier transformation, this gives [cf. eq. (9.75)]

$$C_{\|}(\vec{Q},\omega) = \omega^2 S(\vec{Q},\omega). \tag{9.84}$$

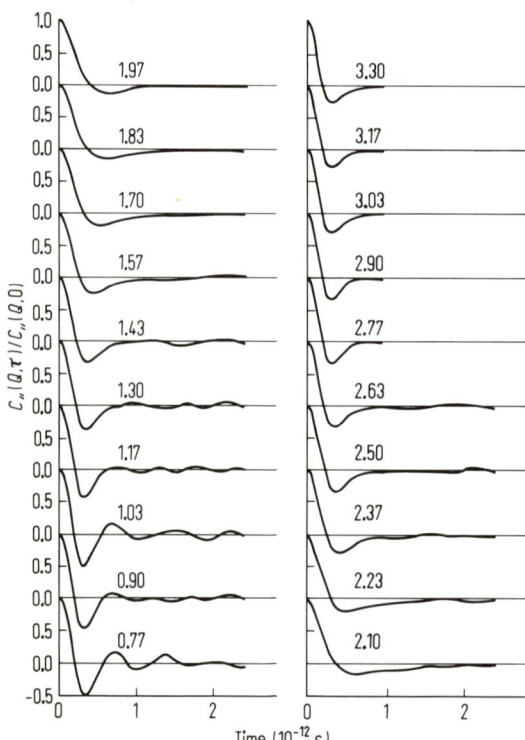

Fig. 9.10. Computer calculated current correlation functions $C_{\|}(Q,\tau)/C_{\|}(Q,0)$ for argon at $T = 76$ K and various values of Q [29].

The computer results on $C_{\|}(Q,\tau)$ — normalized by division through $Q^2 kT/m$ — are shown in fig. 9.10 for various values of Q as parameter. It should be noted that $S(Q)$ has its first and very sharp maximum at $Q = 2.0 \text{Å}^{-1}$ and the next maximum at $Q = 3.6 \text{Å}^{-1}$ (cf. fig. 4.7). The first maximum corresponds to the equilibrium distance between the tetraeder planes of a f.c.c. structure: $d_{111} = 2\pi/Q = R^*\sqrt{2/3}$ with $R^* = 3.82$ Å. Around $Q = 2 \text{Å}^{-1}$, the variation of $C_{\|}(Q,\tau)$ is slower with τ than at all other values of Q, and around $Q = 1 \text{Å}^{-1}$ it is faster. The first case corresponds to an in-phase displacement of immediate neighbours ($\lambda = d_{111}$), the second to an out-of-phase displacement. This is obviously connected to small or big restoring forces, respectively. The characteristics of $C_{\|}(Q,\tau)$ transpose to the Fourier transform $\omega^2 S(Q,\omega)$ such, that $\omega^2 S(Q,\omega)$ is peaked sharply at small ω for $Q \approx 2 \text{Å}^{-1}$ and, of course, for small values of Q. The (normalized) Fourier transforms spread out if the peak height is small. Since the life-time of a fluctuation is inversely related to the width of the frequency response, the peak height of $\omega^2 S(Q,\omega)$ is a measure of the life-time of the corresponding fluctuation.

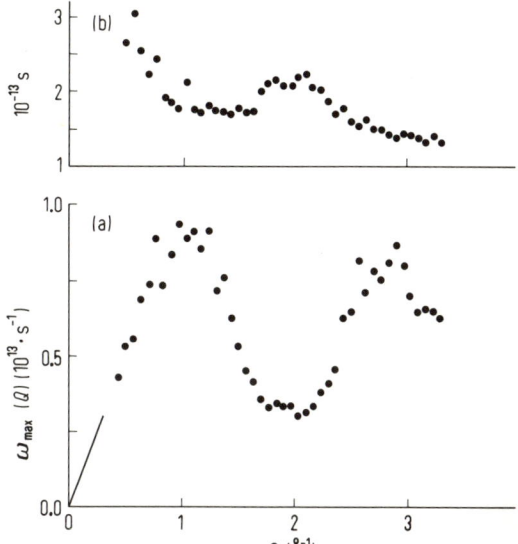

Fig. 9.11. Molecular dynamics calculations for Lennard-Jones-particles (after [29]).
a) Values of ω at the maximum of $C_\|(Q,\omega) = \omega^2 S(Q,\omega)$ (Dispersion for the longitudinal current-current correlation function). The solid line corresponds to the extrapolated sound velocity of argon.
b) Peak height of $C_\|(Q,\omega)/C(Q, t = 0)$ which is a measure of the life-time of the current fluctuations.

The coordinates of the maximum of $\omega^2 S(Q,\omega)$ i.e. ω_{max} and peak height, are shown in fig. 9.11 as a function of Q. It is seen that ω_{max} has its maxima in between the maxima of $S(Q)$, whereas the peak height (i.e. the life-time of the fluctuation) has — superimposed on a general decreasing tendency — a maximum at the maximum of $S(Q)$.

The relation of ω_{max} versus Q (fig. 9.11 a) is called dispersion relation for the current fluctuation. In the hydrodynamic region, where $S(Q,\omega)$ for the Brillouin wing can be approximated by [17]

$$S_{Brill}(Q,\omega) = BQ^2 \frac{2}{(\omega - cQ)^2 + (\Gamma' Q^2)^2} \tag{9.85}$$

(B and Γ' being constants, see next subsection), the maximum of $\omega^2 S(Q,\omega)$ is at

$$\omega_{max} = Qc[1 + (\Gamma' Q/c)^2]. \tag{9.86}$$

For small Q, the width $\Gamma' Q^2$ of the density fluctuation $S(Q,\omega)$ is small compared to the shift cQ of the peak position, and the dispersion of the current and the density fluctuation coincide, i.e. the maxima of $\omega^2 S(Q,\omega)$ are not much different from the maxima of $S(Q,\omega)$. According to eq. (9.86), ω_{max} has an upward curvature at small Q. This is also seen from the limiting slope given by the hypersonic sound velocity and indicated in fig. 9.11a. Around $Q = 1$ Å^{-1}, ω_{max} has a downward curvature. The inflection point (around $Q = 0.5$ Å^{-1}) might be regarded as separating the hydrodynamic region from the microscopic or kinetic region in the liquid.

The sharpening of the $\omega^2 S(Q,\omega)$-function at Q-values corresponding to well populated "planes" in the quasi-structure of the fluids is well understandable as only small restoring forces exist for such a current fluctuation. It is reflected in the scattering law $S(Q,\omega)$ as well, and explains qualitatively the features described in section 9.4.1.

9.4.3. Models

Combination of eq. (1.69), (1.70) for the density fluctuation with the hydrodynamic equations (8.7), (8.8), and (8.9), and the empirical laws of Fourier and Newton (8.1) and (8.3), yields the following relation [17]:

$$S(Q,\omega) = \frac{NkT}{v} \beta_T \left[\frac{C_P - C_v}{C_P} A(Q,\omega) + \frac{C_v}{C_P} B(Q,\omega) \right] \quad (9.87)$$

$$A(Q,\omega) = \frac{D_T Q^2}{\omega^2 + (D_T Q^2)^2}$$

$$B(Q,\omega) = \frac{\Gamma' Q^2 + \delta'(cQ - \omega)}{(\omega - cQ)^2 + (\Gamma' Q^2)^2} + \frac{\Gamma' Q^2 + \delta'(cQ + \omega)}{(\omega + cQ)^2 + (\Gamma' Q^2)^2}.$$

The meaning of the constants D_T, $\Gamma' = \Gamma/2$, and the correction term δ' is as follows:

$$D_T = \frac{\lambda V}{C_P}, \quad \Gamma = \frac{1}{\rho_m}\left(\frac{4}{3}\eta + \phi\right) + \frac{C_P - C_v}{C_v} D_T, \quad (9.88)$$

$$\delta' = \frac{1}{cQ}\left(\Gamma + 2\frac{C_P - C_v}{C_P} D_T\right).$$

With values of the transport coefficients λ, η, and ϕ, and equilibrium properties, the prediction of $S(Q,\omega)$ should be possible. The dotted curves in fig. 9.7 are calculated with eq. (9.87) and (9.88), but fitted values of the transport coefficients. These fitted values were only about half the values of the macroscopic transport coefficients. This discrepancy may be explained either by a beginning breakdown of the hydrodynamic model or by the necessity to assign a proper frequency dependence to the transport coefficients.

In order to fit the experimental results for $S(Q,\omega)$ for higher Q, solid-like models have been developed which treat essentially the liquid as a polycrystalline solid [30]. These models had considerable success. For a review of older models we refer to the literature [25]. Some further improvements have been obtained quite recently [31].

9.5. References

1. A. Rahman, Phys. Rev. A *136*, 405 (1964); A. Rahman, J. Chem. Phys. *45*, 2585 (1966); B. R. A. Nijboer and A. Rahman, Physica *32*, 415 (1966); L. Verlet, Phys. Rev. *159*, 98 (1967); L. Verlet, Phys. Rev. *165*, 201 (1968); D. Levesque and L. Verlet, Phys. Rev. A *2*, 2514 (1970).
2. B. J. Berne and G. D. Harp, Advan. Chem. Phys. *17*, 63 (1970); G. D. Harp and B. J. Berne, Phys. Rev. A *2*, 975 (1970).
3. J. L. Lebowitz, J. K. Percus, and J. Sykes, Phys. Rev. *188*, 487 (1969); A. Z. Akcasu, N. Corngold, and J. J. Duderstadt, Phys. of Fluids *13*, 2213 (1970); A. Z. Akcasu and J. J. Duderstadt, Phys. Rev. *188*, 479 (1969); L. Glass and S. A. Rice, Phys. Rev. *176*, 239 (1968); B. J. Berne, J. P. Boon, and S. A. Rice, J. Chem. Phys. *45*, 1086 (1966); J. Chem. Phys. *47*, 2283 (1967); J. J. Duderstadt and A. Z. Akcasu, Phys. Rev. A *2*, 1097 (1970); N. Corngold and J. J. Duderstadt, Phys. Rev. A *2*, 836 (1970); E. Tong and R. C. Desai, Phys. Rev. A *2*, 2129 (1970); M. Bixon and R. Zwanzig, Phys. Rev. *187*, 267 (1969).
4. S. R. de Groot and P. Mazur, Non Equilibrium Thermodynamics, North Holland, Amsterdam 1962.
5. R. Kubo, in W. E. Brittin and L. G. Dunham (Eds.), Lectures in Theoretical Physics, Vol I, Interscience, New York 1959; R. Kubo, J. Phys. Soc. Japan *12*, 570 (1957).

9.5. References

6. R. Kubo in J. Meixner (Ed.), Statistical Mechanics of Equilibrium and Non-Equilibrium, North Holland, Amsterdam 1965.
7. W. A. Steele in H. J. M. Hanley (Ed.), Transport Phenomena in Fluids, Dekker, New York 1966.
8. L. P. Kadanoff and P. C. Martin, Ann. of Phys. *24*, 419 (1963).
9. C. De Witt and R. Balean (Eds.), Probleme à N Corps-Many Body Physics, Gordon and Breach, New York 1968; B. J. Berne and G. D. Harp, ref. 2.
10. D. Des Cloizeaux, in gen. ref. 14.
11. L. D. Landau and E. M. Lifshitz, Statistical Physics, Pergamon Press, London 1958.
12. H. B. Callen and T. A. Welton, Phys. Rev. *83*, 34 (1951); H. Nyquist, Phys. Rev. *32*, 110 (1928).
13. R. Zwanzig, Ann. Rev. Physic. Chem. *16*, 67 (1965).
14. K. E. Larsson, in D. M. Jović (Ed.), Dynamic and Magnetic Properties of Solid and Liquids, Boris Kidric Institute of Nuclear Sciences, Beograd 1968; K. E. Larsson, in Neutron Inelastic Scattering, Vol. I, International Atomic Energy Agency, Vienna 1968; K. E. Larsson, in P. A. Egelstaff (Ed.), Thermal Neutron Scattering, Academic Press, London 1965.
15. A. Sjölander, in P. A. Egelstaff (Ed.), ref. 14; K. Singwi, in gen. ref. 14; J. E. Enderby, in gen. ref. 12.
16. P. A. Egelstaff, gen. ref. 3.
17. D. McIntyre and J. V. Sengers, in gen. ref. 12; R. D. Mountain, Rev. Mod. Phys. *38*, 205 (1966); J. P. Boon and P. Deguent, Phys. Rev. A *2*, 2542; R. D. Mountain, J. Res. Natl. Bur. Std. *70 A*, 207 (1966).
18. L. van Hove, Phys. Rev. *95*, 249 (1954).
19. V. F. Turchin, Slow Neutrons, Israel Program for Scientific Translations, Jerusalem 1965; A. Sjölander, ref. 15; J. M. Dickey and A. Paskin, Phys. Rev. *188*, 1407 (1969); B. R. A. Nijboer and A. Rahman, ref. 1.
20. J. H. Dymond and B. J. Alder, J. Chem. Phys. *45*, 2061 (1966); B. J. Alder, Ber. Bunsenges. *70*, 968 (1966).
21. B. J. Alder and T. E. Wainwright, Phys. Rev. Letters *18*, 988 (1967); J. Phys. Soc. Japan, Suppl. *26*, 267 (1968); J. H. Dymond and B. J. Alder, Ber. Bunsenges. *75*, 394 (1971).
22. B. J. Alder and T. E. Wainwright, Phys. Rev. A *1*, 18 (1970); R. Zwanzig and M. Bixon, Phys. Rev. A *2*, 2005 (1970); J. R. Dorfman and E. G. D. Cohen, Phys. Rev. Letters *25*, 1257 (1970); M. H. Ernst, E. H. Hange, and J. M. J. van Leeuwen, Phys. Rev. Letters *25*, 1254 (1970).
23. W. M. Lomer and G. G. Low, in P. A. Egelstaff, ref. 14.
24. S. J. Cocking, 34. Physikertagung 1969 Salzburg, B. G. Teubner, Stuttgart 1969; D. Schiff, Phys. Rev. *186*, 151 (1969); for a theoretical explanation cf. L. Bonamy and L. Galatry, Physica *46*, 251 (1970).
25. K. Sköld and K. E. Larsson, Phys. Rev. *161*, 102 (1967); K. E. Larsson, U. Dahlborg, and K. Sköld, gen. ref. 4.
26. W. S. Gornall, G. I. A. Stegeman, B. P. Stoicheff, R. H. Stolen, and V. Volterra, Phys. Rev. Letters *17*, 297 (1966).
27. K. Mika, B. Dorner, and H. H. Stiller, in Neutron Inelastic Scattering, Vol. I, International Atomic Energy Agency, Vienna 1968.
28. B. A. Dasannacharya and K. R. Rao, Phys. Rev. A *137*, 417 (1965).
29. A. Rahman, in Neutron Inelastic Scattering, Vol. I, International Atomic Energy Agency, Vienna 1968.
30. P. A. Egelstaff, Rep. AERE-R 4101 (1962); K. S. Singwi, Phys. Rev. A *136*, 969 (1964); Physica *31*, 1257 (1965).
31. K. N. Pathak and K. S. Singwi, Phys. Rev. A *2*, 2427 (1970); K. S. Singwi, K. Sköld, and M. P. Tosi, Phys. Rev. A *1*, 454 (1970); J. Hubbard and J. L. Beeby, J. Phys. C *2*, 556 (1969); P. Ortoleva and M. Nelkin, Phys. Rev. A *2*, 187 (1970); C. H. Chung and S. Yip, Phys. Rev. *182*, 323 (1969).

10. Polyatomic Molecules

10.1. Introduction

The state of an assembly of N monoatomic molecules is defined by $6N$ coordinates, namely, the positions and the momenta of the molecules. As we have seen in chapter 1.3, the number of coordinates necessary to define an assembly of polyatomic molecules is larger, and comprises the description of rotational motion and intramolecular vibration. However, in many cases these additional contributions are independent of each other, and are not affected by the intermolecular interaction. Then the partition function of a fluid composed of polyatomic molecules can be written as in eq. (1.55)

$$Z = (Z_{\text{trans}})^N (Z_{\text{rot}})^N (Z_{\text{int}})^N Q/v^N, \tag{10.1}$$

where the molecular partition functions are defined as follows:

$$Z_{\text{trans}} = (2\pi m k T/h^2)^{3/2} v; \tag{1.34}$$

$$Z_{\text{rot}} = (8\pi^2 I k T/h^2 \sigma) \quad \text{(linear molecules);} \tag{10.2}$$

$$Z_{\text{rot}} = (8\pi^2 k T/h^2)^{3/2} (I_x I_y I_z \pi/\sigma)^{1/2} \quad \text{(nonlinear molecules);} \tag{10.3}$$

$$Z_{\text{int}} = \prod_i \exp\{(-h v_i/2kT)/[1 - \exp(-h v_i/kT)]\}. \tag{10.4}$$

Here m, I, denote the mass and the moment of inertia, respectively, (I_x, I_y, I_z, principal moments in the three directions x, y, z), and σ is the symmetry number of the molecule. The intramolecular normal frequencies of the molecule are denoted by a set of v_i. In eq. (10.1) all factors except Q/v^N can be evaluated from molecular parameters and determine the properties of dilute gases. The configuration integral (1.56) depends on the total intermolecular energy $E(\vec{r}_1 \ldots \vec{r}_N, \omega_1 \ldots \omega_N)$ which is now a function of the positions of the mass centers ($\vec{r}_1 \ldots \vec{r}_N$), and of the orientations of the molecules (denoted by a set of angular variables $\omega_1 \ldots \omega_N$). In eq. (10.1) it has been assumed that the angular momenta of the molecules are independent of the angular potential energy E. This is analogous to the separation of Z_{trans} from the configuration integral. The classical description of the rotational motion in the fluid may become invalid at low temperatures. In section 10.2 we shall discuss the pair potential between simple non-spherical molecules. In section 10.3 a general method is presented for the evaluation of the configuration integral of fluids composed of certain classes of non-spherical molecules. In 10.4 an extension of the principle of corresponding states to include a parameter characteristic of the kind of molecules is outlined. Section 10.5 is concerned with the influence of molecular shape on the relative stability of orientationally ordered and disordered crystals, and of the liquid, respectively. This will include a discussion of the effect of molecular flexibility.

10.2. Forces between Polyatomic Molecules

All polyatomic molecules have non-spherical symmetry. The non-spherical charge distribution can be expressed by the static moments: all diatomic (and other linear) molecules have a quadrupole moment; hetero diatomic molecules have, in addition, a dipole moment; tetrahedral molecules like CH_4 have an octupole moment; etc. The interaction energy between static multipoles depends on the mutual orientation of the two molecules [see eq. (5.25)]. Also, the polarizability of many polyatomic molecules has a non-central component, which leads to non-central components of the dispersion energy and induction energy.

The short range repulsive forces between polyatomic molecules may vary strongly with their mutual orientation, depending on their shape. The size and shape of molecules is determined by their charge density distribution. This can be obtained theoretically from a suitable set of molecular wave functions. Such calculations have shown that, in general, over 95% of the electronic charge lies within the charge density contour corresponding to 0.002 atomic units and it has been suggested that the dimensions of this contour provide a useful measure of molecular size [1]. The length and width of diatomic molecules can be defined, respectively, as the distance between the intercepts of the 0.002 a.u. contour on the molecular axis and on a line perpendicular to the axis and passing through its mid-point. For molecules like the hydrogen halides a better measure of their width is $2r_x$, where r_x is the "non-bonding radius" of the larger (i.e., the halogen) atom [1]. Non-bonding radii of atoms are also used in the concept of van der Waals radii of molecules which are an experimental measure of molecular dimensions in crystals [2].

Table 10.1 gives the dipole and quadrupole moments, the ratio of the polarizability parallel and perpendicular to the molecular axis, and the ratio of length and width, of some linear molecules.

Table 10.1. Dipole moment μ, quadrupole moment θ, ratio of polarizabilities $\alpha_\parallel/\alpha_\perp$, and ratio length/width from 0.002 a.u. charge density contour, for linear molecules.

Molecule	μ (10^{18} esu cm)	θ (10^{26} esu cm^2)	$\alpha_\parallel/\alpha_\perp$	length/width
H$_2$	0	0.65	1.29	
HCl	1.07		1.31	
HBr	0.79		1.27	
HJ	0.4		1.35	
N$_2$	0	1.5	1.64	1.28
CO	0.13	2.5	1.60	1.37
O$_2$	0	0.4	1.94	1.32
F$_2$	0			1.46
Cl$_2$	0		1.82	
N$_2$O	0.18	3.0		
HCN	2.95		2.04	
C$_2$H$_2$	0		2.11	
CO$_2$	0	4.2	2.08	
CS$_2$	0		2.73	

References: Dipole moments [3]; Quadrupole moments [4];
Polarizabilities [5]; Charge density contours [1].

The influence of the molecular shape on the intermolecular potential can be seen by comparing different isomers of a given formula. Consider, for example, the critical temperatures of the four isomers of pentane, listed in table 10.2. Since the critical temperature is roughly proportional to the depth of the pair potential, we conclude that cyclo-pentane has the largest ε^* and neo-pentane the smallest ε^* of the four pentanes. At the same time cyclo-pentane has the highest

Table 10.2. Properties of the four isomers of pentane.

	n-pentane	iso-pentane	neo-pentane	cyclo-pentane
Critical parameters:				
T_c (K)	469.6	460.4	433.8	511.6
P_c (bar)	33.69	33.81	31.99	45.08
ρ_c (g/cm^3)	0.237	0.236	0.238	0.27
$(PV/RT)_c$	0.262	0.273	0.269	0.276
Saturated liquid:				
T_b/T_c ($P = 1$ atm)	0.659	0.654	0.652	0.630
T/T_c at $P = P_c/50$	0.634	0.629	0.623	0.623
acentric factor ω	0.252	0.227	0.189	
ρ (g/cm^3) at 20 °C	0.6263	0.6196	0.590	0.7454
ρ/ρ_c at $T = 0.57\,T_c$	2.74	2.75		2.76
ρ/ρ_c at $T = 0.68\,T_c$	2.54	2.55	2.48	
η (mP) at 20 °C	2.35	2.25	2.47	4.39
$d\log_{10}\eta/d(1/T)$	0.296	0.364	0.517	0.406
Solid/liquid equilibrium:				
T_t/T_c	0.305	0.246	0.591	0.351
$\Delta S_m/R$	7.0	5.5	1.5	0.4

References: Critical data [6]; viscosity η [7]; the other data are taken from Timmermans [8].

and neo-pentane the lowest critical density. A tentative explanation of this effect is that the disc-like molecules of cyclo-pentane can approach each other more closely, giving rise to a more attractive pair interaction, than the tetrahedral neo-pentane molecules. Note that at a given temperature cyclo-pentane has a considerably higher density than neo-pentane and the other pentanes. Flexible molecules (like n-pentane and iso-pentane) are more complicated because they can assume different shapes depending on temperature and the surrounding molecules. This will be discussed in section 10.5.

10.2.1. Angle-Dependent Pair Potentials

An angle dependent pair potential has the general form

$$\varepsilon_{ij} = \varepsilon(R_{ij}, \vartheta_i, \varphi_i, \vartheta_j, \varphi_j) \qquad (10.5)$$

where R_{ij} is the distance between the mass centres of the pair of molecules i, j, and the angles ϑ and φ are defined with respect to the line joining the centers of the molecules as shown in fig. 5.1. For molecules having axial symmetry only three of these angles are independent, ϑ_i, ϑ_j, and $\varphi_i + \varphi_j$. The pair potential (10.5) has sometimes been written in an abbreviated form as $\varepsilon(R_{ij}, \omega_i, \omega_j)$ [section 1.5, eq. (5.1)].

The interaction energy between hydrogen molecules has been calculated quantum mechanically, using semiempirical molecular wave functions [9]. The total interaction energy has been taken as the sum of exchange, dispersion and electrostatic quadrupole energies. The results obtained by Evett and Margenau are shown in fig. 10.1. The curves in this figure correspond to the

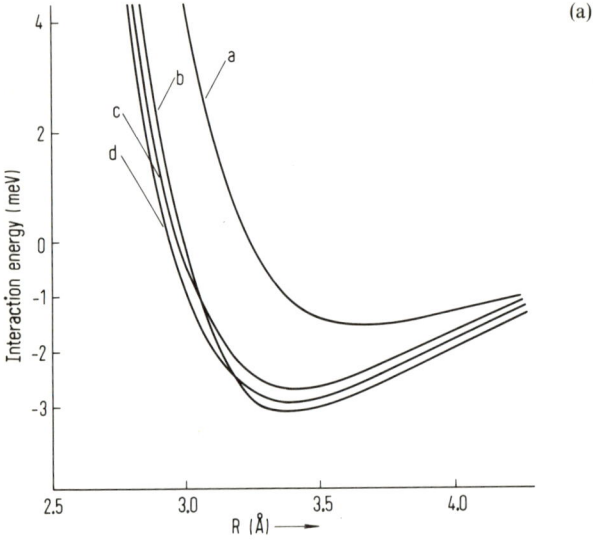

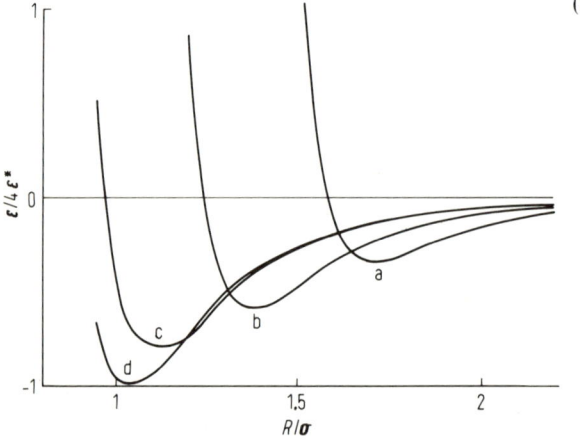

Fig. 10.1. Pair potentials for diatomic molecules at fixed orientations: a) o—o o—o, b) o—o ⌡, c) ⌡ ⌡, d) ⌡ ⌿ (out of plane).
10.1a. Quantum mechanical results for hydrogen molecules (from Evett and Margenau [9]).
10.1b. Diatomic model ($l/\sigma = 0.6$); ε^* and σ denote the minimum and zero of the atomic 12/6 pair potential.

following orientations (relative to $\vec{R}$, the line joining their centers): (a) both molecular axes parallel to $\vec{R}$; (b) one axis parallel, the other perpendicular to $\vec{R}$; (c) both axes perpendicular to $\vec{R}$ and parallel to each other; (d) both axes perpendicular to $\vec{R}$ and perpendicular to each other. In (b), (c), and (d), the centers approach each other more closely and the energy minimum is more attractive than in (a). However, these results are not in agreement with those obtained by de Boer. In his treatment, (d) is the most stable orientation, followed by orientation (b), but again (a) is the least stable orientation.

For molecules other than hydrogen quantum mechanical calculations of the angle dependent pair potential are even more complicated and cannot be made with sufficient accuracy. Therefore, it is useful to study the angle dependence of model pair potentials.

10.2. Forces between Polyatomic Molecules

The intermolecular energy of any pair of molecules of axial symmetry can be expanded in the following form [see eq. (5.25)] [10]:

$$\varepsilon(R_{ij},\vartheta_i,\vartheta_j,\varphi_i,\varphi_j) = \sum_{l=0}^{\infty} \sum_{l'=0}^{\infty} \sum_{m=-l_s}^{l_s} X^{ll'm}(R_{ij}) \, Y_l^m(\vartheta_i,\varphi_i) \, Y_{l'}^m(\vartheta_j,\varphi_j). \tag{10.6}$$

Here, l_s is the smaller of l and l' and

$$X^{ll'm} = X^{ll'-m}. \tag{10.7}$$

The spherical harmonics are defined by

$$Y_l^m(\vartheta,\varphi) = \left[(2l+1)\frac{(l-|m|)!}{(l+|m|)!}\right]^{1/2} P_l^{|m|}(\cos\vartheta) \, e^{im\varphi}, \tag{10.8}$$

where P_l^m is an associated Legendre function defined for positive integers m by

$$P_l^m(x) = (1-x^2)^{m/2} \frac{d^m}{dx^m} P_l(x). \tag{10.9}$$

Eq. (10.6) splits the intermolecular potential into a spherical component, proportional to X^{000} (as Y_0^0 is independent of the angles), and non-spherical components. Due to the properties of Y_l^m each angle dependent term in eq. (10.6) (l or $l' \neq 0$) integrated over all orientations gives zero.
A simple model for the interaction of diatomic molecules consists of a pair of "diatomics", each of which has two Lennard-Jones 12/6 centers of interaction separated by a distance l along the axis. The anisotropy of the molecules is expressed by the parameter l/σ (σ is the distance at which the L.-J. pair potential passes through zero). Sweet and Steele [11] have analysed the "diatomic" potential in terms of the coefficients $X^{ll'm}(R)$ of eq. (10.6). For reasonable values of l/σ the leading term is the spherically symmetric component X^{000}. However, at short separations X^{000} becomes less important, relatively, the greater the anisotropy of the molecules. The most important angle dependent term is that with the coefficient X^{200}, followed by the X^{220} term. For l/σ up to 0.6 the terms with $l,l'=2, m>0$ and all terms with $l,l'>2$ are not important. An estimate of the ratio l/σ of real molecules is obtained by identifying l with the interatomic distance, and σ with the non-bonding diameter of the (larger) atom in the molecule. For the diatomic molecules listed in table 10.1, l/σ estimated in this way does not exceed 0.6 [1]. The diatomic pair potential is compared with the quantum mechanical results for the interaction between a pair of hydrogen molecules in figure 10.1, for the four orientations a–d (see above). The most striking difference between the two potentials is for orientation (a) in which the long axes of the two molecules are in one line: for the "diatomic" this is the most stable orientation at large separations, whereas for hydrogen, according to the quantum mechanical calculations [9], it is the least stable orientation within the range of the calculations.
For tetrahedral, octahedral and other nearly spherical molecules it has been argued that non-central components make only a small contribution to the total pair interaction energy. For such molecules angle independent effective pair potentials have been derived by averaging the pair interaction energy over all orientations of the two molecules, with equal weight given to each orientation [12]. A recent application of this concept is the spherical shell potential [13] which results from the interaction of two spherical surfaces having uniform distribution of L.-J. (12/6) sites. The characteristic parameter of this potential is R^*/d, where R^* is the minimum distance of the L.-J. 12/6 potential, and d is the diameter of the sphere. This pair potential is steeper and more narrow than the L.-J. 12/6 but in the limit $R^*/d \to \infty$ it reduces to the 12/6-potential. The Hamann

potential and the spherical shell potential can be represented adequately by a generalized L.-J. n/m potential [14]. Second virial coefficients of many gases can be fitted over a wide temperature range equally well by the spherical shell or a n/m-potential. For molecules such as CF_4, CCl_4, SF_6, m = 7 and n = 28 (or up to 48) has been used.

The use of angle independent effective pair potentials for the interaction of non-spherical molecules is based on the assumption that the period of time in a molecular encounter is long enough for the molecules to alter their orientation to each other. Hence, the second molecule experiences the potential of the first molecule averaged over all orientations rather than that at a given orientation. The averaging of the pair interaction energy over all orientations of the two molecules, with equal weight given to any orientation, gives the spherical component of the potential:

$$\varepsilon(R_{ij}, \omega_i, \omega_j) \to \varepsilon_0(R_{ij}). \tag{10.10}$$

In chapter 5 another average potential for the interaction of static dipoles has been obtained by weighing each orientation by a Boltzmann factor $\exp[-\varepsilon(R_{ij},\omega_i,\omega_j)/kT]$ before performing the integration [eq. (5.17)–(5.20)]. This leads to a temperature dependent average potential,

$$\varepsilon(R_{ij}, \omega_i, \omega_j) \to \bar{\varepsilon}(R_{ij}, T). \tag{10.11}$$

When calculating the electrostatic contribution to the free energy, $\bar{\varepsilon}/2$ is the appropriate contribution to the pair potential [see eq. (10.24) and (10.25)].

10.3. Configuration Free Energy

Pople [10] has derived a formalism to show the effect of non-central forces on the configuration free energy of a fluid. By splitting the pair potential into a spherically symmetric component ε_0 and a small non-spherical term ε_1, the intermolecular energy of the fluid is given by

with
$$E = E_0 + E_1 \tag{10.12}$$

$$E_0 = \sum_{i>j} \varepsilon_0(R_{ij}), \tag{10.13}$$

$$E_1 = \sum_{i>j} \varepsilon_1(R_{ij}, \omega_i, \omega_j). \tag{10.14}$$

E_1 is considered as a (small) perturbation. Each term ε_1 can be expanded into spherical harmonics [see eq. (10.6)]. By inserting eq. (10.12) into the configuration integral and expanding the factor $\exp(-E_1/kT)$ into powers of E_1, and remembering the properties of the spherical harmonics,

$$\iint (\varepsilon_1)_{ij} \, d\omega_i \, d\omega_j = 0, \tag{10.15}$$

one obtains

$$Q = \frac{1}{N!} \int \ldots \int e^{-E_0/kT} d\vec{r}_1 \ldots d\vec{r}_N$$

$$+ \left(\frac{1}{4\pi}\right)^N \frac{1}{N!} \frac{1}{2(kT)^2} \int \ldots \int E_1^2 \, e^{-E_0/kT} d\vec{r}_1 \ldots d\vec{r}_N \, d\omega_1 \ldots d\omega_N$$

$$= Q_0 + Q_2. \tag{10.16}$$

E_1^2 represents a sum of terms of the following types: $(\varepsilon_1)_{ij}^2$, $(\varepsilon_1)_{ij}(\varepsilon_1)_{jk}$, and $(\varepsilon_1)_{ij}(\varepsilon_1)_{kl}$. The integrals over terms of the latter type are factorized into products of integrals of the type (10.15). By inserting the number of terms $(\varepsilon_1)_{ij}^2$ and $(\varepsilon_1)_{ij}(\varepsilon_1)_{jk}$ one obtains

10.3. Configuration Free Energy

$$Q_2 = \frac{1}{N!} \frac{N!}{2!(N-2)!} \frac{1}{2(kT)^2} \int \ldots \int e^{-E_0/kT} \left[\frac{1}{(4\pi)^2} \int\int (\varepsilon_1)_{ij}^2 \, d\omega_i \, d\omega_j \right] d\vec{r}_1 \ldots d\vec{r}_N$$

$$+ \frac{1}{N!} \frac{N!}{(N-3)!} \frac{1}{2(kT)^2} \int \ldots \int e^{-E_0/kT}$$

$$\times \left[\frac{1}{(4\pi)^3} \int\int\int (\varepsilon_1)_{ij} (\varepsilon_1)_{ik} \, d\omega_i \, d\omega_j \, d\omega_k \right] d\vec{r}_1 \ldots d\vec{r}_N . \tag{10.17}$$

In the first term on the r.h.s. of this equation we have

$$\int\int (\varepsilon_1)_{ij}^2 \, d\omega_i \, d\omega_j = \int\int \left[\sum_{l,l',m} X^{ll'm}(R_{ij}) \, Y_l^m(\omega_i) \, Y_{l'}^m(\omega_j) \right]^2 d\omega_i \, d\omega_j = (4\pi)^2 \sum_{l,l',m} (X^{ll'm})^2 , \tag{10.18}$$

where the sum starts with either l or $l' > 0$. The latter identity follows from the orthogonality of spherical harmonics. In the second term on the r.h.s. of eq. (10.17)*),

$$\int\int\int (\varepsilon_1)_{ij} (\varepsilon_1)_{ik} \, d\omega_i \, d\omega_j \, d\omega_k = (4\pi)^3 \sum_{l>0} X^{l00}(R_{ij}) \, X^{l00}(R_{ik}) \, P_l(\cos \vartheta_{jik}) , \tag{10.19}$$

where ϑ_{jik} is the angle between the centers of the molecules j, i, and k. The configuration free energy is then

$$f = -kT \ln Q = -kT \ln Q_0 - kT \frac{Q_2}{Q_0} - \cdots = f_0 + f_2 + \cdots . \tag{10.20}$$

f_2 represents the first term of an expansion of the configurational free energy about that of a central force fluid. By inserting eq. (10.18) and (10.19) into (10.17):

$$f_2 = -\frac{1}{4kT} \{ \int\int n_0^{(2)}(\vec{r}_1, \vec{r}_2) \sum_{l,l',m} [X^{ll'm}(R_{12})]^2 \, d\vec{r}_1 \, d\vec{r}_2 \tag{10.21}$$

$$+ 2 \int\int\int n_0^{(3)}(\vec{r}_1, \vec{r}_2, \vec{r}_3) \sum_{l>0} [X^{l00}(R_{12})] [X^{l00}(R_{13})] \, P_l(\cos \vartheta_{213}) \, d\vec{r}_1 \, d\vec{r}_2 \, d\vec{r}_3 \} .$$

In this equation, $n_0^{(2)}$ and $n_0^{(3)}$ denote the molecular distribution functions of the unperturbed (central force) assembly [cf. eq. (4.3)]:

$$n_0^{(h)} = \frac{1}{Q_0 (N-h)!} \int \ldots \int e^{-E_0/kT} d\vec{r}_{h+1} \ldots d\vec{r}_N . \tag{10.22}$$

The three-molecule distribution function $n^{(3)}$ is, in general, not known. In eq. (10.21) the term containing $n^{(3)}$ vanishes if coefficients of the type X^{l00} do not appear in the pair potential function (10.5). In this particular case eq. (10.21) may be rewritten as

$$f_2 = -\frac{N^2}{2v} \int [\sum (X^{ll'm})^2 / 2kT] \, g_0(R_{12}) \, 4\pi R_{12}^2 \, dR_{12} . \tag{10.23}$$

---

* The l.h.s. of eq. (10.19) consists of a sum of terms, each containing an integral of the type

$$\int\int\int Y_l^m(\omega_i^j) \, Y_{l'}^m(\omega_j) \, Y_{l^*}^{m*}(\omega_i^k) \, Y_{l'^*}^{m*}(\omega_k) \, d\omega_i \, d\omega_j \, d\omega_k$$

$$= \int Y_l^m(\omega_i^j) \, Y_{l^*}^{m*}(\omega_i^k) \, d\omega_i \int\int Y_{l'}^m(\omega_j) \, Y_{l'^*}^{m*}(\omega_k) \, d\omega_j \, d\omega_k .$$

Here ω_i^j and ω_i^k define the orientation of molecule i relative to the molecules j and k, respectively. The double integral gives $(4\pi)^2$ when $l' = l^{*'} = m = m^* = 0$, and else zero. The integral over ω_i is zero when $l^* \neq l$ and else it gives $P_l(\cos \vartheta_{jik})$.

This result also follows, by an expansion like eq. (10.16), from a first order perturbation treatment of a central-force assembly with a temperature dependent perturbation energy $E'_1 = \sum_{i>j} \varepsilon'_{ij}$, where

$$\varepsilon'_{ij} = -\sum (X^{ll'm})^2/2kT. \tag{10.24}$$

On the other hand the Boltzmann-weighted average of $(\varepsilon_1)_{ij}$ for $\varepsilon_1 \ll kT$ is given by

$$\bar{\varepsilon}_{ij} = \frac{1}{(4\pi)^2} \iint \varepsilon_1 e^{-\varepsilon_1/kT} d\omega_i d\omega_j = \frac{1}{(4\pi)^2} \iint \varepsilon_1 \left(1 - \frac{\varepsilon_1}{kT} + \cdots\right) d\omega_i d\omega_j \cong -\sum (X^{ll'm})^2/kT \tag{10.25}$$

which is twice the value of ε'_{ij} in eq. (10.24).

10.3.1. Ideal Dipoles

For molecules with point dipoles the orientation energy is given by (cf. table 5.2)

$$\sum (X^{ll'm})^2 = 2(X^{111})^2 + (X^{110})^2 = 6\frac{\mu^4}{9R^6}. \tag{10.26}$$

The induction energy caused by point dipoles in polarizable molecules may be written [cf. eq. (5.21)]

$$(\varepsilon_{ij})_{\text{ind}} = -\frac{\alpha_i \mu_j^2 + \alpha_j \mu_i^2}{R_{ij}^6} - \frac{\alpha_i \mu_j^2}{\sqrt{5 R_{ij}^6}} (3\cos^2 \vartheta_j - 1) \sqrt{\frac{5}{4}} \tag{10.27}$$

$$- \frac{\alpha_j \mu_i^2}{\sqrt{5 R_{ij}^6}} (3\cos^2 \vartheta_i - 1) \sqrt{\frac{5}{4}} = X^{000} + X^{200} Y_2^0(\vartheta_j) + X^{020} Y_2^0(\vartheta_i).$$

X^{000} is the central part of the induction energy. From eq. (10.27), $[(X^{200})^2 + (X^{020})^2]/2kT$ is normally small compared with X^{000}. Thus, the non-central contribution of the induction energy to f_2 can be neglected in the first term, and probably also in the second term, of eq. (10.21). To that approximation, by inserting eq. (10.26) into (10.23), the following expression is derived for ideal dipoles:

$$f_2 = -\frac{N^2}{2v} \int \frac{\mu^4}{3R_{12}^6 kT} g_0(R_{12}) 4\pi R_{12}^2 dR_{12}. \tag{10.28}$$

If the central component of the pair potential (which includes the contribution of the induction energy) is represented by a 12/6-potential, the overall potential compatible with eq. (10.28) is

$$\varepsilon = 4\varepsilon_0^* \left[\left(\frac{\sigma_0}{R}\right)^{12} - \left(\frac{\sigma_0}{R}\right)^6 (1 + 2\chi(T)) \right], \tag{10.29}$$

with

$$\chi(T) = \mu^4/(24 \varepsilon_0^* \sigma_0^6 kT). \tag{10.30}$$

This result is important as it permits to compare the properties of liquids composed of weak dipole molecules with the properties of central force liquids. The comparison is based on the principle of corresponding states, and the deviations from central forces are expressed by the attractive parameter $\chi(T)$. The potential (10.29) conforms with the 12/6-potential if the parameters ε_0^* and σ_0 are replaced by

$$\sigma = \sigma_0 [1 - \chi(T)/3], \quad \varepsilon^* = \varepsilon_0^* [1 + 4\chi(T)]. \tag{10.31}$$

10.3.2. Molecules Having a Non-Spherical Shape

In the "diatomic" model for the interaction of a pair of diatomic molecules (section 10.2.1) we have seen that the X^{200} terms comprise the most important part of the angle dependence of this potential (at not too small separations). Considering eq. (10.21) we may conclude that a non-spherical shape of the molecules affects the triplet distribution function (and, probably, higher distribution functions) in a characteristic way. For the diatomic model, with a 12/6-interaction between the atoms, X^{200} is given in the limit of low anisotropy by $X^{200} = X^{020} = AR^{-14} - BR^{-8}$ [11]. On the basis of (10.21) this leads to an effective pair potential with perturbation terms proportional to R^{-28} and R^{-16}.

Rowlinson [15] has represented the non-central part of the potential by an arbitrary function such that the resulting perturbation conforms with the 12/6-potential:

$$(\varepsilon_1)_{ij} = R_{ij}^{-6} \sum_{\substack{l \text{ or } l' > 0 \\ m}} \xi^{ll'm} Y_l^m(\omega_i) Y_{l'}^m(\omega_j), \tag{10.32}$$

where the coefficients ξ are constants. Inserting (10.32) into (10.21) gives

$$f_2 = -\frac{N^2}{2v} \int \left[\sum_{ll'm} (\xi^{ll'm})^2 / (2 R_{ij}^{12} kT) \right] g_0(R_{ij}) 4\pi R_{ij}^2 d R_{ij}. \tag{10.33}$$

The overall pair potential compatible with eq. (10.33) can then be expressed by

$$\varepsilon = 4\varepsilon_0^* \left[\left(\frac{\sigma_0}{R}\right)^{12} \{1 - 2\delta(T)\} - \left(\frac{\sigma_0}{R}\right)^6 \right], \tag{10.34}$$

with
$$\delta(T) = \sum (\xi^{ll'm})^2 / (16 \varepsilon_0^* \sigma_0^{12} kT). \tag{10.35}$$

The parameters of the perturbed 12/6-potential corresponding to eq. (10.34) are

$$\sigma = \sigma_0 \{1 - \delta(T)/3\},$$
$$\varepsilon^* = \varepsilon_0^* \{1 + 2\delta(T)\}. \tag{10.36}$$

The potential eq. (10.34) is an empirical function because the form of the perturbation term ε_1 in eq. (10.32) has no theoretical justification. Furthermore, from figure 10.1 b it is clear that any perturbation treatment will break down for small separations R_{ij}, i.e., at high densities of the fluid. From that point of view, the pair potential eq. (10.34) is remarkably successful as will be discussed below.

10.4. Extension of the Theorem of Corresponding States

According to this theorem [cf. eq. (6.3)] the configuration free energy of any assembly whose pair potential conforms to a reference assembly is given by

$$f(v, T) = ef_0(v/s^3, T/e) - 3NkT \ln s, \tag{10.37}$$

where
$$e = \varepsilon^*/\varepsilon_0^*, \quad s = \sigma/\sigma_0. \tag{10.38}$$

The subscript refers to the reference assembly.

For an assembly of molecules interacting by a 12/6 potential with superimposed point dipoles, as eq. (10.29), $e = (1 + 4\chi)$, $s = (1 - \chi/3)$. Neglecting terms higher than first order in χ, the configuration free energy of such a fluid is given by

$$f(v, T) = (1 + 4\chi) f_0 [v(1 + \chi), T(1 - 4\chi)] + \chi N k T. \tag{10.39}$$

If the equation of state of the reference substance is

$$\varphi_0(P, v, T) = 0, \tag{10.40}$$

then for an assembly with point dipoles

$$\varphi(P, v, T) = \varphi_0 [P(1 - 5\chi), v(1 + \chi), T(1 + 4\chi)] = 0, \tag{10.41}$$

with $\chi = \chi(T)$ [cf. eq. (10.30)].

For an assembly of molecules interacting by a 12/6 potential with a modified repulsive coefficient, as eq. (10.34), the expressions analogous to eq. (10.39) and (10.41) are

$$f(v, T) = (1 + 2\delta) f_0 [v(1 + \delta), T(1 - 2\delta)] + N k T, \tag{10.42}$$

$$\varphi(P, v, T) = \varphi_0 [P(1 - 3\delta), v(1 + \delta), T(1 - 2\delta)] = 0, \tag{10.43}$$

with $\delta = \delta(T)$ [cf. eq. (10.35)]. All other thermodynamic functions can be derived from eq. (10.42) and (10.43). The most important differences between a perturbed and an unperturbed 12/6 liquid are: a lower reduced saturated vapour pressure P/P_c (i.e. a higher reduced boiling point); a higher reduced density ρ/ρ_c (or V_c/V, in terms of the molar volume); and an enlarged heat capacity of the liquid conforming to a perturbed 12/6 potential. This is in agreement with the observed deviations from the principle of corresponding states of most molecular liquids relative to liquid argon. For typical substances with non-spherical molecular shape the quantities $\ln(P/P_c) - \ln(P/P_c)_{argon}$; $V_c/V - (V_c/V)_{argon}$; and the residual heat capacity are plotted vs. the reduced temperature in figure 10.2; 10.3; and 10.4, respectively. From these deviations the parameter $\delta(T_c) = \delta(T) T/T_c$ can be calculated. In most cases, the deviations in vapour pressure give a higher value of $\delta(T_c)$ than the deviations in density or heat capacity.

The attempt to cover the deviations from the theorem of corresponding states by a single parameter (δ or χ) is related to other empirical treatments which are again successful at low and moderate densities. One of them, which is due to Riedel and to Pitzer, uses the critical point as a reference, neglecting deviations of the critical coefficient from the value for the rare gases. Another correlation, suggested by Holleran, uses the Boyle temperature as the reference, i.e. the temperature at which the second virial coefficient is zero at zero pressure. Holleran's correlation is more fundamental but more difficult to employ than the former.

Pitzer's and Riedel's treatment [21] is based on a comparison of the compressibility factor of different substances at the same reduced temperature $T_r = T/T_c$, and reduced pressure $P_r = P/P_c$:

$$(PV/RT) = Z(T_r, P_r, \omega) \tag{10.44}$$

For the parameter ω, which is characteristic for a given substance, Pitzer suggested the name acentric factor. The equation of state is expressed as

$$Z = Z^{(0)} + \omega Z^{(1)}, \tag{10.45}$$

where $Z^{(0)}$, $Z^{(1)}$ are each functions of T_r and P_r. Extensive tables have been computed empirically for these two functions. The acentric factor is defined by

$$\omega = -\log P_r - 1 \quad \text{at } T_r = 0.7. \tag{10.46}$$

10.4. Extension of the Theorem of Corresponding States

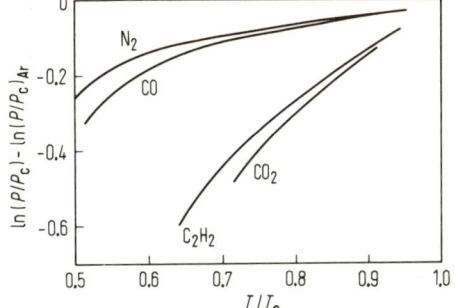

Fig. 10.2. Deviations of the reduced vapour pressure from the reduced vapour pressure of argon. N_2 and CO from ref. [18], CO_2 and C_2H_2 from ref. [16].

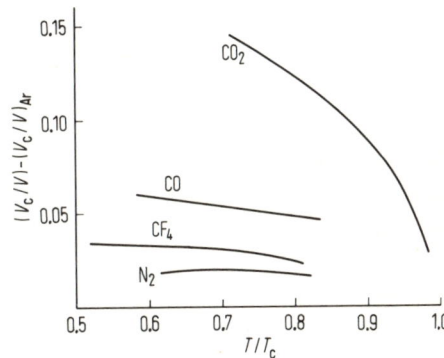

Fig. 10.3. Deviations of the reduced orthobaric density from the reduced orthobaric density of argon. N_2, CO, and CF_4 from ref. [17], CO_2 from ref. [16]; critical densities from Kudchadker et al. [6].

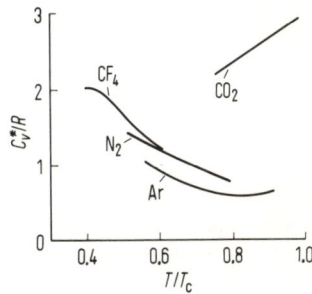

Fig. 10.4. Residual heat capacity $C'_v = C_v(\text{liquid}) - C_v(\text{ideal gas})$ of N_2 [15], CF_4 [17, 19], CO_2 [18, 20], and argon (cf. table 6.3).

It is nearly zero for the rare gases. Riedel has independently suggested an expression equivalent to eq. (10.45) with a characteristic parameter defined as [cf. eq. (6.6)]

$$\alpha_k = (d \ln P/d \ln T) \quad \text{at } T_c. \tag{10.47}$$

The parameters ω, α_k and $\delta(T_c)$ are numerically related to each other by

$$\omega = 2.40 \delta(T_c); \quad \alpha_k = 5.808 + 11.8 \delta(T_c). \tag{10.48}$$

Holleran [22] suggests an equation of state of the form

$$Z = 1 + k_B Y(T_\rho, V_\rho) \tag{10.49}$$

where Y is a function of a reduced temperature, $T_\rho = T/T_B$, and a reduced volume, $V_\rho = V/V_0$, and k_B, V_0, and T_B (the Boyle temperature) are three characteristic parameters for a given substance. This treatment is based on an empirical linear relationship among those temperatures and molar volumes at which $Z = 1$,

$$(T_\rho + 1/V_\rho)_{Z=1} = 1. \tag{10.50}$$

The parameter $1/V_0$ is thus obtained by extrapolating the experimental densities $1/V_\rho$ along the states with $Z = 1$ to the limit of zero temperature; hence $1/V_0$ is effectively the zero point density. Holleran has shown, that the third parameter k_B can be related to the second and third virial coefficient at $T = T_B$, and thus to the form of the pair potential.

10.5. Melting

In the previous section we have found that many properties of fluids at moderate and low density are not sensitive to details of the molecular shape. This is reflected in the various versions of the principle of corresponding states. By contrast, for most substances the principle of corresponding states is not applicable to melting properties such as the melting point, the entropy change and the volume change at melting. Consider the properties of the isomers of pentane (table 10.2): the equilibrium properties[*] of the saturated liquids approximately follow the principle of corresponding states, i.e., the reduced temperatures corresponding to a given reduced pressure (e.g., $P_c/50$) and the reduced densities at a given reduced temperature are similar for all four isomers. The acentric factor [eq. (10.46)] increases in the order neo-pentane, iso-pentane, n-pentane. On the other hand, the melting properties of these substances, the reduced triple points, the enthalpies and entropies of melting differ widely from each other. The least symmetrical molecule, iso-pentane, has the lowest reduced triple point and the tetrahedral neo-pentane has the highest reduced triple point. In the following, the relationship between molecular shape and the stability of ordered and disordered phases will be discussed.

10.5.1. Rigid Molecules

Eucken, and Ubbelohde, have first pointed at the different melting properties of (a) substances which undergo phase transitions associated with orientational randomization in the solid state; and (b) substances for which orientational randomization is part of the melting phenomenon [23]. Substances of the first group consist of molecules with a small non-central part of the pair potential. These substances form a high temperature phase of plastic crystals [24], which show no (or only low) optical birefrigence and have low entropies of fusion.
Pople and Karasz [25] have developed a simple model for melting of rigid non-spherical molecules. This theory is remarkably successful in predicting qualitatively the trends in the melting behaviour of such substances. The theory is an extension of the two lattice model of

---

[*] As an example of a non-equilibrium property table 10.2 lists the viscosities of the four pentanes. The derivative $\partial \log \eta/\partial(1/T)$, which is proportional to the energy of activation of viscous flow, differs widely between the four liquids, and hence the similarity of some values of η at 20° is a coincidence [7]. A comparison of the reduced viscosities is shown in fig. 8.9.

melting of monoatomic liquids by Lennard-Jones and Devonshire, who assumed that the molecules can be accommodated on lattice sites α or β. The α and β-sites form two equivalent lattices such that each α-site is surrounded by z β-sites and vice versa. At low temperatures the free energy of the assembly is least when the particles are distributed mainly over sites of one kind such that the assembly displays long range order and corresponds to a solid. Conversely, at high temperatures both types of sites are equally populated, the system is disordered and corresponds to the liquid. Pople and Karasz, to allow for different orientations of non-spherical molecules in the crystal, assume that each molecule can have two orientations, 1 or 2, on each site, such that there are, all together, four possibilities for any molecule: $\alpha_1, \alpha_2, \beta_1, \beta_2$. A repulsive energy w for molecules on neighbouring α and β sites is introduced to account for the higher energy of interstitial sites (β) in a mainly α lattice. Similarly, a repulsive energy w' is introduced to allow for the higher energy of molecules in unfavourable relative orientations, i.e., molecules in neighbouring α_1 and α_2, or β_1 and β_2 positions. Thus, zw is the energy required to move one molecule from a α-site to a β-site when all other molecules are situated on α sites, and zw' is the energy necessary to turn one molecule from α_1 to α_2 when the rest of the molecules is in α_1 orientation. The ratio

$$v = w'/w \tag{10.51}$$

is a measure of the relative energy barriers against orientational and positional disorder. It is the essential parameter of this model. The partition function of the system can be written (on the basis of an Einstein model)

$$Z = (Z_0)^N Q, \tag{10.52}$$

$$Q = \sum \exp\left[-(N_{\alpha\beta} w + N_{\alpha_1\alpha_2} w' + N_{\beta_1\beta_2} w')/kT\right], \tag{10.53}$$

where Z_0 is the partition function of a single molecule. Z_0 is taken by Pople and Karasz from the spherically smoothed three-shell cell theory by Wentorf, Buehler and Hirschfelder. For a state of perfect order, $Z = Z_0^N$. In eq. (10.53), N_{ij} denotes the number of nearest neighbour contacts between molecules in states i, j, and the sum extends over all arrangements and orientations of particles on $\alpha_1, \alpha_2, \beta_1, \beta_2$ sites. The terms with $N_{\alpha_1\beta_2}$ and $N_{\beta_1\alpha_2}$ are omitted in the exponential of eq. (10.53): it is thought that re-orientation of a single molecule (while all surrounding molecules remain orientationally ordered) causes a strain in the vicinity of that site which results in an increased tendency for (neighbouring) molecules to move onto interstitial sites. This is taken into account in the simplest possible way by assuming that β-sites near a misorientation in a mainly α lattice are favoured by not experiencing the repulsive w' term. By introducing the parameters q and s, representing the degree of positional and orientational order (i.e., the fraction of molecules on α sites or in orientation 1), respectively, the zeroth order (or random mixing) approximation for $Q(q, s)$ is

$$Q(q,s) = g(q,s) \exp\left\{-[zNwq(1-q) + zNw's(1-s)(1-2q+2q^2)]/kT\right\}, \tag{10.54}$$

where $g(q, s)$ is the number of ways of arranging the particles for given values of q, s:

$$g(q,s) = \left\{\frac{N!}{[Nq]![N(1-q)]!}\right\}^2 \frac{[Nq]!}{[Nqs]![Nq(1-s)]!}$$
$$\times \frac{[N(1-q)]!}{[N(1-q)s]![N(1-q)(1-s)]!}. \tag{10.55}$$

Maximizing the partition function Q with respect to q and s yields the conditions

$$\log \frac{q}{1-q} = \left[\frac{zw}{2kT} - \frac{zw'}{kT} s(1-s)\right](2q-1), \tag{10.56}$$

$$\log \frac{s}{1-s} = \frac{zw'}{kT}(1 - 2q + 2q^2)(2s-1). \tag{10.57}$$

The equilibrium values of q and s are functions of zw/kT and the parameter v. For $v < 0.325$, orientational disorder occurs at lower T than positional disorder. This situation corresponds to a plastic crystal exhibiting a rotational transformation in the solid state. For larger v, orientational and positional disorder occurs at the same T. At very large v ($v > 2$), positional disorder occurs at lower T than orientational disorder. The physical analog to this situation are substances forming liquid crystals [26]. To obtain the thermodynamic functions for the above model it is necessary to specify the volume dependence of the repulsive energies w and w'. These are taken proportional to R^{-12} (or V^{-4}),

$$w = w_0 (V_0/V)^4 \qquad w' = w'_0 (V_0/V)^4 \tag{10.58}$$

where V_0 is the molar volume of the ordered crystal when the distance between nearest neighbours is R^* of the L.-J. potential. Both w_0 and w'_0 are proportional to kT/ε^*, but the ratio v is independent of temperature and volume. The reduced melting temperature, kT_m/ε^*, the entropy of melting, ΔS_m, and the relative volume change at melting, $\Delta V/V_s$ (where V_s is the volume of the solid at T_m), as a function of the parameter v are shown in figure 10.5.

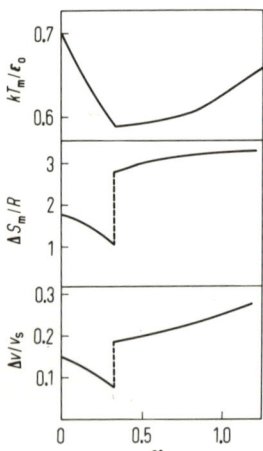

Fig. 10.5. Reduced melting point kT_m/ε^*, entropy of melting ΔS_m, and relative volume change of melting $\Delta V/V_s$, calculated as a function of the parameter v (After Pople and Karasz [25]).

For some linear molecules the reduced triple point T_t/T_c (instead of kT_m/ε^*), the entropy of melting, and the relative volume change of melting are summarized in table 10.3. Phase transitions in the solid state are indicated and the total entropy change at these transitions is also given. In general, the trends predicted by the model of Pople and Karasz are well reproduced: with increasing molecular asymmetry one observes first a decrease and then an increase in T_t/T_c; much higher entropies of melting are observed for such "dumbbell" or linear molecules which have no solid state transitions. The few reliable data on $\Delta V/V_s$ are also in agreement with the model.

Table 10.3. Melting properties of linear molecules.

Substance	$T_t(K)$	T_t/T_c	$\Delta S_m/R$	$\Delta V/V_s$	Solid-Solid-Trans. No.	$\sum \Delta S_{trans}/R$	Ref.
HCl	158.9	0.490	1.51		1	1.45	27
DCl	158.4	0.490	1.50		1	1.53	27
HBr	186.3	0.513	1.55		3	2.05	27
DBr	185.6	0.513	1.56		2	2.18	27
HJ	222.3	0.524	1.55		2	2.49	27
DJ	221.2	0.524	1.56		2	2.66	27
CO	68.1	0.513	1.47		1	1.24	18
N_2	63.2	0.500	1.37	0.087	1	0.78	18
O_2	54.4	0.351	0.98	0.038	2	2.52	28
F_2	53.5	0.372	1.15	0.041	1	1.96	28, 29
Cl_2	172.2	0.413	4.48		0		29
Br_2	265.9	0.452	4.78		0		29
J_2	386.7	0.472	4.91		0		29
N_2O	182.3	0.589	4.31		0		30
C_2H_2	192.4	0.623	2.55	0.2	0		18
CO_2	216.6	0.712	4.86	0.28	0		18
COS	134.3		4.23	0.12	0		23
CS_2	161.1	0.293	3.28	0.07	0?		23, 31

Critical Temperatures: Reference [6].

Note the difference in the melting properties of CO_2 and CS_2 for which there is no explanation as yet. The heat capacity of solid CS_2 has a kink near 80 K indicating a higher order phase transition in the solid [31]. Also note that there are no significant differences between the melting properties of polar and non-polar substances. Melting properties of some tetrahedral molecules are summarized in table 10.4. Tetrahedral molecules have attained much interest due to their

Table 10.4. Melting properties of tetrahedral molecules.

Substance	$T_t(K)$	T_t/T_c	$\Delta S_m/R$	$\Delta V/V_s$	Solid-Solid-Trans. No.	$\sum \Delta S_{trans}/R$	Ref.
CH_4	90.6	0.474	1.24	0.087	1	0.39	27
CD_4	89.8		1.21	0.087	2	1.54	27
SiH_4	88.5	0.328	0.9		1	1.17	27
GeH_4	107.3		0.94		2	1.75	27
CF_4	89.6	0.394	0.96	0.022	1	2.36	19
CCl_4	250.4	0.450	1.21	0.053	1	2.47	33, 34
CBr_4	364		1.31		1	2.51	33
$C(CH_3)_4$	256.4	0.591	1.52		1	2.2	23
SiF_4	182.9	0.706	4.7				30
$SiCl_4$	205.5	0.405	4.56	0.124			23, 34
$SnCl_4$	239.9		4.58	0.119			32, 34
$TiCl_4$	250.0		4.53	0.124			32, 34
SF_6	222.5	0.698	2.71	0.37	1	2.05	23

Critical Temperatures: Reference [6].

high degree of symmetry [32]. The melting properties of these substances resemble those of linear molecules. Methane, SiH_4, and GeH_4, like the hydrogen halides, exhibit rotational phase transitions in the solid state. Phase transitions involving rotational randomization are also observed for CF_4, CCl_4, CBr_4, and CMe_4, but not for the tetrahalides of Si, Sn, and Ti. These latter substances, like the more asymmetrical linear molecules, have large entropy changes of melting, and somewhat larger volume changes.

The deuterated compounds have similar melting properties as the "light" molecules, but orientational randomization in the solid state occurs at distinctly higher temperatures and involves a larger entropy change. This has been explained by the smaller zero-point energy and the somewhat higher packing density of the deuterated compounds [27]. For the onset of molecular rotation it is necessary to overcome the energetical interlocking with the neighbouring molecules. With increasing temperature this will occur the sooner the higher the zero-point energy of the molecules and the lower the packing density of the solid.

10.5.2. Flexible Molecules

Here we consider the influence of an equilibrium among different rotational isomers of a molecule on the thermodynamic properties of the fluid [*]. The simplest molecules to exhibit rotational isomerism are 1,2-disubstituted ethanes like n-butane. For such molecules it is well established that only "staggered" conformations are stable and, moreover, the *trans* conformation is more stable than the two *gauche* conformations [35]. Hence the coordinate ϕ which measures the angle of twist about the central C—C bond is associated with a potential of the form (see fig. 10.6a):

$$E(\phi) = E_0(1 - \cos 3\phi) + E_1(1 - \cos \phi) \tag{10.59}$$

and the corresponding factor in the classical partition function of the molecule is

$$Q_{(\phi)} \propto \int_0^{2\pi} \exp\left[-E(\phi)/kT\right] \mathrm{d}\phi. \tag{10.60}$$

In order to simplify the evaluation of eq. (10.60) Pitzer [36] used the following approximation

$$Q_{(\phi)} \propto [1 + 2\exp(-\Delta E/kT)] \int_0^{2\pi} \exp\left[-E_0(1 - \cos 3\phi)/kT\right] \mathrm{d}\phi, \tag{10.61}$$

where ΔE is the potential energy difference between the minima of the potential wells of *gauche* and *trans* and $E_0(1 - \cos 3\phi)$ is the potential for hindered internal rotation in ethane which can be evaluated when the barrier height E_0 is known. When E_0 is a purely molecular constant, the contribution due to hindered internal rotation is the same in the solid and liquid phase and does not influence the relative stability of the two phases. According to this model the *trans-gauche* conformational equilibrium in liquid 1,2-substituted ethanes manifests itself in the first factor on the r.h.s. of eq. (10.61) whereas this factor is near unity in the solid.

For longer hydrocarbon chains, by analogy with butane, Pitzer assumed that there should be one *trans* (t) and two *gauche* (g,g') conformations for each C—C bond except for the first and the last in the chain. However, some of the resulting configurations would lead to an overlapping

---

[*] The possible influence of the internal rotation of methyl groups will not be considered.

of different parts of the molecule and are therefore impossible. Pitzer estimated the number of possible isomers by excluding all configurations which involve consecutive gg' or g'g conformations. For the remaining chain configurations, each g or g' conformation was assumed to be associated with an energy ΔE in excess to a *trans* conformation. With this simplification the factor in the molecular partition function accounting for rotational isomerization is

$$Q_{r.i.} = \sum_{i=0}^{r} m_i \exp(-i\Delta E/kT), \qquad (10.62)$$

where i is the number of *gauche* conformations in an isomer, m_i is the weight factor for that isomer, and $r = n - 3$, where n is the number of carbon atoms per molecule. In an assembly of N molecules the equilibrium distribution numbers N_i among the $(r + 1)$ energy states is given by

$$x_i = N_i/N = (m_i/Q_{r.i.}) \exp(-i\Delta E/kT). \qquad (10.63)$$

Assuming that ΔE is independent of temperature, the entropy and heat capacity associated with rotational isomerisation is

$$s_{r.i.} = Nk\left[\ln Q_{r.i.} + (\Delta E/kT)\sum_i i x_i\right], \qquad (10.64)$$

$$c_{r.i.} = Nk(\Delta E/kT)^2 \sum_{j>i} (j - i)^2 x_j x_i. \qquad (10.65)$$

For some n-alkanes the entropy $S_{r.i.}$ from eq. (10.64) is shown in table 10.5. For the energy difference between *trans* and *gauche* Pitzer used a value $\Delta E = 800$ cal mol^{-1} but later Person and Pimentel [36] suggested a lower value, $\Delta E = 500$ cal mol^{-1}. $S_{r.i.}$ accounts for only a fraction of the observed entropy of melting ΔS_m, but the increment in $S_{r.i.}$ between two alkanes accounts for a substantial fraction of the increment in ΔS_m. A further contribution to the entropy of melting of flexible molecules is the entropy of mixing of the various rotational isomers in the liquid. This is a problem in the theory of polymer solutions [38]. The alternation in the increments of ΔS_m between subsequent pairs of n-alkanes (see table 10.5) is a consequence of the alternation of the increments in the melting temperature T_m of even and odd numbered alkanes. This is caused by the influence of molecular symmetry on the crystal structure [39].

Table 10.5. Entropies of melting ΔS_m and entropies of rotational isomerization $S_{r.i.}$ of n-alkanes. $S_{r.i}$ is given for $\Delta E = 500$ and 800 cal mol^{-1} [eq. (10.64)] and also for the energy levels given by Scott & Scheraga [40].

n-alkane	T_m(K)	$\Delta S_m/R$	$S_{r.i.}/R$ 500	$S_{r.i.}/R$ 800	$S_{r.i.}/R$ S & S
Butane	134.9	4.16	0.71	0.37	—
Pentane	143.4	7.06	1.37	0.79	1.17
Hexane	177.82	8.85	2.28	1.60	2.04
Heptane	182.55	9.25	2.96	2.05	2.21
Octane	216.38	11.53	3.97	3.07	—
Nonane	219.66	8.47*)	4.77	3.81	—
Decane	243.51	14.18	5.69	4.76	—

* Nonane undergoes a solid state transition at 217.19 K with an entropy $\Delta S_{trans}/R = 3.48$; hence $(\Delta S_m + \Delta S_{trans})/R = 11.95$ [37].

More recently several authors have calculated the conformational potential energy of normal hydrocarbon molecules in a more rigorous manner [40]. According to the treatment of Scott and Scheraga it is not justified for longer-chain alkanes to assume that each C—C bond can

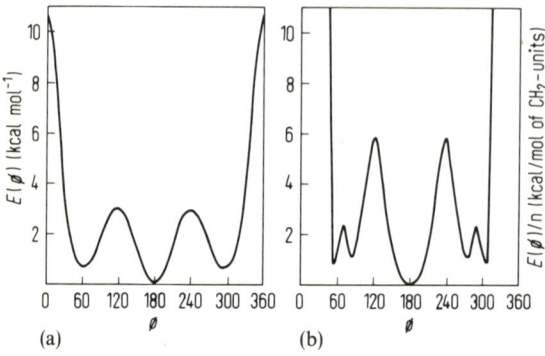

Fig. 10.6. Angular potentials of hindered rotation about single bonds (according to Scott and Scheraga [40]).
10.6a. Energy vs. angle of twist ϕ for rotation about the central C—C bond in butane.
10.6b. Energy per monomer unit as a function of ϕ in a long hydrocarbon chain.

exist in only three rotationally isomeric states (t, g, g'), but as many as nine states might be required. For a long hydrocarbon chain they obtain five isomeric states per CH_2 unit (cf. fig. 10.6b). The entropy of isomerization at the melting point which follows from this theory is included in table 10.5.

A conformational equilibrium can be affected by transferring the molecules from the gas into the liquid phase. For 1,2-dihalogenethanes the polar *gauche* form is considerably more stable in the (pure) liquid and in polar solvents than in nonpolar solvents or in the gas [41]. For longer chain paraffins the more extended forms may be stabilized by a cooperative effect. Indeed, for liquid long-chain paraffins there is some spectroscopic evidence that *gauche* conformations are relatively scarce and are located preferably near the ends of the chains [42].

Another question concerns the molecular partition functions of *gauche* and *trans* conformations. An analysis of the spectra of liquid n-butane, over a range of temperatures, enables the assignment of the individual lines to either the *trans* or *gauche* isomer [43]. Most Raman lines of the *gauche* isomer appear at lower wave numbers than the same vibrations of the higher symmetrical *trans* isomer. In the Raman spectrum of solid butane only the lines of the *trans* isomer are present and their frequencies are precisely the same as in the liquid. Similar results were found for liquid and solid pentane, hexane and heptane [43]. This implies that (1) the molecular partition function of the lowest energy isomer is not altered on melting and (2) the internal partition functions of different rotational isomers are different, on account of the different internal vibrational frequencies. Therefore, in the equations (10.61)–(10.63) ΔE is to be replaced by ΔG^0, the Gibbs free energy difference between *trans* and *gauche* forms. In principle ΔG^0 can be obtained from any experimental method which *directly* measures the concentrations of the rotational isomers. For the alkanes butane to heptane in the gas phase, $\Delta G^0 = 610$ cal mol^{-1} has been obtained from the pair distribution curves by electron diffraction [35]. The enthalpy difference between the rotational isomers, ΔH^0, obtained from the temperature dependence of the Raman intensities of the different rotational isomers, is distinctly higher for butane ($\Delta H^0 = 760$ cal mol^{-1}) than for pentane and hexane ($\Delta H^0 = 450$ and 500 cal mol^{-1}, respectively) [43]. The difference in volume occupied by the *gauche* and *trans* isomers of butane has been determined indirectly from ultrasonic relaxation in the liquid [44], and was found to be small ($\Delta V/V < 0.3\%$).

10.6. References

1. R. F. W. Bader, W. H. Henneker, and P. E. Cade, J. Chem. Phys. *46*, 3341 (1967); R. F. W. Bader, I. T. Keaveny, and P. E. Cade, ibid, *47*, 3381 (1967); R. F. W. Bader and A. D. Bandrauk, ibid, *49*, 1653 (1968).
2. L. Pauling, The Nature of the Chemical Bond, Cornell University Press, Ithaca, N. Y. 1960, 3rd ed., p. 257–264; W. E. Streib, T. H. Jordan, H. W. Smith, and W. N. Lipscomb, J. Chem. Phys. *37*, 2962 (1962); *41*, 756, 760 (1964).
3. A. L. McClellan, Tables of Experimental Dipole Moments, Freeman, San Francisco 1963.
4. A. D. Buckingham, Advances in Chem. Phys. *12*, Interscience, New York 1967, p. 107–142; see also G. Birnbaum, ibid, p. 487–548.
5. H. A. Stuart, Molekülstruktur, Springer, Berlin 1967.
6. A. P. Kudchadker, G. H. Alani, and B. J. Zwolinski, Chem. Rev. *68*, 659 (1968); K. A. Kobe and R. E. Lynn, Chem. Rev. *52*, 117 (1953); Selected Values of Properties of Chemical Compounds, Texas A & M University, 1968; J. A. Burriel Lluna, C. B. Cragg, and J. S. Rowlinson, Quimica, *B*, *64*, 1 (1968); see also J. F. Mathews, Chem. Rev. *72*, 71 (1972).
7. M. K. Phibbs, J. Chem. Phys. *19*, 1420 (1951).
8. J. Timmermans, Physico-Chemical Constants of Pure Organic Compounds, Elsevier, Amsterdam 1950 (Vol. 1), 1965 (Vol. 2).
9. J. de Boer, Physica *9*, 363 (1942); H. Margenau, Phys. Rev. *63*, 131, 385 (1943); A. A. Evett and H. Margenau, ibid, *90*, 1021 (1953); F. R. Britton and D. T. W. Bean, Can. J. Phys. *33*, 668 (1955); V. Magnasco and G. F. Musso, J. Chem. Phys. *48*, 2657 (1968), and previous papers.
10. J. A. Pople, Proc. Roy. Soc. (London) *A 221*, 498 (1954).
11. J. R. Sweet and W. A. Steele, J. Chem. Phys. *47*, 3022 (1967).
12. M. Atoji and W. N. Lipscomb, J. Chem. Phys. *21*, 1480 (1953); S. D. Hamann, W. J. McManamey, and J. F. Pearse, Trans. Faraday Soc. *49*, 351 (1953); S. D. Hamann and J. A. Lambert, Austral. J. Chem. *7*, 1 (1954).
13. A. G. de Rocco and W. G. Hoover, J. Chem. Phys. *36*, 916 (1962).
14. R. N. Lichtenthaler, Ber. Bunsenges. *73*, 1041 (1969).
15. J. S. Rowlinson, Trans. Faraday Soc. *50*, 647 (1954); see also Rowlinson, Liquids and Liquid Mixtures, gen. ref. 11.
16. A. Michels, T. Wassenaar, T. N. Zwietering, and P. Smits, Physica *16*, 501 (1950); D. Ambrose and R. Townsend, Trans. Faraday Soc. *60*, 1025 (1964).
17. M. J. Terry, J. T. Lynch, M. Bunclark, K. R. Mansell, and L. A. K. Staveley, J. Chem. Thermodyn. *1969*, 413.
18. F. Din, Thermodynamic Functions of Gases, Volume 1, 2, 3; Butterworth, London 1956, 1961.
19. J. H. Smith and E. L. Pace, J. Phys. Chem. *73*, 4232 (1969); R. A. Aziz, C. C. Lim, and D. H. Bowman, Can. J. Chem. *45*, 1037 (1967); Ch. E. Decker, A. G. Meister, and F. F. Cleveland, J. Chem. Phys. *19*, 784 (1951).
20. A. Eucken and F. Hauck, Z. Phys. Chem. *134*, 161 (1928).
21. M. L. Riedel, Chem. Ing. Techn. *26*, 83, 259, 679 (1954); K. S. Pitzer, D. Z. Lippman, R. F. Curl, C. M. Huggins, and D. E. Petersen, J. Am. Chem. Soc. *77*, 3433 (1955); K. S. Pitzer and R. F. Curl, ibid, *79*, 2369 (1957).
22. E. M. Holleran, J. Chem. Phys. *47*, 5318 (1967); J. Phys. Chem. *72*, 1230 (1968), *73*, 167, 3700 (1969); E. M. H. and G. J. Gerardi, ibid. *72*, 3559 (1968).
23. A. Eucken, Z. angew. Chem. *55*, 163 (1942); A. R. Ubbelohde, Quart. Rev. *4*, 356 (1950); See also Ubbelohde, Melting and Crystal Structure, Clarendon Press, Oxford 1965.
24. J. Phys. Chem. Solids, entire volume *18* (1961); J. Chim. Phys. *63* (1966), entire issue No. 1; L. Meyer, Advances in Chem. Phys. *16*, Interscience, New York 1969.
25. J. A. Pople and K. E. Karasz, J. Phys. Chem. Solids *18*, 28 (1961); ibid. *20*, 294 (1961).
26. S. Chandrasekhar, R. Shashidhar, and N. Tara, Mol. Cryst. *10*, 337–358 (1970).
27. K. Clusius, Z. Phys. Chem. *B 23*, 213 (1933); K. C. and K. Weigand, *B 46*, 1 (1940); K. S. and L. Popp, *B 46*, 63 (1940); K. C. and G. Faber, *B 51*, 352 (1942); K. C. and G. Wolf, Z. Naturforsch. *2 a*, 495 (1947); F. I. A. T. Review, The Physics of Solids I, p. 193.
28. J. A. Jahnke, J. Chem. Phys. *47*, 336 (1967).

29. Landolt-Börnstein, Zahlenwerte und Funktionen, Band 2/4, Springer, Berlin 1961.
30. Selected Values of Chemical Thermodynamic Properties, Circular Nat. Bur. Stand. *500*, 1952.
31. O. L. I. Brown and G. G. Manov, J. Am. Chem. Soc. *59*, 500 (1937).
32. J. H. Hildebrand, J. Chem. Phys. *15*, 727 (1947).
33. G. B. Guthrie and J. P. McCullough, J. Phys. Chem. Solids *18*, 53 (1961).
34. H. Sackmann and G. Kloos, Z. Phys. Chem. *209*, 319 (1958).
35. L. S. Bartell and D. A. Kohl, J. Chem. Phys. *39*, 3097 (1963).
36. K. S. Pitzer, J. Chem. Phys. *8*, 711 (1940); W. B. Person and G. C. Pimentel, J. Am. Chem. Soc. *75*, 532 (1953).
37. H. L. Finke, M. E. Gross, G. Waddington, and H. M. Huffman, J. Am. Chem. Soc. *76*, 333 (1954).
38. H. Tompa, Polymer Solutions, Butterworth, London 1956.
39. H. Sackmann and P. Venker, Z. Phys. Chem. *199*, 100 (1952).
40. R. A. Scott and H. A. Scheraga, J. Chem. Phys. *44*, 3054 (1966); L. Radom and J. A. Pople, J. Am. Chem. Soc. *92*, 4786 (1970).
41. A. Neckel and H. Volk, Z. Elektrochemie (Ber. Bunsenges.) *62*, 1104 (1958).
42. R. G. Snyder, J. Chem. Phys. *47*, 1316 (1967); R. G. Snyder and J. H. Schachtschneider, Spectrochimica Acta *19*, 85 (1963).
43. G. J. Szasz, N. Sheppard, and D. H. Rank, J. Chem. Phys. *16*, 704 (1948); N. Sheppard and G. J. Szasz, J. Chem. Phys. *17*, 86 (1949).
44. J. E. Piercy and M. G. S. Rao, J. Chem. Phys. *46*, 3951 (1967); but see K. R. Crook and E. Wyn-Jones, J. Chem. Phys. *50*, 3445 (1969).

11. Molecular Re-Orientation in Liquids

11.1. Introduction

The rotational motion of molecules [1] can be studied by any method which yields the time correlation function of some orientation dependent molecular parameter $\vec{u}$ [2]. The ensemble average of this correlation function, $\langle \vec{u}(0)\cdot\vec{u}(t)\rangle$, is the Fourier transform of the spectral density, or line shape, $S(\omega)$, of the scattered radiation, which is experimentally accessible. In this chapter, we will present a short introduction to the analysis of IR and Raman spectra and of neutron scattering (section 11.2), to the interpretation of dielectric absorption and of Rayleigh scattering of laser beams (section 11.3), to the calculation of relaxation times in n.m.r. (section 11.4), and to ultrasonic relaxation in fluids (section 11.5). The last method is not directly applicable to the study of rotational motion of the molecules, but rather to the study of different energy states of molecules as given, e.g., by conformational equilibria.

11.2. Absorption Spectra and Angle Correlation Functions

Fig. 11.1 shows the far-infrared spectrum of a dilute solution of H_2 in liquid argon [3]. The molecular quadrupole moment of H_2 induces a dipole in the surrounding liquid. This induced dipole is modulated as H_2 rotates and translates in the liquid solution, which explains the infrared absorption. The peaks correspond to the first rotational transitions ($J = 0 \rightarrow J = 2$ and $J = 1 \rightarrow J = 3$), which are the same as for the unperturbed gas-phase molecules indicated on the energy level diagram in fig. 11.1. Similar observations are made in a D_2-argon solution. Thus the molecules H_2 and D_2 are free rotors in this liquid state system. Profile analysis of the absorption lines revealed further the translational wings indicated by the dotted lines in fig. 11.1. Purely trans-

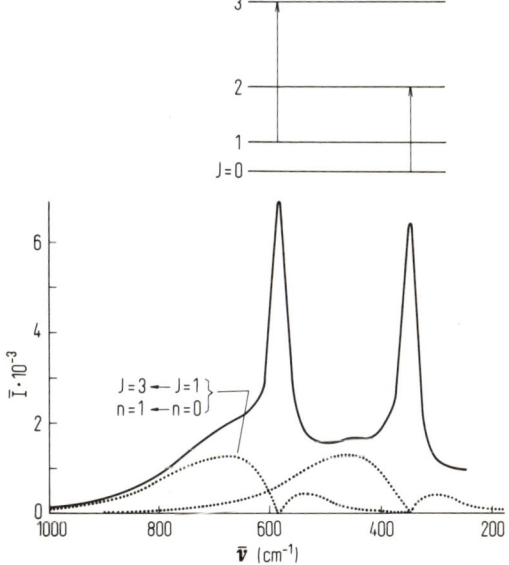

Fig. 11.1. Rotation-translation spectrum of H_2 in liquid argon (1 mol %, $T = 87$ K). The energy level diagram above indicates the pure rotational transitions of H_2. The dotted curves show the contribution of translational fine structure to the experimental absorption. After Ewing [3].

lational absorption has been also observed in H_2 and D_2 solutions in liquid argon. Here also a mode appears (at 38 cm^{-1}) which has been assigned to the translational mode of the solvent argon, on which the solute molecules induce absorption.

Whereas these modes can probably be directly compared to some features in the frequency spectrum $\tilde{z}(\omega)$ obtained by neutron scattering, the situation is more difficult in mixtures of rare gases, where translational modes have also been observed [4]. Here the absorption is obviously caused by the dipole induced by the quadrupole moment of a collision complex, and appears at higher frequency as can be expected for a single translational mode. If translational spectra appear as satellites on other spectra, as in the case of fig. 11.1, there is some preference for them on the high frequency side, i.e., translational excitation goes parallel with rotational or other excitation.

The hydrogen isotopes also provide the best example of coupling between rotation and translation [3], in the case of induced rotational absorption of HD in liquid argon. The HD rotational energy levels in liquid argon are shifted from their unperturbed gas phase values. For the molecular interaction of HD, the electrical center, located at the midpoint between the two atoms, is essential. But rotation occurs around the center of mass, which is nearer to the D atom. Thus rotation brings the HD molecule out of its potential well in the liquid, so that a translational motion is necessary to restore equilibrium. A simplified perturbation treatment of a rotor in a spherically symmetric cavity explains quite nicely the perturbed energy levels of HD.

The hydrogen molecules are unique in showing free rotational motion in the liquid state. The reasons for this are the small size of the molecule, the weak intermolecular interaction, and the small moment of inertia. The latter causes a relatively wide spacing of the rotational levels, so that the spacing is large in comparison to possible energy barriers. That the reverse situation applies already to CH_4 follows from an infrared absorption study of CH_4 in solid argon at 4 K [3]. The fine structure of the ν_3-absorption band is very pronounced but completely different from the gas phase transition. Calculation of the rotational energy level diagram for a rigid tetrahedron (CH_4) in a cubic field (solid argon) with the barrier height as parameter shows that the observed spectrum corresponds to a barrier height between 1.1 and 1.5 kcal/mol. This barrier height is above the first eight unperturbed rotational levels of methane. The perturbed levels display very pronounced splitting. In view of this complication and the fluctuating local environment of a molecule in the liquid it is clear that no distinct rotational levels can be observed in the liquid state. Only band contours can be resolved. Therefore, the detailed analysis of the absorption bands and the construction of appropriate angle correlation functions is of great importance.

If there is no coupling between translational and rotational motion of the molecule, the van Hove self-correlation function for a characteristic part of the molecule (say an H atom) can be written [5] in terms of the correlation functions for the center-of-gravity motion $G_{\text{tr}}(\vec{r}',t)$ and for the angle dependent relative motion $G_i(\vec{r} - \vec{r}',t)$ (the subscripts tr and i are for translational and internal, resp.):

$$G_s(\vec{r},t) = \int G_{\text{tr}}(\vec{r}',t) G_i(\vec{r} - \vec{r}',t) \, d\vec{r}'. \tag{11.1}$$

The Fourier transforms can then be written

$$F_s(\vec{Q},t) = F_{\text{tr}}(\vec{Q},t) F_i(\vec{Q},t), \tag{11.2}$$

and

$$S_s(\vec{Q},\omega) = \int S_{\text{tr}}(\vec{Q},\omega - \omega') S_i(\vec{Q},\omega') \, d\omega'. \tag{11.3}$$

11.2. Absorption Spectra and Angle Correlation Functions

Rather than formulating the angle dependent correlation function $G_i(\vec{r} - \vec{r}', t)$, it is more convenient to introduce the probability $G(\Omega_0, \Omega, t) d\Omega_0 d\Omega$ that $\vec{b} = \vec{r} - \vec{r}'$ lies in the solid angle element $d\Omega_0$ at time zero and in $d\Omega$ at time t. This quantity does not contain the magnitude of b. For isotropic liquids $G(\Omega_0, \Omega, t)$ depends only on the angle ϑ between Ω_0 and Ω, and can be separated into an angle dependent part and a time dependent part:

$$G(\Omega_0, \Omega, t) = \frac{1}{4\pi} \sum_{l=0}^{\infty} \sum_{m=-l}^{+l} F_l(t) Y_l^{m*}(\Omega_0) Y_l^m(\Omega) = \frac{1}{16\pi^2} \sum_l (2l+1) F_l(t) P_l(\cos\vartheta). \tag{11.4}$$

In eq. (11.4), the Y_l^m are the spherical harmonics, and by their addition theorem, the sum over m can be expressed in terms of the Legendre polynomials [6]. The number l can be readily identified as proportional to the change of angular momentum. With the help of eq. (11.4), the internal contribution to the scattering law can be written [5]:

$$S_i(Q, \omega) = \sum_{l=0}^{\infty} (2l+1) [j_l(Qb)]^2 S_l(\omega), \tag{11.5}$$

where $j_l(Qb)$ is a spherical Bessel function and

$$S_l(\omega) = \frac{1}{2\pi} \int_{-\infty}^{\infty} F_l(t) e^{-i\omega t} dt. \tag{11.6}$$

The unknowns in the contribution of $S_i(\vec{Q}, \omega)$ to the self-scattering law $S_s(\vec{Q}, \omega)$ are the time dependent functions $F_l(t)$. Whereas the entire set of these functions is of importance in neutron scattering data of, say, hydrogen containing molecules, the function $F_1(t)$ is accessible from infrared spectra and the function $F_2(t)$ from Raman measurements [7]. Again, the detailed connections are beyond the scope of this book, but the essential idea should be indicated. The fraction $A(\omega)$ of electromagnetic radiation absorbed per unit thickness of the sample can be shown to be [2]

$$A(\omega) = \frac{2\pi\omega}{3\hbar c} S(\omega),$$

$$S(\omega) = \frac{1}{2\pi} \int_{-\infty}^{+\infty} e^{-i\omega t} \langle \vec{\mu}(0) \cdot \vec{\mu}(t) \rangle dt. \tag{11.7}$$

Here $\langle \vec{\mu}(0) \cdot \vec{\mu}(t) \rangle$ is the ensemble average of the dipole moment of the whole system projected on the dipole moment which the system had the time interval t before, and multiplied with the moment at time zero. In purely rotational spectra, $\vec{\mu}$ is the permanent dipole moment, in vibrational and electronic spectra, it is the transition moment. If the dipoles of different molecules are uncorrelated, then

$$\langle \vec{\mu}(0) \cdot \vec{\mu}(t) \rangle = N \langle \vec{\mu}_i(0) \cdot \vec{\mu}_i(t) \rangle, \tag{11.8}$$

where $\vec{\mu}_i$ is the dipole moment of an individual molecule. Eq. (11.8) will not hold for rotational spectra, where a correlation of the dipole moments certainly exists, but it will hold for vibrational spectra, as the transition moments will be usually uncorrelated. In the case of purely rotational spectra it would be necessary to go to dilute solutions. Apart from multiplicative constants, the dipole correlation function is equal to $\langle \cos\vartheta(t) \rangle$, the ensemble average of the cosine of the angle between the two orientations at time t and at time zero. On the other hand, it follows from the orthogonality of the Legendre polynomials that [cf. eq. (11.4)]

$$\int P_l(\cos\vartheta) G(\Omega_0, \Omega, t) \frac{1}{2} \sin\vartheta \, d\vartheta = \frac{1}{16\pi^2} (2l+1) F_l(t) \frac{2}{2l+1}. \tag{11.9}$$

Here $\sin\vartheta\,d\vartheta/2$ is the differential between two orientations, and $2/(2l+1)$ is the integral over $P_l^2(x)$. Therefore, $F_l(t)$ is proportional to the ensemble average of $P_l(\cos\vartheta(t))$. Recognizing that $P_1(\cos\vartheta) = \cos\vartheta$, we have

$$\langle\cos\vartheta(t)\rangle \propto F_1(t). \tag{11.10}$$

Whereas the infrared spectrum is caused by changes of the dipole moment, the Raman spectrum is caused by changes of electric polarizability. In a vibration-rotation Raman band, the dipole correlation function in eq. (11.7) has to be replaced by $\mathrm{Tr}\langle\tilde{\alpha}^v(0)\cdot\tilde{\alpha}^v(t)\rangle$, where $\tilde{\alpha}^v$ is the vibrational matrix element of the anisotropy of the polarizability, taken between two vibrational states involved in the transition and the trace (Tr) is over the three spatial indices of $\tilde{\alpha}^v$. For a totally symmetric vibration of a symmetric top or a linear molecule there is

$$\mathrm{Tr}\langle\tilde{\alpha}^v(0)\cdot\tilde{\alpha}^v(t)\rangle = \mathrm{const}\cdot\langle 3\cos^2\vartheta(t) - 1\rangle \propto F_2(t), \tag{11.11}$$

as $P_2(\cos\vartheta) = (3\cos^2\vartheta - 1)/2$. Inverse Fourier transformation of $S(\omega)$ [eq. (11.7)] thus gives $F_1(t)$ from the band shape of infrared absorption spectra and $F_2(t)$ from the band shape of Raman spectra. In general, it is advisable to confine the integration interval to the band, e.g.

$$F(t) = \int_{\text{band}} S(\omega)\cos\omega t\,d\omega \Big/ \int_{\text{band}} S(\omega)\,d\omega. \tag{11.12}$$

Such a normalization also helps to eliminate dielectric effects which modify the simple relation between the local electric field and the field due to the radiation. This treatment of spectra does

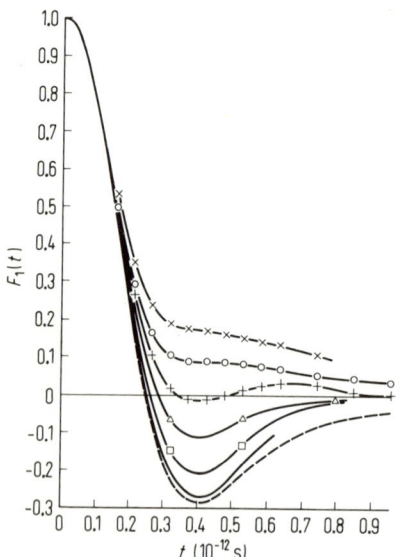

Fig. 11.2. The dipole-dipole-correlation function of CO in various environments. Full curves refer (from top to bottom) to liquid CHCl$_3$, CCl$_4$, n-heptane, and gaseous argon of a density of 510, 270, and 66 amagat^{-1}, resp. The dashed curve is calculated for freely rotating CO. After Ewing [3].

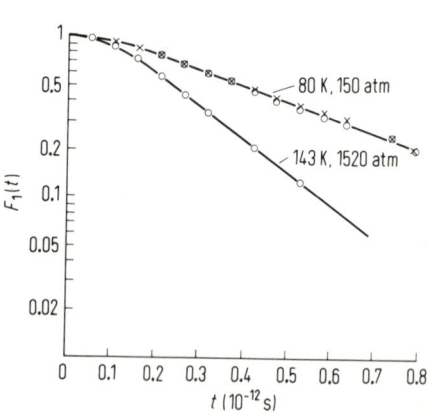

Fig. 11.3. Dipole-dipole-correlation function for liquid CO from infrared spectra. Crosses refer to the fundamental vibration, circles to the second overtone. After Gordon [7].

not involve induced absorption (as the H_2 absorption in fig. 11.1), and is limited to non-overlapping bands. This limitation rules out all more complicated molecules.

Fig. 11.2 and 11.3 show the dipole correlation function for CO in various media. For small time intervals the cosine of eq. (11.12) can be expanded into a power series ($\bar{S}_1(\omega)$ denoting the normalized absorption):

$$F_1(t) = \sum_{n=0}^{\infty} \frac{(-1)^n t^{2n}}{(2n)!} \int \omega^{2n} \bar{S}_1(\omega) \, d\omega = 1 - \frac{M_2}{2!} t^2 + \frac{M_4}{4!} t^4 \ldots \tag{11.13}$$

The first few moments of $\bar{S}_1(\omega)$ have been calculated [7], with the result that

$$M_2 = 2kT/I \tag{11.14}$$

(I being the moment of inertia), and M_4 contains the mean square torque on a molecule due to the other molecules. With M_2 containing only the kinetic energy, $F_1(t)$ starts always parabolically, the width of the parabola being determined by the moment of inertia of the molecule. The bigger the moment of inertia, the wider the starting parabola. The bigger M_4, i.e. the mean square torque or the hindering of the rotation, the more remains $F_1(t)$ above the correlation function of the free rotor. This is seen nicely in fig. 11.2. The sequence of rotational hindrance in solutions is apparently given by the molar volume of the various solvents ($V_{CHCl_3} : V_{CCl_4} : V_{n-C_7H_{16}} = 81 : 97 : 147$): The larger the quasi-lattice of the solvent, the more free the rotation of the small solute molecule. Whereas even in chloroform a bit of the wave-like behaviour of the correlation function is preserved, in pure liquid CO there is a quick transition to an exponential decay of $F_1(t)$ (fig. 11.3). Such an exponential decay should be expected if rotation is strongly hindered, i.e. occurs only in steps of small angles. This leads to a kind of "random walk reorientation" which satisfies a differential equation of the kind of a diffusion equation:

$$D_r \nabla^2 G(\Omega_0, \Omega, t) = \partial G(\Omega_0, \Omega, t)/\partial t, \tag{11.15}$$

where D_r is termed rotational diffusion coefficient. The solutions of eq. (11.15), using eq. (11.4), are

$$F_l(t) = \exp\{-l(l+1) D_r t\}. \tag{11.16}$$

Fourier transformation gives a Lorentzian shape of the spectra:

$$S_l(\omega) = \frac{1}{\pi} \frac{(l+1) D_r}{l^2(l+1)^2 D_r^2 + \omega^2}, \tag{11.17}$$

where $l(l+1) D_r$ gives the half width of the line.

Fig. 11.4 shows $F_1(t)$ and $F_2(t)$ for methane. Though there is a certain time interval for free rotation (corresponding to a relatively large change in $\cos \vartheta$), both correlation functions change to exponential decay at the same time. The values of the correlation functions where the change occurs are of course different, as in the free rotor case $F_2(t)$ declines much faster than $F_1(t)$. This causes quite different line widths in the Raman and infrared spectra, which once led to discordant conclusions about freedom of rotation of CH_4. From an observation of the Raman bands, the rotation was thought to be almost free, whereas from a study of infrared bands, the rotation was considered to be highly hindered. This discrepancy is completely resolved by the detailed analysis shown in fig. 11.4. A hindrance of rotation in liquid CH_4 follows also from the results of neutron scattering [8].

More recent studies on N_2O in various solvents [9], on methyl iodide [10], and on methylene chloride [11], have shown the same qualitative features as exhibited by CO in solvents and by

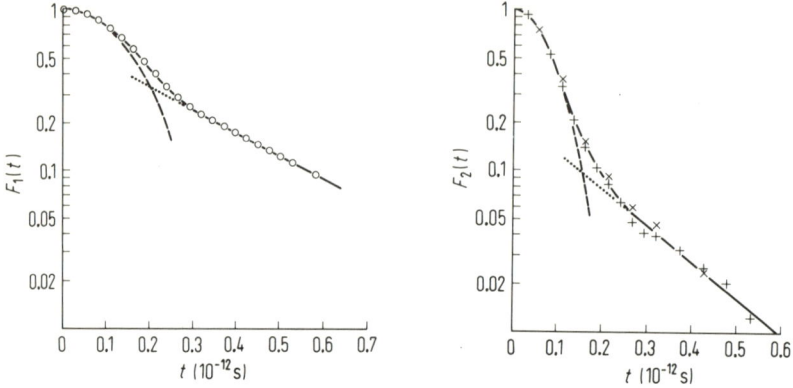

Fig. 11.4. The correlation functions $F_1(t)$ and $F_2(t)$ for liquid CH_4 from infrared data at 98 K and Raman band shapes at 95 K, resp. After Gordon [7].

CH_4, except that the rotation is more hindered. From the time interval, where the correlation function changes from a parabolic to an exponential behaviour, one may conclude in a very simplified fashion, that methyl iodide rotates freely for an angle of less than 30° in the end-over-end rotation, and less than 60° in the rotation about the symmetry axis. The corresponding values of the rotational diffusion constant are given in table 11.1. Unfortunately, the relation between the breakpoint in the correlation function and the rotational diffusion constant is not always as clear as in the case of methyl iodide (cf. [9]).

Finally, a short remark should be made about neutron scattering. The formalism starting with eq. (11.1) applies to a characteristic part of the molecule, so in neutron scattering summation over all scattering centers is necessary. It has not been possible yet to evaluate the $F_l(t)$ from neutron scattering data. Instead, several models for molecular rotation have been developed [5, 8, 12, 13], and have been used in combination with spectral data to predict neutron scattering results, thus testing the existent data. In the case of CH_4 [8, 14], the partial disagreement may be explained either by the importance of higher $F_l(t)$-terms or by the existence of higher molecular complexes. It is interesting in this connection that evidence for a H_2-Ar complex has been found [15]. A reported discrepancy between neutron scattering results and results of Raman and Rayleigh scattering of cyclohexane [13, but cf. 11] has now been resolved by new Raman scattering results [45]. The interpretation of Rayleigh scattering in cyclohexane is complicated due to other processes.

11.3. Dielectric Relaxation and Rayleigh Scattering

Dielectric relaxation is the decay with time of the polarization in a dielectric when an external electric field is removed. The classical theory of orientation polarization of polar fluids has been developed by Debye [16]. When an external electrical field of angular frequency ω is applied to a dielectric the resulting polarization $\vec{P}$,

$$4\pi \vec{P} = \vec{D} - \vec{E} = (\varepsilon - 1)\vec{E}, \tag{11.18}$$

11.3. Dielectric Relaxation and Rayleigh Scattering

is in general not in instantaneous equilibrium with the field $\vec{E}$. The displacement $\vec{D}$ lags behind $\vec{E}$ by a phase difference φ:

$$\vec{E} = \vec{E}_0 e^{i\omega t} \tag{11.19}$$

$$\vec{D} = \vec{D}_0 e^{i(\omega t - \varphi)} = \vec{D}_0 e^{i\omega t} (\cos \varphi - i \sin \varphi). \tag{11.20}$$

By introducing formally a complex permittivity*$^{)}$,

$$\varepsilon = \varepsilon' - i\varepsilon'', \tag{11.21}$$

we define the frequency dependent quantities

$$\varepsilon' = (\vec{D}_0/\vec{E}_0) \cos \varphi, \tag{11.22}$$

$$\varepsilon'' = (\vec{D}_0/\vec{E}_0) \sin \varphi, \tag{11.23}$$

where ε' is the measured permittivity in a condenser and ε'' is the loss factor. At sufficiently low frequencies the phase difference between $\vec{D}$ and $\vec{E}$ can be neglected, and ε becomes equal to the static permittivity ε_0. At very high frequencies (corresponding to the far IR) only that part of $\vec{D}$ remains which can follow $\vec{E}$ instantaneously, so that $\vec{D} = \varepsilon_\infty \vec{E}$ (ε_∞ is the "optical" permittivity resulting from the atomic and electronic polarizability). Hence the relaxing part of the displacement is $(\vec{D} - \varepsilon_\infty \vec{E})$. The Debye theory assumes that the attainment of equilibrium is exponential, i.e., the rate of relaxation of $(\vec{D} - \varepsilon_\infty \vec{E})$ is proportional to the deviation of this quantity from its equilibrium value:

$$\frac{d}{dt}(\vec{D} - \varepsilon_\infty \vec{E}) = \frac{1}{T_0}\left[(\varepsilon_0 - \varepsilon_\infty)\vec{E} - (\vec{D} - \varepsilon_\infty \vec{E})\right]. \tag{11.24}$$

$(1/T_0)$ is the reciprocal macroscopic decay time. Performing the differentiation [using eq. (11.19) and (11.20)] and rearranging gives

$$\varepsilon = \varepsilon_\infty + \frac{\varepsilon_0 - \varepsilon_\infty}{1 + i\omega T_0}, \tag{11.25}$$

or, by separating the real and imaginary parts,

$$\varepsilon' = \varepsilon_\infty + \frac{\varepsilon_0 - \varepsilon_\infty}{1 + \omega^2 T_0^2}, \tag{11.26}$$

$$\varepsilon'' = \frac{\varepsilon_0 - \varepsilon_\infty}{1 + \omega^2 T_0^2} \omega T_0. \tag{11.27}$$

From eq. (11.26) and (11.27), $\varepsilon' = \varepsilon_0$ for zero frequency, and $\varepsilon' = \varepsilon_\infty$ for very high frequencies. The loss factor ε'' approaches zero at both low and high frequencies and has a maximum for $\omega = 1/T_0$, which is called the critical frequency ω_c. For an analysis of experimental results it is convenient to plot ε in the complex plane, i.e. to plot ε'' vs. ε' (Cole-Cole plot) [17]. In such a plot eq. (11.26) and (11.27) which are based on a single decay time T_0 give a semicircular arc. A considerable number of simple rigid molecules, either in the pure liquid or dissolved in non-polar solvents show such a behaviour. Molecules with internal rotation usually exhibit a more complicated relaxation behaviour than rigid molecules. In some cases a Cole-Cole plot exhibits two well separated semicircular arcs corresponding to two separate relaxation processes:

---

* The term "(relative) permittivity" is used in the modern literature instead of "dielectric constant".

molecular and group relaxation. Some results for substituted benzene and naphthalene molecules have been summarized by Smyth [18]. Relaxation times of molecules in non-associated liquids near room temperature are usually between 0.5 and $5 \cdot 10^{-11}$ s, depending on molecular size and shape. It is frequently assumed that group relaxation times are shorter than those of the whole molecule (but see the end of section 11.5).

The macroscopic relaxation time T_0 can be related to the decay of the dipole autocorrelation function of the molecular dipole moment $\vec{\mu}_i$ [51],

$$\gamma(t) = \langle \vec{\mu}_i(0) \cdot \vec{\mu}_i(t) \rangle / \langle \vec{\mu}_i^2(0) \rangle . \tag{11.28}$$

The difficulty is to express the total moment of the fluid in terms of $\vec{\mu}_i(t)$. Glarum, and Cole [19], on the basis of the Kirkwood-Onsager model, derived an expression relating $\gamma(t)$ to $\varepsilon(\omega)$. An exponential decay of $\gamma(t)$ with a rate constant $1/\tau$, where τ may be called the molecular relaxation time, gives eq. (11.25)–(11.27) with

$$T_0 = \frac{3\varepsilon_0}{2\varepsilon_0 + \varepsilon_\infty} \tau . \tag{11.29}$$

Some molecular relaxation times determined in such a way are listed in table 11.1. A distribution of molecular relaxation times (each for an exponential decay) corresponds to a distribution of macroscopic relaxation times [19]. Such a distribution of relaxation times has been proposed for many liquids on the basis of Cole-Cole plots which show a 'depressed circular arc' or a non-symmetrical 'skewed arc'.

The autocorrelation function of the molecular dipole moment, eq. (11.28), has been studied for a two-dimensional array of rotators by the method of molecular dynamics [20]. It was found that when the rotators interact only by dipole-dipole orientation energies [eq. (5.17)], then the correlation function $\gamma(t)$ has a time dependence proportional to $\exp -(t/t_c)^2$, where t_c is the mean period for rotation. It was concluded that dipolar forces play little part in dielectric relaxation and do not lead to an exponential decay law. Angular potentials other than those resulting from dipole interaction are much more effective in transforming the free rotator autocorrelation function into a function with exponential decay as demanded by experiment.

An important development has been the observation of a second dispersion region in liquids at frequencies higher than the Debye frequency [21]. Whereas dipole relaxation usually has a dispersion broader than corresponding to a Debye single relaxation time, these absorptions, which are in the 20–100 cm^{-1} region, are sharper than the Debye type. This second dispersion is believed to arise from 'quasilattice modes' of the liquid [22] or from collisional complexes [23, 46]. A comparison of the absorption spectrum of liquid and solid chlorobenzene is shown in fig. 11.5. The liquid exhibits a much broader band than the crystal but the absorption maximum is in the same region. Below 20 cm^{-1} the relaxation process arising from reorientation of dipoles is dominant in the liquid. In the crystal where molecules cannot reorient themselves this relaxation is absent [*].

Rayleigh light scattering is caused by fluctuations in polarizability, i.e. fluctuations of induced moments. The spectrum of scattered light is given by the Fourier transform of the time correlation function of the electric field, which can be related to the moments induced in the molecules,

---

[*] Pseudolattice absorption in the 20–100 cm^{-1} region has been observed also for non-polar liquids, with a smaller integrated intensity. Substances with a high polarizability, such as CCl_4, CS_2, or benzene, show much stronger absorption than e.g. cyclohexane [21].

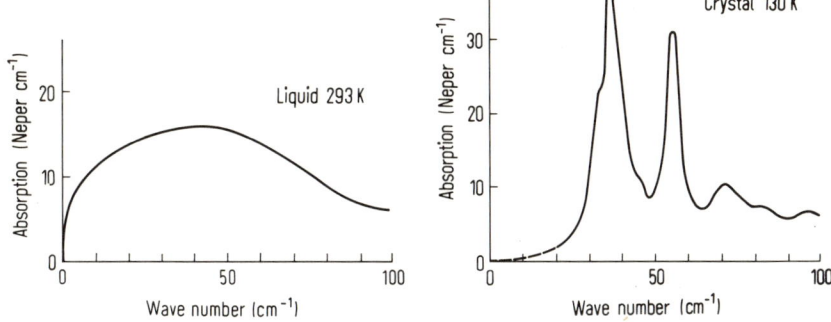

Fig. 11.5. The submillimeter absorption of chlorobenzene (solid and liquid) vs. the wave number. After Chantry and Gebbie [21].

their positions and orientations. Fluctuations of induced moments are caused by density fluctuations (isobaric fluctuations in Rayleigh scattering) and by orientational fluctuations of anisotropic molecules. The latter contribution, which frequently amounts to only a small fraction of the total scattering, can be singled out by observing depolarized scattering [24]. Assuming that only changes in the orientation of a single molecule are correlated, the spectrum of depolarized Rayleigh scattering can essentially be described by eq. (11.6) and (11.11).

Some results obtained from dielectric relaxation and from Rayleigh scattering are collected in table 11.1. A further comparison is shown in fig. 11.6. It can be seen that depolarized Rayleigh scattering gives a relaxation time increasing regularly with the chain length of the molecules (characteristic of end-over-end rotation), whereas dielectric relaxation in long molecules is caused by other reorientation processes (group reorientation or reorientation around the long axis of the chain molecule).

Orientational relaxation in Rayleigh scattering as well as dielectric relaxation gives the time correlation functions of macroscopic quantities, which can be traced back to the time dependence

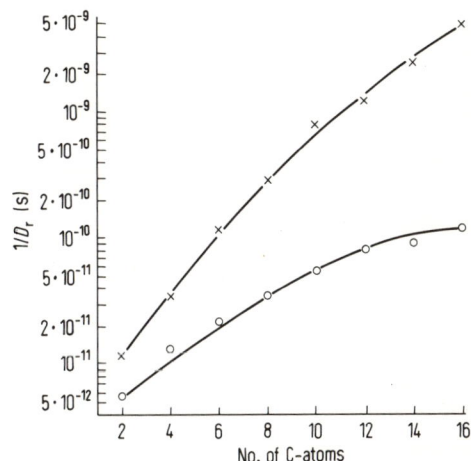

Fig. 11.6. The reciprocal rotational diffusion constant for the n-alkyl bromides from measurements of dielectric relaxation (circles) [37], and Rayleigh scattering (crosses) [24], at 25°C.

of molecular correlation functions by the use of models. In table 11.1 and fig. 11.6 we have assumed an exponential behaviour on the basis of a rotational diffusion model.

Finally, it should be mentioned that the same kind of information which is obtained from short wavelength dielectric relaxation (absorption due to quasi-lattice vibrations or collisional complexes) can be derived from the detailed shape of the wings of the depolarized Rayleigh line [23, 46]. Such investigations are important for our knowledge of translational motions in liquids.

11.4. Nuclear Magnetic Resonance Relaxation [25]

N.m.r. experiments are related to the molecular motions in liquids through the nuclear relaxation times T_1 and T_2. The spin-lattice (or longitudinal) relaxation time T_1 is a measure for the time required by the nuclei to exchange energy with their surroundings in order to achieve an equilibrium population among the allowed energy levels in the external magnetic field. The spin-spin (or transversal) relaxation time T_2 measures the time necessary for the nuclei to exchange energy among themselves by an adiabatic process.

When a nucleus of spin 1/2 is placed in a magnetic field $\vec{H}_0$ in the z direction, the magnetic moment $\vec{\mu}$ of the nucleus has two possible orientations, corresponding to two energy levels separated by an energy difference μH_0. Transitions between the two energy levels can be induced by an oscillating magnetic field in the x-y plane, of a frequency $\omega_n = \gamma H_0$, where γ is the nuclear magnetogyric constant.

Now consider an assembly of nuclear spins 1/2. In the absence of an external field the population of the two spin states is equal. When an external field $\vec{H}_0$ is now applied suddenly, the system will tend to relax to the new equilibrium populations. The oscillating magnetic field of frequency ω_n, which is needed to induce these transitions, is supplied by the rapid random motions of the molecules in the liquid. Let N_m and N_n be the number of nuclei in the upper and lower states, respectively, and $\Delta N = N_m - N_n$. A single transition changes the excess number ΔN by two. When the relaxation process is exponential in time then the change in ΔN is proportional to the deviation of ΔN from its equilibrium value ΔN^*:

$$d\Delta N/dt = -2P_{mn}(\Delta N - \Delta N^*) . \tag{11.30}$$

The transition probability P_{mn} is related to the spin-lattice relaxation time by

$$T_1 = \frac{1}{2P_{mn}} . \tag{11.31}$$

For diamagnetic liquids T_1 is usually between 10^{-2} and 10^2 s. For techniques of its measurement we refer to the literature [26].

We turn to the relationship between the transition probability P_{mn} of the nuclear spin and the random motions of the molecules in the liquid, which produce the fluctuating local magnetic field. These random motions can be characterized by correlation functions of the form

$$Z(\tau) = \langle F(0) F(\tau) \rangle / \langle F^2(0) \rangle , \tag{11.32}$$

where $F(0)$ and $F(\tau)$ are the values of some dynamical variable (translational or angular, according to the relaxation mechanism) at time zero and τ, respectively. The Fourier transform of $Z(\tau)$

11.4. Nuclear Magnetic Resonance Relaxation

gives the intensity of the fluctuations as a function of frequency ω, that is the spectral density $J(\omega)$,

$$J(\omega) = \frac{1}{2\pi} \int_{-\infty}^{+\infty} Z(\tau) e^{-i\omega\tau} d\tau. \tag{11.33}$$

Relaxation will be the more effective the larger the Fourier component at the resonance frequency ω_n. Hence P_{mn} is proportional to $J(\omega)$ [27]. When the correlation function $Z(\tau)$ decays exponentially in time, with a correlation time τ_c,

$$Z(\tau) = Z(0) \exp(-\tau/\tau_c), \tag{11.34}$$

then its Fourier transform $J(\omega)$ is given by

$$J(\omega) \propto 2\tau_c/(1 + \omega^2 \tau_c^2). \tag{11.35}$$

For sufficiently low values of $\omega\tau_c$ the function $J(\omega)$ is linearly related to τ_c and is independent of ω. Now, for liquids τ_c is in the order of 10^{-10} s. For static magnetic fields $\vec{H}_0$ of the order of 10^4 Gauss the resonance frequency ω_n is near 10^7 s^{-1}, so that indeed $\omega_n^2 \tau_c^2 \ll 1$. Thus, $1/T_1 = 2P_{mn}$ becomes proportional to τ_c, and independent of the resonant frequency ω_n.

In liquids containing nuclei of spin 1/2 the most important mechanisms producing n.m.r. relaxation are the dipole-dipole interactions between the nuclear magnetic moments, and the spin-rotational interaction between a nuclear magnetic moment and the rotation of electric charges distributed in the molecule. Nuclear dipole-dipole interaction leads to both intra- and intermolecular relaxation: intramolecular relaxation may occur as a consequence of molecular rotation, which causes a fluctuation in the local field $\vec{H}'$ at the position of one nucleus due to another nucleus of the same molecule. The intermolecular contribution to T_1 arises from the diffusive motion of one molecule relative to another, which causes a fluctuation of $\vec{H}'$ at one nucleus due to a nucleus in another molecule. In the case of proton-proton interaction it is possible to separate the intramolecular contribution from the intermolecular contribution to T_1 by dissolving the molecular species in a fully deuterated solvent and extrapolating to zero concentration[*]. It is more difficult to separate then $(1/T_1)_{intra}$ into the dipole-dipole part and the spin-rotational part. With some assumptions such a separation is possible by studying the temperature dependence of $(1/T_1)_{intra}$ [28].

Another important relaxation mechanism, which is dominant in liquids with molecules of nuclear spin larger than 1/2, is the interaction between the nuclear quadrupole moment and an electric field gradient at the nucleus, which varies with time due to molecular rotation. Quadrupolar relaxation is thus specific for rotational motions by which the electric field gradient is changed. E.g., in CD_3CN the electric field gradient is in the C–N direction, so that the ^{14}N-relaxation is caused by the end-over-end rotation, whereas the D-relaxation has also a contribution from rotation around the symmetry axis.

Table 11.1 presents some relaxation times measured by different methods. The relaxation times have been converted into rotational diffusion coefficients, in order to eliminate the factor $l(l + 1)$ of eq. (11.16), which is 2 for IR spectra and dipolar relaxation, and 6 for Raman spectra, Rayleigh scattering, and n.m.r. relaxation. An exponential decay[**] of the correlation function has been assumed on the basis of a rotational diffusion model. This model seems to be justified for

---

[*] The proton-deuteron interaction is much weaker than the proton-proton interaction.
[**] For non-exponential decay the n.m.r. relaxation time τ_c would be the integral over the correlation function [25].

relatively small and simple molecules (constituting "collision-limited" liquids [45]) by spectral investigations of correlation functions, which indicate that free rotation usually occurs through angles smaller than 30° (but see also [11]). The situation may be different for the rotation of methane and of methyl groups. An alternative model to rotational diffusion is reorientation by large jumps, followed by periods of residence in the cage of the neighbouring molecules. For this model the correlation functions decay also exponentially with time. In the rotational diffusion model, due to the factor $l(l + 1)$ in the exponent, the exponential decay of $F_2(t)$ is 3 times faster than of $F_1(t)$. In the jump model the decay of both correlation functions is the same. A jump model may be appropriate for large and complicated molecules (constituting "sterically hindered structure-limited" liquids [45]), where the reorientation process is connected with the breakup of the surrounding cage. Possibly C_4H_9Br and C_6H_5Cl are examples for this class. The discrepancies between dielectric $1/D_r$-values and those from other measurements in table 11.1 may be caused by the inappropriate multiplication of the relaxation times with $l(l + 1)$.

11.5. Ultrasonic Relaxation in Fluids

The propagation of sound in a fluid, in the absence of absorption, is described by the equation

$$c^2 = \left(\frac{\partial P}{\partial \rho_m}\right)_s = \frac{1}{\rho_m \beta_s} = \frac{1}{\rho_m \beta_T} \frac{C_P}{C_v}, \tag{11.36}$$

where c is the velocity of sound. The phase velocity can be expressed by the magnitude of the wave vector k and the angular frequency ω as $c = \omega/k$. If the density wave lags behind the pressure wave, absorption occurs. It is then convenient to define a complex wave vector as

$$k^* = k - i\alpha, \tag{11.37}$$

$$\left(\frac{1}{c^*}\right)^2 = \left(\frac{1}{c} - i\frac{\alpha}{\omega}\right)^2 = \rho_m \beta_T \left(\frac{C_v}{C_P}\right)^*, \tag{11.38}$$

where α is the absorption coefficient per unit length. The complex parameters are labelled by an asterisk. The absorption is given by the Stokes-Kirchhoff equation,

$$\alpha = \frac{2\omega^2}{3c^3 \rho_m}\left[\eta + \frac{3}{4}\phi + \frac{3}{4}(\lambda/C_P)(C_P/C_v - 1)\right], \tag{11.39}$$

where η, ϕ, and λ denote the shear and bulk viscosity, and the thermal conductivity of the fluid, respectively. Eq. (11.39) with $\phi = 0$ gives the "classical" absorption. Non-zero values of ϕ arise when the energy transfer between the translational kinetic energy in the direction of the wave and other degrees of freedom of the molecules needs time. For a single relaxation process (contributing α_1 to the total absorption α) the heat capacity may be written as a sum of two terms,

$$C_v = C_e + C_r(\omega), \tag{11.40}$$

where C_e is the constant external contribution and $C_r(\omega)$ is the "reaction heat capacity". The latter is caused by the temperature variation of the distribution of molecules among those energy levels which are responsible for the relaxation process. When the rate of energy transfer is proportional to the deviation from the equilibrium distribution (exponential decay in time) then

$$C_r(\omega) = \frac{C_r(0)}{1 + i\omega\tau}, \tag{11.41}$$

Table 11.1. Orientational relaxation times from different experimental methods, expressed as $1/D_r$, in 10^{-12} s. End-over-end rotation (reorientation of the long axis or the sixfold axis of the ring molecules) is denoted by $\perp$, rotation around the symmetry axis is denoted by $\parallel$. In CH_2Cl_2, a line parallel to both chlorine atoms corresponds to the symmetry axis [11]. Temperatures of measurements are 95–98 K for methane, 303–311 K for CH_2Cl_2, and 295–303 K for all other substances. The dielectric data are converted to molecular relaxation times using eq. (11.29).

Substance	infrared $(1/D_r = 2\tau_1)$		Raman $(1/D_r = 6\tau_2)$		N.m.r. $(1/D_r = 6\tau_2)$ dipole-dipole		quadrupole		dielectric $(1/D_r = 2\tau_1)$	Rayleigh scattering $(1/D_r = 6\tau_2)$	τ_1/τ_2 after [45]	References
	$\perp$	$\parallel$	$\perp$	$\parallel$	$\perp$ (a)		$\perp$	$\parallel$	$\perp$	$\perp$		
CH_4		0.8		1.1							—	b
CH_3J	2.1	0.5	9.0		9.5					9.6	—	10, 29, 45, 28, 45
CD_3J	3.0	0.6									—	10
CH_2Cl_2	2.2				3.4						—	11
$CHCl_3$			9.0			11			9.4		2.7	45, 46, 30, 31
$CDCl_3$						10.8	5.5				—	c, 32, 33
$CHBr_3$			32								2.7	45
CH_3CN	7.4		9.0		7.8				5.2	10.8	2.9	50, 45, 29, 34, 35, 47
CD_3CN							7.4	0.8			—	34, 36
$CBrCl_3$			17								—	45
CCl_4				10.8				10.2			—	45, 38
CS_2			9.0			8.4				12.0	—	45
C_2H_5Br									5.6	12	—	37, 24
C_4H_9Br									13.4	35	—	37, 24
C_6H_6			15.6	19	10.4					15.6	—	45, 48, 49
C_6F_6			42								—	45
C_6H_5Cl						25			18.5	57	1.1	45
c-C_6H_{12}	10.0		10.2								3	50, 45

a) The values for the dipole-dipole n.m.r. relaxation represent a combination of $\perp$ and $\parallel$, but correspond numerically almost to $\perp$.
b) Deduced from fig. 11.4.
c) Deduced from fig. 4 of ref. 11.

where τ is the relaxation time typical for that energy transfer (cf. the analogous eq. (11.25) for dielectric relaxation). By substituting eq. (11.40) and (11.41) into (11.38) and rearranging [41],

$$\left(\frac{1}{c} - i\frac{\alpha_1}{\omega}\right)^2 = \left(\frac{1}{c(0)}\right)^2 \left(1 - A\frac{i\omega\tau'}{1 + i\omega\tau'}\right), \tag{11.42}$$

with

$$A = \frac{C_r(C_P - C_v)}{C_v(C_P - C_r)}; \quad \tau' = \frac{C_P - C_r}{C_P}\tau. \tag{11.43}$$

From eq. (11.42), by separating real and imaginary parts, and neglecting $(\alpha_1/\omega)^2$ against $(1/c)$ one obtains

$$\left(\frac{c(0)}{c}\right)^2 = 1 - A\frac{\omega^2\tau'^2}{1 + \omega^2\tau'^2}, \tag{11.44}$$

$$\alpha_1 = \frac{1}{2}\frac{c}{c(0)^2}A\tau'\frac{\omega^2}{1 + \omega^2\tau'^2}. \tag{11.45}$$

At low frequencies ($\omega^2\tau'^2 \ll 1$) eq. (11.44) gives $c = c(0)$, and eq. (11.45) shows that α_1 is proportional to ω^2 or

$$\frac{\alpha_1}{v^2} = \frac{2\pi^2}{c(0)}A\tau'. \tag{11.46}$$

At very high frequencies ($\omega^2\tau'^2 \gg 1$)

$$\left(\frac{c(0)}{c(\infty)}\right)^2 = 1 - A, \tag{11.47}$$

and α_1/v^2 becomes zero. A useful parameter is the absorption per wavelength μ,

$$\mu = \alpha_1\lambda = \left(\frac{c}{c(0)}\right)^2 \pi A\frac{\omega\tau'}{1 + \omega^2\tau'^2}, \tag{11.48}$$

which has a bell-shape like other absorption curves (see fig. 11.7). The maximum of μ is at $\omega^2\tau'^2 = 1/(1 - A)$ and the maximum value is

$$\mu_{max} = \frac{\pi}{2}\frac{A}{\sqrt{1 - A}}. \tag{11.49}$$

It is seen that the increase in the velocity from $c(0)$ to $c(\infty)$ [eq. (11.47)] and the maximum absorption per wavelength [eq. (11.49)] are given by the parameter A, whereas at low frequencies α_1/v^2 [eq. (11.46)] is determined by $A\tau'$.

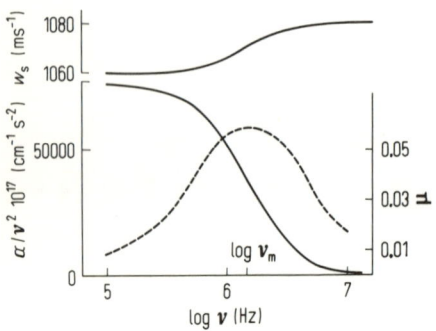

Fig. 11.7. Ultrasonic velocity (denoted here by w_s) (upper curve), absorption α/v^2 (lower curve), and absorption per wavelength μ (dashed curve) in the relaxation region. The scale refers to acetic acid at 40°C (cf. chapter 12, ref. 40).

Above we have considered a single relaxation process which may be caused by vibrational excitation, conformational equilibria of flexible molecules, or association equilibria. Furthermore, we have assumed that the "reaction" is not associated with a volume change. Otherwise the compressibility β_T in eq. (11.38) is also a complex quantity depending on frequency. Such a volume change may be revealed by investigating the pressure dependence of the relaxation phenomena.

In the present context, the application of ultrasonic absorption to the study of conformational equilibria is of interest [42]. Relaxation due to *trans-gauche* equilibria has been observed in 1,2-dichloro- and dibromoethane, with a relaxation frequency of 20–40 MHz at −80 °C [43]. Similarly, *trans-gauche* conversion causes a relaxation in n-butane, with $2\pi/\tau' = 20$ MHz at −120 °C, and in n-pentane, with $2\pi/\tau' = 15$ MHz at −120 °C [44]. An energy barrier of 3.4 and 3.9 kcal/mol, respectively, has been deduced. At room temperature, this would correspond to a relaxation time of 10^{-9} s. This is considerably higher than dielectric relaxation times for group relaxation. Further work will be necessary in order to arrive at a clear interpretation.

11.6. References

1. Discussion Meeting of Deutsche Bunsengesellschaft at Herrenalb, Oct. 1970, Ber. Bunsenges. *75* (1971), No. 3/4.
2. cf. W. A. Steele, in H. J. M. Hanley, Transport Phenomena in Fluids, Dekker, New York 1969.
3. G. E. Ewing, Accounts of Chemical Research *2*, 168 (1969).
4. P. Marteau, R. Granier, H. Vu, and B. Vodar, Compt. Rend. *265 B*, 685 (1967).
5. K. E. Larsson, Neutron Inelastic Scattering, Proceed. Copenhagen Symp., Internatl. Atom. Energ. Agency, Vienna 1968, Vol. I, p. 397; V. F. Sears, Can. J. Phys. *44*, 1279 and 1299 (1966).
6. G. Arfken, Mathematical Methods for Physicists, Academic Press, New York 1966.
7. R. G. Gordon, J. Chem. Phys. *43*, 1307 (1965), and previous papers.
8. V. F. Sears, Can. J. Phys. *45*, 237 (1967).
9. J. Vincent-Geisse, J. Soussen-Jacob, Nguyen-Tan Tai, and D. Descout, Can. J. Chem. *48*, 3918 (1970).
10. C. E. Favelukes, A. A. Clifford, and B. Crawford, Jr., J. Phys. Chem. *72*, 962 (1968); T. Fujiyama and B. Crawford, Jr., J. Phys. Chem. *73*, 4040 (1969).
11. W. G. Rothschild, J. Chem. Phys. *53*, 3265 (1970); *53*, 990 (1970).
12. K. E. Larsson, Phys. Rev. *167*, 171 (1968).
13. P. A. Egelstaff, J. Chem. Phys. *53*, 2590 (1970).
14. O. Eder and P. A. Egelstaff, p. 223, Vol. II of Neutron Inelastic Scattering, ref. 5.
15. A. Kudian, H. L. Welsh, and A. Watanabe, J. Chem. Phys. *43*, 3397 (1965).
16. P. Debye, Polare Molekeln, Hirzel, Leipzig 1929; C. P. Smyth, Dielectric Behaviour and Structure, McGraw-Hill, New York 1955; V. V. Daniel, Dielectric Relaxation, Academic Press, New York 1967.
17. K. S. Cole and R. H. Cole, J. Chem. Phys. *9*, 341 (1941).
18. C. P. Smyth, Ann. Rev. Phys. Chem. *17*, 433 (1966).
19. S. H. Glarum, J. Chem. Phys. *33*, 1371 (1960); R. H. Cole, J. Chem. Phys. *42*, 637 (1965).
20. A. Bellemans, M. Köhler, and M. Gancberg, J. Chem. Phys. *51*, 2578 (1969).
21. G. W. Chantry and H. A. Gebbie, Nature *208*, 378 (1965); G. W. Chantry, H. A. Gebbie, B. Lassier, and G. Wyllie, Nature *214*, 163 (1967); S. K. Garg, J. E. Bertie, H. Kilp, and C. P. Smyth, J. Chem. Phys. *49*, 2551 (1968).
22. N. Hill, Proc. Phys. Soc. (London) *82*, 723 (1963).
23. T. A. Litovitz, in ref. 1.
24. D. A. Pinnow, S. J. Candau, and T. A. Litovitz, J. Chem. Phys. *49*, 347 (1968).
25. M. D. Zeidler, in ref. 1.
26. H. Strehlow, Magnetische Kernresonanz und chemische Struktur, Steinkopff, Darmstadt 1968.

27. J. A. Pople, W. G. Schneider, and H. J. Bernstein, High Resolution Nuclear Magnetic Resonance, McGraw-Hill, New York 1959.
28. E. U. Franck, H. G. Hertz, and C. Rädle, Z. Phys. Chem. NF *73*, 18 (1970).
29. E. v. Goldammer and M. D. Zeidler, Ber. Bunsenges. *73*, 4 (1969).
30. A. A. Antony and C. P. Smyth, J. Am. Chem. Soc. *86*, 152 (1964).
31. S. Mallikarjun and N. E. Hill, Trans. Faraday Soc. *61*, 1389 (1965).
32. W. T. Huntress, J. Phys. Chem. *73*, 103 (1969).
33. D. L. Hogenboom, D. E. O'Reilly, and E. M. Peterson, J. Chem. Phys. *52*, 2793 (1970).
34. D. E. Woessner, B. S. Snowden, and E. T. Strom, Mol. Phys. *14*, 265 (1968).
35. K. Mansingh and A. Mansingh, J. Chem. Phys. *41*, 827 (1964).
36. T. T. Bopp, J. Chem. Phys. *47*, 3621 (1967).
37. E. J. Hennelly, W. M. Heston, Jr., and C. P. Smyth, J. Am. Chem. Soc. *70*, 4102 (1948).
38. D. E. O'Reilly and G. E. Schacher, J. Chem. Phys. *39*, 1768 (1963).
39. R. Figgins and M. Rhodes, Mol. Phys. *17*, 669 (1969).
40. S. K. Garg and C. P. Smyth, J. Chem. Phys. *42*, 1397 (1965).
41. K. F. Herzfeld and T. A. Litovitz, Absorption and Dispersion of Ultrasonic Waves, Academic Press, New York 1959.
42. cf. also R. A. Pethrick and E. Wyn Jones, Quart. Rev. *23*, 301 (1969); R. A. Pethrick, Sci. Progr. *58*, 563 (1970).
43. J. E. Piercy, J. Chem. Phys. *43*, 4066 (1965).
44. J. E. Piercy and M. G. S. Rao, J. Chem. Phys. *46*, 3951 (1967).
45. F. J. Bartoli and T. A. Litovitz, J. Chem. Phys. *56*, (1972), in press.
46. J. A. Bucaro and T. A. Litovitz, J. Chem. Phys. *55*, 3585 (1971).
47. H. Dardy (private communication).
48. R. Kosfeld, B. B. Gross, and W. Dietrich, Z. Naturforsch. *25 A*, 40 (1970).
49. D. A. Jackson and Simic-Glavaski, Mol. Phys. *18*, 393 (1970).
50. A. V. Rakov, Opt. i Spektr. *13*, 369 (1962).
51. For a recent review on the use of the dipole correlation function in dielectric relaxation see G. Williams, Chem. Rev. *72*, 55 (1972).

12. Associated Liquids

12.1. Introduction

Modern theories of fluids have led to the conclusion that the structure of simple liquids near their triple point is determined primarily by the repulsive forces between the molecules, whereas the attractive forces are responsible mainly for the configurational energy of the fluid. There are, however, cases in which the attractive energy between a pair of molecules at a certain orientation is much larger than for any other orientation of the two molecules. At low temperatures these favourable orientations may have a dominant influence on the structure of the assembly. Such liquids are usually termed associated liquids. The best-known examples are hydrogen bonded liquids[*]. It is instructive to compare the pair distribution functions $g(R)$ of liquid argon [1] and of water [2]. Near the triple point the average number of nearest neighbours of a molecule is about 10 for argon and about 4 for water. Furthermore, with increasing temperature (along the saturation curve) the first peak in $g(R)$ moves to shorter R for argon and to larger R for water. The behaviour of liquid argon has been explained by the fact that the repulsive part of the pair potential is not infinitely steep and hence the effective size of argon atoms decreases with increasing temperature. In water the nearest neighbour distance is determined by $O-H...O$ hydrogen bonds, which are weakened when the temperature rises [3].

An associated complex may be defined [4] as a group of molecules with a definite orientation to each other, for a period of time long compared with the period of molecular vibrations (which is in the order of 10^{-13} s). On such a time scale an associated complex behaves like a composite molecule with translational, rotational, and internal (vibrational) modes. Spectroscopic studies (Raman and IR) can give valuable information about such motions. The life time of an associated complex is determined by the frequency of reorientations of the molecules to each other. This can often be investigated by dielectric and ultrasonic relaxation measurements. In many liquids different associated complexes may exist simultaneously and it is difficult to correlate the experimental results with a certain association mechanism.

The above definition of an associated complex is particulary useful in the study of association in the gas phase or in dilute solutions. In pure dense liquids it may break down in two ways: Firstly, it is difficult to determine the exact structure of the associated complex. For example, in pure acetic acid there is evidence for a specific attractive interaction between more than two molecules, but the orientation of these further molecules relative to the acetic acid dimer is unknown. Secondly, in some hydrogen bonded liquids molecular reorientation is a very rapid process, comparable with the average period of molecular vibrations. This seems to be the case for water.

In section 12.2 we will present some aspects of hydrogen bonds. The thermodynamic stability of associates will be discussed in section 12.3. In section 12.4 we will review briefly some methods which are used to study association equilibria, and in section 12.5 we will bring some examples of associated liquids.

---

[*] Other examples are charge transfer complexes, and some molten salts.

12.2. Hydrogen Bonds

A hydrogen bond between two molecules can be represented as A−H...B, where H is a hydrogen atom bonded to an atom A, and B is an atom with a lone electron pair orbital[*]. A and B may be atoms of the same element as for example oxygen in water or alcohols. The character of the hydrogen bond varies depending on the two atoms A and B, but also on the entire molecules.

In the electrostatic model the two molecules are represented by multipoles with charge centers located near the surface of the molecules. Consider two dipoles, e.g.

each consisting of two electronic charges e^+ and e^- separated by the distance l (fig. 12.1). The potential energy of a pair of such dipoles, for different orientations (described by the angles ϑ_1, ϑ_2) in plane, is shown in fig. 12.2 for two ratios R_{12}/l. It is seen that a purely electrostatic inter-

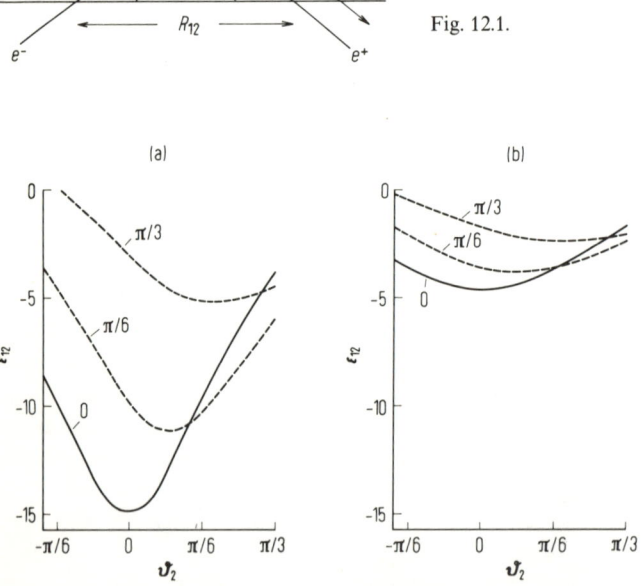

Fig. 12.2. Potential energy ε_{12} of a pair of non-ideal dipoles as a function of the angle ϑ_2 (see fig. 12.1), with $\vartheta_1 = 0, \pi/6, \pi/3$ as parameter;
a) $R_{12}/l = 1.5$; b) $R_{12}/l = 2$. The ordinate scale is in units of 10^5 J per mole of pairs, if each dipole consists of two unit charges separated by a distance $l = 1$ Å ($\mu = 4.8$ D).

---

[*] We will not consider hydrogen bonds in boron compounds.

action leads to a strong attractive energy for a relatively narrow range of orientations around $\vartheta_1 = \vartheta_2 = 0$ when R_{12} is not much greater than l. More sophisticated electrostatic models of H-bonding have been developed by considering a more realistic charge distribution in the molecules. In the *equivalent orbital representation* of a molecule the total electronic charge of each localized orbital (either bonding or a lone pair orbital) is assumed to be concentrated at the centroid of the orbital, such that the model gives the correct dipole moment of the molecule [5]. This model has been used successfully to calculate the H-bond energy in liquid water [6] and other substances.

The electrostatic model, while being useful for the estimation of thermodynamic properties, is not a realistic representation of hydrogen bonds. For a quantitative treatment of hydrogen bonds it is necessary to consider in addition to the electrostatic energy, the exchange energy, the induction and dispersion energy, and also the charge transfer from one molecule to the other. For relatively weak H-bonds the interaction energy can be calculated by a perturbation approach, based on an extension of London's theory of intermolecular forces to the region of non-zero intermolecular overlap. By such a treatment [7] it has been shown that the bond A−H need not be polar for the formation of a stable H-bond, but the relative stability of two H-bonds differing in A−H bond polarity may be explained by the difference in dipolar attraction. For strong H-bonds the charge transfer energy becomes very large. In such cases the H-bond is treated as a weak chemical bond within a composite molecule, using a variational approach. The H-bond energy is then obtained as the difference between the calculated energy of this complex and of its components [8].

It is generally accepted that the linear arrangement of the three atoms A−H...B is energetically most favourable. Perhaps the strongest evidence for this is the structure of ice where each molecule is surrounded by only four nearest neighbours to give four linear H-bonds. Further evidence for the stereospecific nature of H-bonds has been reviewed by Luck [9]. In the electrostatic dipole-dipole model (fig. 12.1 and 12.2) the linear arrangement (both angles ϑ_1 and ϑ_2 zero) has the lowest energy for any given value of R_{12}/l. The variation of either ϑ_1 or ϑ_2 (while the other is zero) gives a symmetric potential function. By a variation of both angles further (relative) energy minima are obtained. In this model the angle stability of the "bond" is enhanced by a decrease of the relative separation R_{12}/l. Ab initio self consistent field molecular orbital calculations of the H-bond energy have been made for three plausible structures of the H_2O dimer [8, 10]:

linear   cyclic   bifurcated

The linear form has a higher stabilization (H-bond) energy than either of the other forms, despite the fact that these have two hydrogen bonds. On the basis of these calculations it has been suggested that the energy minimum of the H-bond is determined by the balance of the H...O attraction and the O...O electrostatic repulsion. The colinearity of O−H...O would thus be preferred because for a given H...O distance the two oxygen atoms are farthest apart.

12.3. Thermodynamic Stability of Associates

The thermodynamic stability of an associate is determined by two factors: (1) the strength of the bond between the constituent molecules; (2) the change in entropy connected with the formation of the complex. Consider an equilibrium between monomers (M) and dimers (D)

$$2M \rightleftharpoons D. \qquad (12.1)$$

The equilibrium constant for this reaction, expressed in terms of the activities a_i,

$$K = a_D/(a_M)^2, \qquad (12.2)$$

is related to the change in standard Gibbs free energy of the reaction, $\Delta G = G_D - 2G_M$, by

$$\Delta G = -RT \ln K. \qquad (12.3)$$

Let us consider the conditions under which dimerization will occur in the gas and the liquid phase. The Gibbs free energy of either species $i(M,D)$ is

$$G_i = E_{0i} - kT \ln Z_i + PV_i \qquad (12.4)$$

where E_{0i} is the energy at zero temperature and Z_i is the partition function per mole as defined by eq. (10.1). The translational, rotational, and internal partition functions, given by eq. (1.34), (10.2)–(10.4), can be evaluated from molecular parameters. Consider the simple molecular models of a monomer and a dimer as shown in fig. 12.3. The molecular parameters used in the model calculation are summarized in table 12.1. By the formation of a dimer three translational and three rotational modes are replaced by six internal modes associated with the dimer bond,

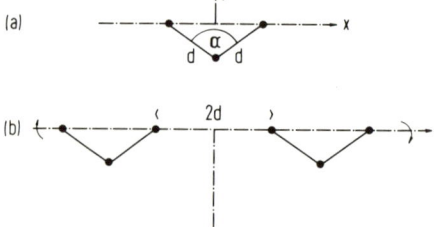

Fig. 12.3. Simplified model of a monomer and a dimer (which may apply to CH_3CH_2OH);
a) monomer consisting of three equal mass centers, $d = 1.54$ Å, $\alpha = 109°$;
b) linear dimer with a O–H...O distance $2d$. Cf. table 12.1.

Table 12.1. Molecular parameters used in the model calculation of dimerization equilibrium. The moments of inertia are calculated according to figure 12.3.

	Monomer	Dimer
Mass, g mol^{-1}	50	100
I_x, 10^{-40} g cm^2	23.8	47.6
I_y, 10^{-40} g cm^2	85.6	1441.7
I_z, 10^{-40} g cm^2	109.4	1489.3
I_{red}, 10^{-40} g cm^2	–	10.7
$\bar{\nu}_{stretching}$, cm^{-1}	–	200 (1)
$\bar{\nu}_{bending}$, cm^{-1}	–	100 (4)

namely, a stretching mode in the x direction; internal rotation (or torsional oscillation) about the x-axis; two vibrations in the xy plane and two vibrations in the xz plane [11]. In our model we assume that the internal rotation is non-hindered. The reduced moment of inertia for internal rotation, and the corresponding contribution to the free energy have been calculated using the standard formulae [12]. The internal partition function of the monomeric units is usually not significantly affected by the dimerization. The contributions to the function $\Delta(G - E_0)/RT$ as obtained from the parameters in table 12.1 are given in table 12.2.

Table 12.2. Contributions to the function $(E_0 - G)/RT$ at $T = 300$ K and $P = 1$ bar. For the gases the molar volume $V = RT/P$, for the liquids the translational volume is 6 cm^3 mol^{-1} for monomers and 12 cm^3 mol^{-1} for dimers.

	M	D	D-2M
Translation gas	16.5	17.5	−15.5
liquid	9.1	10.9	−7.3
Rotation	9.7	12.8	−6.6
Internal rotation	−	2.8	
Stretching vibration	−	0.5	+7.2
Bending vibrations	−	3.9	
$\Delta(E_0 - G)/RT$ gas			−14.9
liquid			−6.7

For reaction (12.1) in the gas phase, from table 12.2, $\Delta G - \Delta E_0 = 14.9\,RT$. ΔE_0 is the energy of formation of the dimer from the monomers at zero temperature. It is related to the enthalpy of formation, ΔH, at a given temperature T, by

$$\Delta H = \Delta E_0 + RT^2 \Delta(\partial \ln Z/\partial T)_P. \tag{12.5}$$

For translational and rotational modes, $T(\partial \ln Z/\partial T) = 1/2$ per degree of freedom (classical limit). For vibrational modes, $T(\partial \ln Z/\partial T)$ is between zero ($h\nu \gg kT$) and unity ($h\nu \ll kT$). Hence, for an association equilibrium $\Delta H - \Delta E_0$ has a value between $-3RT$ and $+3RT$ (in most cases between 0 and $+2RT$). In our model, for the vibrational modes $T(\partial \ln Z/\partial T) = 0.779$ (for $\nu = 100$ cm^{-1}) and 0.596 (for $\nu = 200$ cm^{-1}), respectively, at 300 K, so that $\Delta H - \Delta E_0 = 1.2\,RT$. Thus we find that $\Delta G = \Delta H + 13.7\,RT$. The condition for a reasonable concentration of dimers, from eq. (12.2) and (12.3), is $\Delta G < 0$ which requires that ΔH of the dimerization equilibrium is more negative than $-13.7\,RT$. Since the enthalpy of forming a hydrogen bond between two (non-ionic) molecules is usually about $-8\,RT$ it is obvious that there will be no appreciable association via hydrogen bonds in the gas phase. Carboxylic acids form a cyclic dimer by two strong hydrogen bonds and are known to be associated in the gas phase, in accordance with our simple estimate.

Estimation of the stability of the associate in the liquid state requires additional assumptions. In the translational partition function [eq. (1.34)] the volume is replaced by the "free volume" in the liquid which is taken to be one tenth of the total volume. The molar volume of the dimer V_D is taken as twice the molar volume of the monomer V_M (for the perfect gas at constant pressure $V_D = V_M$). Furthermore, it is assumed that $(Q_D/v_D^N) = (Q_M/v_M^N)^2$. With these assumptions the standard free energy of dimerization in the liquid, from table 12.2, is $\Delta G = \Delta E_0 + 6.7 RT = \Delta H + 5.5 RT$. Hence the condition for a reasonable degree of association in the liquid is that the enthalpy ΔH is about $-6RT$, which is often the case [13]. Thus, our model calculation offers an explanation why many hydrogen bonded associates are stable in the liquid phase but not in the gas phase. Note that the difference in stability is a consequence of the large difference (by a factor 10^3 to 10^4) in the translational free volume of the molecules in the two phases.

12.4. Study of Association in Liquids

It is much easier to determine the structure of an associated complex in the gas phase or in dilute solutions than in the pure liquids. Usually it is inferred that in the pure liquid there exist essentially the same types of associates (i.e., chain or/and cyclic multimers, etc.) as in dilute solutions. When the nature of these associates has been established from the work on dilute solutions, measurements on concentrated solutions and the pure associating liquid are used to evaluate the *concentration* of the associated species as a function of mole fraction. Dielectric and spectroscopic methods, n.m.r. chemical shifts, and various thermodynamic measurements have been used for this purpose. However, this approach has severe shortcomings which cast some doubt on much of the earlier work. It is known that association mechanisms depend on the dielectric permittivity and the local structure of the liquid. For example, polar associates are stabilized in polar solvents and non-polar (cyclic) associates are favoured in non-polar media. As a consequence, the mole fraction equilibrium constant of an association equilibrium can vary (by several powers of ten) from dilute solutions to the pure associating liquid. Accordingly, different association mechanisms can be dominant in different concentration regions. For a proper analysis of experimental results in terms of association equilibrium constants it would be necessary to know the activity coefficients of all associated species. Furthermore, the kinetics of association may change from dilute solutions to pure liquids, so that a certain complex may be well defined in a certain concentration region and not so in another.

In this section we will discuss dielectric and ultrasonic methods which are sensitive to association equilibria and association kinetics. Further methods will be mentioned in section 12.5. For some basic kinetic methods we refer to the literature [14].

12.4.1. Dielectric Methods

Dielectric methods are used in the study of association in two ways: by a measurement of the static dielectric permittivity and of the dielectric relaxation time [15].

The (relative) permittivity ε is a macroscopic quantity resulting from the time-averaged structure of the fluid. Fröhlich [16] has derived the following relationship:

$$\frac{(\varepsilon - 1)(2\varepsilon + 1)}{3\varepsilon} = \frac{4\pi N}{3v} \frac{\vec{\mu}\cdot\vec{\mu}^*}{kT}. \tag{12.6}$$

$\vec{\mu}$ is the molecular dipole moment under the polarizing influence of the surroundings, and $\vec{\mu}^*$ is the vector sum of the dipole moment of that molecule and the moments of the neighbouring molecules. Kirkwood, on the basis of a quasicrystalline model, defined a correlation parameter g as

$$g = \frac{\vec{\mu}\cdot\vec{\mu}^*}{\mu^2} = 1 + \sum_{i=1}^{\infty} z_i \langle \cos \gamma_i \rangle, \tag{12.7}$$

where z_i is the number of neighbours to the central molecule in the i-th coordination shell, and $\langle \cos \gamma_i \rangle$ is the average cosine of angle γ formed by the dipole moments of molecules in the i-th shell with the dipole of the central molecule. Both quantities, g and μ, depend on the structure of the liquid[*]. For "normal" liquids $g \simeq 1$ but for associated liquids g may be considerably larger than unity. However, g has rarely been calculated rigorously.

In the theory of Onsager correlations between the molecules are neglected, and the central molecule is treated as a point dipole at the center of a spherical cavity of radius a in a continuous dielectric medium. This leads to the expression

$$\frac{(\varepsilon - 1)(2\varepsilon + 1)}{3\varepsilon} = \frac{4\pi N}{v} \left\{ \frac{\alpha}{1 - f\alpha} + \frac{\mu^2}{3kT} \frac{1}{(1 - f\alpha)^2} \right\}, \tag{12.8}$$

where f is the reaction field factor [17],

$$f = (1/a^3)(2\varepsilon - 2)/(2\varepsilon + 1). \tag{12.9}$$

As eq. (12.8) neglects molecular correlations we may expect that it will not apply to associated liquids. In order to compare the permittivity of a real liquid with that resulting from eq. (12.8), the polarizability α may be calculated from the refractive index n_D using the Lorentz-Lorenz formula,

$$(4\pi/3) N \alpha = v(n_D^2 - 1)/(n_D^2 + 2), \tag{12.10}$$

and the radius of the cavity in eq. (12.9) is taken as the effective molecular radius by

$$4\sqrt{2} N a^3 = v, \tag{12.11}$$

assuming a f.c.c. packing of the liquid. Some results are shown in table 12.3. For "normal" liquids the calculated dielectric permittivity and its temperature dependence is in reasonable agreement with the experimental values. On the other hand, for associated liquids like carboxylic acids, alcohols, and water the calculated value is usually very different from the observed permittivity.

Under normal working conditions (field strengths up to 100 Vcm^{-1}) the polarization of a fluid is strictly proportional to the applied field E, in other words ε is independent of E. At very large field strengths ($E > 30000$ Vcm^{-1} or 100 esu cm^{-2}) deviations from this linear response occur as a consequence of a beginning saturation (complete orientation of dipoles). This dielectric saturation effect is usually expressed by $\Delta\varepsilon/E^2$, where $\Delta\varepsilon = \varepsilon - \varepsilon_0$ (ε_0 ... dielectric permittivity at low field strength). For molecules with nearly ideal dipoles a small negative dielectric saturation is observed ($\Delta\varepsilon/E^2 = -0.3 \cdot 10^{-8}$ esu^{-2} cm^4 for ethyl ether) (cf. Böttcher [17], Piekara [19]). Some polar liquids exhibit a large positive $\Delta\varepsilon$ (for nitrobenzene $\Delta\varepsilon/E^2 = 26 \cdot 10^{-8}$ esu^{-2} cm^4). This has been explained by the existence of molecular pairs with nearly anti-parallel orientation; this pairwise orientation is partly destroyed by a strong external electric field. If such liquids are dissolved in non-polar solvents a normal (i.e. negative) saturation prevails [18]. For alcohols a large negative dielectric saturation has been observed, but by diluting with non-polar solvents $\Delta\varepsilon$ becomes remarkably positive. This behaviour seems to be typical for some hydrogen bonded liquids [19]. The positive effect in solutions has been explained by a "field stimulated proton

---

[*] Note that for liquids with a high dielectric permittivity $\varepsilon \propto g\mu^2$.

Table 12.3. Observed and calculated dielectric permittivity ε of polar and associated liquids. μ is the molecular dipole moment in the gas phase.

liquid	μ(D)	t(°C)	ε		$-10^3 \dfrac{1}{\varepsilon} \dfrac{d\varepsilon}{dT}$		Literature		
			obs.	calc.	obs.	calc.	μ	ε	ρ, n_D
$(C_2H_5)_3N$	0.66	25	2.42	2.4			21	24	31
$(C_2H_5)_2O$	1.15	20	4.34	3.6	5.1	3.7	21	24	31
SO_2	1.61	−10	16.5	13.7			21	25	25
C_6H_5Cl	1.70	20	5.69	8.9	3.1	4.6	21	26	31
C_6H_5Br	1.70	25	5.40	9.1			21	24	31
$(CH_3)_2CO$	2.88	20	21.20	22.1	4.7	5.6	21	24	31
CH_3NO_2	3.44	20	38.6	44.6	4.5	5.4	21	27	31
$HCON(CH_3)_2$	3.80	20	38.0	42.2	4.2	5.0	22	22	22
HCOOH	1.51	25	56.5	12.4	5.5	4.9	21	28	32
CH_3COOH	1.91	25	6.35	12.9	−1.1	4.8	23	28	31
C_2H_5COOH	1.75	25	3.23	9.0	−1.4	4.7	21	28	31
C_6H_5OH	1.45	45	10.92	7.4	7.0	3.9	21	24	31
CH_3OH	1.70	20	33.62	13.5	6.2	5.1	21	29	31
H_2O	1.85	20	80.36	34.7	4.6	3.7	21	24	32
H_2O	1.85	100	55.3	25.5	4.7	3.9	21	30	32
H_2O	1.85	400*⁾	16.5	8.8			21	30	30

* at 1000 bar (corresponding to a density of 0.69 g cm⁻³).

shift": the hydrogen bond proton, under the influence of the strong external field, is shifted towards the "base" molecule:

$$\begin{array}{c} R \\ \diagdown \\ O-H \cdots O \\ \diagup \\ H \quad R \end{array} \longrightarrow \begin{array}{c} R \\ \diagdown \\ O^{\ominus} \cdots H-O^{\oplus} \\ \diagup \\ H \quad R \end{array}$$

This shift produces a considerable increase of the dipole moment in short chain associates (particularly in dimers), but not so much in longer linear multimers (which seem to exist in pure alcohols, see section 12.5.2) [19] *⁾.

Dielectric relaxation measurements in many associated liquids have shown the existence of two (or more) relaxation times of dipole reorientation which are understood as reorientation of "free" molecules and of molecules engaged in association complexes, respectively. An important quantity is the activation enthalpy $\Delta H^{\#}$ for reorientation. As $1/T_0$ is proportional to the rate of molecular reorientation [eq. (11.29)], the activation enthalpy can be obtained from the temperature dependence of $1/T_0$ by means of an Arrhenius-type expression:

$$1/T_0 = A \exp(-\Delta H^{\#}/RT). \tag{12.12}$$

Some examples of dielectric relaxation measurements are given in section 12.5.

---

* The formation of cyclic trimers in non-polar surroundings [20] seems to offer an alternative explanation.

12.4.2. Ultrasonic Absorption

An ultrasonic field sets up periodic variations of pressure and temperature, which affect the equilibrium state of a reaction (cf. section 11.5). Consider the dimerization equilibrium, eq. (12.1). Let n_M and n_D denote the number of moles of monomer and dimer, respectively, so that the total amount of substance is $n = n_M + 2n_D$. The relative concentration of monomers, expressed as mole fraction, is

$$x = \frac{n_M}{n_M + n_D} = \frac{2n_M}{n + n_M}, \tag{12.13}$$

and hence the dissociation constant of the dimer is

$$K_d = \frac{x^2}{1 - x} \psi, \tag{12.14}$$

where ψ contains the activity coefficients. For an alteration of the equilibrium we define the degree of advancement by $d\xi = \frac{1}{2} dn_M = -dn_D$. The overall reaction rate is

$$\frac{1}{n} \frac{d\xi}{dt} = k(1 - x) - k' x^2, \tag{12.15}$$

where k and k' are the rate constants of the dissociation and association reaction, respectively. If the dissociation of a dimer does not involve a volume change, the variation of pressure in the sonic wave influences the equilibrium only through the variation of temperature. The reaction heat capacity is related to the dissociation enthalpy ΔH_d and ξ by

$$C_r = \Delta H_d \left(\frac{1}{n} \frac{\partial \xi}{\partial T} \right)_P = R \left(\frac{\Delta H_d}{RT} \right)^2 \frac{x(1-x)}{(2-x)^3} \psi'. \tag{12.16}$$

In eq. (12.16), $(\partial \xi / \partial T)$ is obtained by differentiating eq. (12.14). The factor ψ' is a correction containing the activity coefficients. We now want to know the conditions under which the reaction causes a strong ultrasonic absorption. From eq. (11.47) and (11.49) the total dispersion and the maximum of the absorption per wavelength is determined by the parameter A [eq. (11.43)], which depends on the equilibrium properties ΔH_d and x. The low frequency absorption [eq. (11.46)] depends also on the relaxation time, which characterizes the kinetics near the equilibrium. Since C_r is usually only a small fraction of the total heat capacity of the fluid we may say, from eq. (11.43), that A is proportional to C_r. Furthermore, when the equilibrium concentration of monomers is small we may write $x \simeq \exp(-\Delta G_d/2RT) = \exp(-\Delta H_d/2cRT)$, when $c = \Delta H_d / \Delta G_d$. Hence,

$$A \propto C_r \propto \left(\frac{\Delta H_d}{RT} \right)^2 \exp(-\Delta H_d/2cRT). \tag{12.17}$$

From this expression, A approaches zero both for very small and very high values of $\Delta H_d/RT$ and it has a maximum when $\Delta H_d/RT = 4c$ or $\Delta G_d = 4RT$. Under the assumptions of the previous section, where we have considered association reactions, $\Delta G_d = \Delta H_d - 5.5 RT$. As ΔH_d for hydrogen bonds is often near $8RT$ we have $\Delta G_d = 2RT$ to $3RT$. Hence, we note that dissociation equilibria involving hydrogen bonds may have a high reaction heat capacity and a high value of A.

The relaxation time is related to the reaction rate by

$$1/\tau = k' x^2 \varphi, \tag{12.18}$$

where φ is a function of the concentration and the stoichiometric coefficients [33]. At equilibrium, from eq. (12.15), $k' x^2 = k(1 - x) \simeq k$ (for $x \ll 1$) so that the relaxation time is inversely proportional to the dissociation rate constant. This rate constant differs widely for different types of associates. Dissociation of a cyclic dimer like the acetic acid dimer requires the simultaneous breaking of two strong H-bonds and will thus have a small rate constant and long relaxation time. This is a plausible explanation for the large value of absorption observed in acetic acid. On the other hand, the low ultrasonic absorption in alcohols implies a large dissociation rate constant, which indicates a preference for linear association. Thus, in strongly associated liquids the measurement of ultrasonic absorption is selective for strong hydrogen bonds and low dissociation rate constants.

Experimentally, the parameter A is obtained by a measurement of the maximum of the absorption per wavelength. When A is known for two temperatures then the dissociation constant and ΔH_d can be evaluated. The function ψ is either neglected [34] or must be estimated independently.

12.5. Some Examples

In this final section we deal in more detail with some liquids which are generally considered as associated liquids. In particular we shall discuss the following topics: (1) the evidence for the existence (or non-existence) of certain associated species; (2) the concentration of associates (association equilibrium constants); (3) the kinetics of the association and dissociation process from which the mean life time of the various species can be deduced.

12.5.1. Carboxylic Acids

Formic acid and acetic acid in the gas phase form cyclic dimers held together by two strong hydrogen bonds [35]. The dimer in its lowest vibrational state has a plane structure and hence zero dipole moment. This is supported indirectly by microwave measurements on mixtures of carboxylic and fluoro-carboxylic acid vapours: microwave spectra of the (polar) heterodimers but not of the (non-polar) homodimers have been observed [36]. In the crystal, formic acid and acetic acid form zig-zag chains in which each molecule is bonded to two others by H-bonds (see figure 12.4) [37].

Fig. 12.4. Acetic acid molecules in the crystal, linked to infinite chains by hydrogen bonds − − − −. After [37].

There is only indirect evidence about the structure of carboxylic acids in the liquid state. In spectroscopic studies of association it has been difficult to assign the various bands to specific species (monomer, cyclic dimer, open dimer, chain polymer) of the acids. This difficulty can now be matched by investigating the low frequency infrared and Raman spectra. The acetic acid monomer has no fundamentals below 500 cm^{-1} (except for the internal rotation of the methyl group which can be identified by the frequency shift on deuteration). All Raman lines observed in this low frequency region can be attributed to out-of-plane vibrations of hydrogen bonded associates [38]. In hydrocarbon solutions of acetic acid only lines assignable to cyclic dimers were observed. In the pure liquid at room temperature, from the total intensity of the spectrum, the fraction of molecules present as cyclic dimers was obtained to be 85% or more (i.e., the mole fraction of monomers, defined by eq. (12.13), is 0.26 or less).

Ultrasonic absorption measurements on acetic and propionic acid have been analysed in terms of the reaction: 2 monomers $\rightleftharpoons$ cyclic dimer [34]. The dimerization constants and specific rate constants for dissociation are summarized in table 12.4. From the mole fraction dimerization constant in acetic acid the mole fraction of monomers can be calculated and is shown in table 12.5*[). Also given in this table is a rough estimate of the thermodynamic activity of monomers in the liquid. The estimated activity**[) is less than the mole fraction by a factor of about 10. This indicates that the mixture of monomers and dimers exhibits strong negative deviations from Raoult's law [23]. A strong attractive interaction between monomers and dimers could mean that higher associates (trimers etc.) exist in liquid acetic acid. It has also been suggested that a strong inductive attraction may exist between the highly polarizable hydrogen bonds of the cyclic dimer and the bond dipole moments of the monomers. This could lead to molecular aggregates similar to hydrogen bonded trimers but without a definite orientation of the third molecule with respect to the dimer [23].

The low frequency dielectric permittivity of acetic acid and propionic acid increases with increasing temperature [28]. Qualitatively, this can be explained by an increase of the concentra-

Table 12.4. Mole fraction dimerization constant K_x and rate constant for dimer dissociation k, from ultrasonic absorption (From Freedman [34]).

$t(°C)$	Acetic Acid		Propionic Acid	
	K_x	$10^{-4} k(s^{-1})$	K_x	$10^{-4} k(s^{-1})$
10			2720	3.7
20	62	0.110	1550	8.1
30	44	0.218	920	17
40	31	0.422	560	34
50	23	0.808	370	61
60	17	1.49		

---

* There has been some argument in the literature about Freedman's evaluation [39]. In our opinion the only shortcoming of Freedman's treatment is the neglect of activity coefficients for the monomer-dimer interaction. Accordingly, the monomer mole fractions in table 12.5 are somewhat too large [40].
** The activity of component i in a mixture is given by $a_i =$ (partial pressure of i in the mixture)/(saturation pressure of i). The partial pressure of acetic acid monomers has been calculated from the saturation pressure of acetic acid and the dimerization constant in the vapour, assuming ideal behaviour of the gas mixture. The saturation pressure of a hypothetical liquid composed of pure monomeric acetic acid has been identified with the saturation pressure of acetone (which has a CH_3 group instead of OH).

Table 12.5. Mole fraction x_M and estimated activity a_M of monomers in pure liquid acetic acid. K_x and K_P is the dimerization constant in the liquid and in the vapour, respectively, P is the vapour pressure, P_M the partial pressure of monomers. P^* is the vapour pressure of acetone (as a model liquid of pure monomeric acetic acid).

$t(°C)$	K_x	x_M	$K_P(\text{Torr}^{-1})$	$P(\text{Torr})$	$P_M(\text{Torr})$	$a_M = P_M/P^*$
20	62	0.12	3.16	11.6	1.8	0.010
30	44	0.14	1.29	20.5	3.6	0.013
40	31	0.16	0.562	34.9	7.1	0.017
50	23	0.19	0.263	57.2	13.0	0.021
60	17	0.21	0.126	90.5	23.1	

tion of polar monomers. However, to account quantitatively for the observed permittivity either the concentration of monomers must be higher than given in table 12.5 or the dipole moment of the monomer must be higher than 1.9 D. It is also possible that the cyclic dimer in the liquid has a non-zero effective dipole moment [41].

The permittivity of formic acid is much higher than that of acetic and the other carboxylic acids, and also much higher than to be expected on the basis of eq. (12.8) for a "normal" polar liquid of the same dipole moment, refractive index, and density as formic acid (see table 12.3). This indicates the existence of highly polar associates in this liquid. Unlike the higher carboxylic acids the permittivity of formic acid has a normal (negative) temperature coefficient. The dielectric absorption reveals two relaxation processes, I and II, with critical frequencies at 2.08 (I) and 40 (II) GHz. Acetic acid has three relaxation regions, at 1.19 (I), 6.5 (II), and 70 (III) GHz, whereas for the higher carboxylic acids only the relaxations II and III are found [28]. A relaxation process analogous to I is also observed for methyl and ethyl alcohol, and for ethyl glycol. This relaxation is probably determined by the life time of H-bonded chain associates. The relaxation mechanism of absorption region II and III is likely to be the reorientation of cyclic dimers (opening of one bond?) and of free monomers, respectively. It can be concluded that chain association is dominant in formic acid and cyclic dimerization in the higher carboxylic acids.

Note the small dissociation rate constants of acetic acid and propionic acid dimers (table 12.4) in comparison to the dielectric relaxation region II. E.g., for acetic acid we have $k = 2 \cdot 10^3 \text{ s}^{-1}$ (ultrasonic) and $v_c = 7 \cdot 10^9 \text{ s}^{-1}$ (dielectric), at room temperature. Probably the ultrasonic relaxation involves the breaking of both hydrogen bonds.

12.5.2. Alcohols

The (high temperature) crystal structure of methyl alcohol consists of chains of molecules, parallel to each other, in which each molecule is hydrogen bonded to two other molecules [42]:

From the X-ray diffraction patterns of liquid methyl alcohol Zachariasen in 1935 concluded that the time averaged structure of the liquid consists of similar chains of hydrogen bonded molecules [43].

Alcohols and phenol have a high dielectric permittivity at low temperatures and a large negative temperature coefficient $\left(\dfrac{1}{\varepsilon}\dfrac{d\varepsilon}{dT}\right)$ (see table 12.3). This supports the concept of highly polar chain associates, which break up when the temperature increases. In contrast, the permittivity of dilute solutions of alcohols in non-polar solvents has indicated the existence of associates of low resulting dipole moment [44]. The effective dipole moment μ of various dissolved normal alcohols, plotted vs. concentration, shows a minimum which has been ascribed to cyclic associates of low resultant moment. At higher concentrations μ increases steadily and at mole fraction 0.9 it reaches the value of the pure alcohols. At a mole fraction near 0.3, μ is independent of temperature. At lower concentrations μ increases with temperature (probably caused by the breaking up of cyclic associates) whereas at higher concentrations μ decreases with increasing temperature (probably caused by the breaking up of linear associates) [44]. There are also other hints of the existence of cyclic trimers in non-polar surroundings [20].

To identify the various associated species infrared spectroscopic studies have been made. In most cases the O—H stretching mode of the alcohols in solution has been studied. By a matrix isolation technique (rapid cooling of a mixture of nitrogen and methanol vapour to 20 K), with varying concentrations of methanol, a number of narrow bands between 3660 and 3250 cm^{-1} have been assigned to methanol monomers, dimers, trimers, tetramers, and higher multimers. It was concluded that the dimers and trimers had a cyclic structure and the higher multimers an open structure [45]. Infrared measurements on simple alcohols in liquid solutions (usually carbon tetrachloride as the solvent) have revealed three bands and it is generally agreed that these bands correspond, respectively, to monomers, dimers, and various more highly associated species [46]. An attempt to distinguish between cyclic and open dimers has been made by a combination of infrared and cryoscopic measurements [47]. The multimer band is farthest shifted from the monomer band, indicating a higher energy of the hydrogen bonds in these species than in dimers. It probably arises from hydrogen bonds either in long chain multimers such as proposed by Zachariasen, or from cyclic multimers, or from a mixture of all these forms. With increasing concentration of the alcohol the polymer band increases both in intensity and in half-width: for example, for n-heptanol in carbon tetrachloride at a mole fraction of 0.5 a broad and intense multimer band is the most important feature, but the monomer and dimer bands still appear as separate sharp peaks [46]*). Measurements of the dielectric saturation effect of n-alcohols in various solvents and in the pure state have also been explained by an increasing degree of chain association with increasing concentration of the solutions**).

The dynamical nature of the association equilibria in liquid alcohols has been studied by relaxation measurements. Generally, alcohols have a low ultrasonic absorption coefficient, indicating the absence of relatively slow relaxation processes. The absorption coefficient decreases with

---

* For certain "globular" alcohol molecules such as pentamethylethanol no multimer band is observed. This can be understood from steric considerations since linear hydrogen bonded chains of these molecules cannot be formed [46].
** For pure hexanol Malecki [19] assumed the existence of only trimers, tetramers, and pentamers of a specified chain structure, and obtained the following species mole fractions: trimers 0.15; tetramers 0.28; pentamers 0.57.

increasing temperature, in contrast to non-associated liquids [48]. For alcohols dissolved in non-polar solvents the absorption coefficient vs. concentration displays a maximum at low alcohol concentration. It is likely that this absorption is caused by cyclic multimers for which longer relaxation times than for linear multimers can be expected. Similar results are also found for phenol [49].

Dielectric absorption measurements on alcohols reveal a strong absorption with relaxation times in the order $\tau_1 = 10^{-9}$ to 10^{-10} s, and two further absorption regions at higher frequencies, with relaxation times τ_2 near $3 \cdot 10^{-11}$ s and $\tau_3 = 2 \cdot 10^{-12}$ s, respectively (at room temperature). It is believed that τ_1 is related to the breaking of hydrogen bonds in the various multimers, that τ_2 results from a reorientation of single molecules, and τ_3 is interpreted as intramolecular OH rotation [50]. A characteristic feature of the relaxation process is that τ_1 (like τ_2 and τ_3) is single valued (very low distribution factor) for almost all alcohols investigated. This leads to the conclusion that τ_1 cannot result from a reorientation of the entire associated complexes (since the various multimers should have different relaxation times). The activation enthalpy of this relaxation ranges from 5–8 kcal/mol for the normal alcohols. From these values it is assumed that the breaking of a hydrogen bond in an associated complex is the initial step of this relaxation process. However, for the normal alcohols τ_1 increases regularly with increasing number of carbon atoms per molecule. Hence a molecular reorientation must be involved in this relaxation process to account for the increase of τ_1 with molecular size. The relaxation time τ_1 is therefore attributed to the breaking of hydrogen bonds in molecular associates followed by a rotation of a single molecule [50]. A somewhat different relaxation mechanism has also been proposed [51]. The activation enthalpy of the second relaxation process increases with molecular size and for the normal alcohols ranges from 1.5 to 3.0 kcal/mol. This relaxation is interpreted either as resulting from bonded molecules near the ends of chain associates [52] or from the reorientation of free alcohol molecules. The latter argument is based partly on a comparison of τ_2 for alcohols with the relaxation time of the corresponding (non-associated) alkyl bromide molecules [50].

12.5.3. Water [53]

In ice each H_2O molecule is bonded to four neighbouring molecules by four strong hydrogen bonds. In liquid water, from the pair distribution function obtained by X-ray diffraction measurements, the average number of nearest neighbours to any given molecule is about 4.4 over a wide temperature range [2]. It is likely, therefore, that liquid water also comprises a three dimensional network of hydrogen bonded molecules. However, since each molecule can form only four hydrogen bonds, either the network of H-bonds is considerably distorted (such as to bring some molecules which are not directly linked by a hydrogen bond into the range of nearest neighbour distances) or some of the molecules are not incorporated into the network (non-bonded molecules in interstitial cavities). Both these concepts have been proposed to explain the time averaged structure of water. At 4 °C the nearest neighbour oxygen-oxygen distance is distributed around a value of 2.82 Å and increases with increasing temperature. The nearest neighbour oxygen-oxygen distance in ice at 0 °C is 2.76 Å. Since the oxygen-oxygen distance is a measure of the strength of the hydrogen bond it follows that the average strength of the hydrogen bonds in water is somewhat less than in ice. The same conclusion follows from the shift in the O–H stretching frequency in ice and liquid water (see below). Other properties characteristic of the time averaged structure of water are the static dielectric permittivity, the density, the heat capacity and other thermodynamic properties. The large dielectric permittivity of water [54] is evidence for a strongly correlated mutual orientation of the molecules [6], which persists even at supercritical temperatures [30]. The temperature dependence of the molar volume of water (with a density maximum at 4 °C) is usually explained by two competing effects: (a) an increase in the average number of neighbours to a molecule due to the breakdown of the voluminous ice I structure; (b) an increase of the O—H...O distance due to distortion of hydrogen bonds and increasing amplitude of inharmonic intermolecular vibrations. The configurational heat capacity of water is very high: $C'_v = C_v - C_v$ (id. gas) $\simeq 6R$ at 20 °C. The configurational C'_v arising from the six intermolecular modes (three translational and three librational) must be below $3R$, as is the case in ice [53]. Hence $3R$ or more arises from hydrogen bond breaking or distortion with increase of temperature. A different approach to the explanation of a large configurational heat capacity has been given in section 10.3 on the basis of a large non-central contribution to the pair potential.

Several authors have argued that water consists of a mixture of distinctly different structural units; e.g., monomers, dimers, tetramers[*] and octomers [55]; or monomers and groups of 20 hydrogen bonded molecules forming open pentagonal dodecahedra like gas hydrates [56]. If such different units do exist their bonds should have distinctly different molecular environments and hence different energies, and this can be investigated by spectroscopic methods. The O—D stretching vibration of dilute solutions of HDO in H_2O is particulary well-suited for such an investigation, one reason being that it is only weakly coupled to the vibrations of the neighbouring molecules. The IR-absorption of the O—D stretching shows a simple band shape with one single maximum [58]. It has been emphasized that such a distribution is inconsistent with a mixture model of a small number of distinctly different structural units[**]. However, the Raman band of the O—D stretching has a small shoulder on the high frequency side indicating the existence of two or more forms of water molecules of not very different energy. This view is supported by the temperature dependence of the Raman O—D stretching band [60], and by its variation with the density in the supercritical region [61]. These observations are also consistent with earlier conclusions drawn from overtone vibrations observed in the near IR [68]. Hence, the spectroscopic investigations mentioned above suggest that water consists of a mixture of two (or more) forms of not very different energy. On the other hand, the continuum model [62] cannot be ruled out definitely. Similar conclusions are reached from the study of the intermolecular modes in the $50-1000$ cm^{-1} range [53].

The dielectric absorption of water can be described by a single relaxation time $\tau = 1 \cdot 10^{-11}$ s (at 20 °C). The activation enthalpy $\Delta H^\ddagger$ calculated from the temperature dependence of τ is 4.5 kcal/mol at 5 °C and 2.8 kcal/mol at 67.5 °C [63]. Although the mechanism of the relaxation process is not known it is very unlikely that a reorientation of associated complexes is involved since this should lead to a rather wide distribution of relaxation times (see the similar argument for τ_1 of alcohols). The single valued relaxation time has been explained by a cooperative model of hydrogen bond breaking involving a relatively large number of molecules ("flickering clusters") [64], but this model is difficult to reconcile with the low value of $\Delta H^\ddagger$ [65]. Furthermore, water near room temperature has a low small angle scattering intensity. This is taken as an indication

---

[*] Water tetramers seem to play some role under various other conditions [57].
[**] An example for the sensitivity of the O-D streching frequency to different local environments presents ice II in which there exist four different nearest neighbour separations, each of which gives a separate peak in the uncoupled O-D stretching band [59].

that large fluctuations of the local density as necessitated, for example, by the assumption of ice-like clusters [66] do not exist in liquid water [67].

A new and promising approach to the understanding of the structure of water are computer experiments on samples of water-like molecules, using the Monte Carlo method [69] and the molecular dynamics method [70]. An orientation dependent effective pair potential function has been used, consisting of the electrostatic interaction of four point charges (two positive and two negative) per molecule, plus a Lennard-Jones 12/6 function [71]. This interaction model neglects quantum mechanical refinements and, in particular, the non-additivity of H-bond energies [8, 10, 72]. Rahman and Stillinger [70] present a detailed analysis of their molecular dynamics calculations, discussing both the static (time averaged) structure and kinetic properties of their model. An important result of this calculation is that the distribution function of the effective pair interaction energy has a wide flat region in the range from -4.5 to -2.0 kcal mol^{-1}, revealing a wide class of moderately strong bonds. This fact makes it difficult to distinguish between bonded and non-bonded molecules. The dynamical behaviour of the model liquid water is characterized by an interconversion from one highly strained H-bond configuration to another. This interconversion is a continual rather than a discrete process. There is no evidence for a hopping process between alternative stable positions. Instead, diffusion proceeds via a continual reconstruction of the labile random H-bonded network [70]. This feature of the model is consistent with the absence of large angle jumps in reorientation processes, as revealed by a comparison of dielectric and n.m.r. relaxation times [73]. The ratio of these relaxation times is about 3, which corresponds to a rotational diffusion model (cf. end of section 11.4).

We have considered here the structure of liquid water at low temperatures and pressures. It is interesting that water at supercritical temperatures under high pressure (at densities $\rho_m > 1$ g cm^{-3}) is strongly dissociated into ions and has several properties similar to ionic melts [74].

12.6. References

1. C. J. Pings, Disc. Faraday Soc. *43*, 89 (1967).
2. A. H. Narten, M. D. Danford, and H. A. Levy, Disc. Faraday Soc. *43*, 97 (1967).
3. J. S. Rowlinson, Disc. Faraday Soc. *43*, 243 (1967).
4. A. Eucken, Z. Elektrochem. *52*, 255 (1948); F. Kohler, Monatsh. Chem. *100*, 1151 (1969).
5. J. A. Pople, Proc. Roy. Soc. (London) *A 202*, 323 (1950); A. B. F. Duncan and J. A. Pople, Trans. Faraday Soc. *49*, 217 (1953); W. G. Schneider, J. Chem. Phys. *23*, 26 (1955).
6. J. Lennard-Jones and J. A. Pople, Proc. Roy. Soc. (London) *A 205*, 155 (1951); J. A. Pople, ibid. 163.
7. F. B. van Duijneveldt and J. N. Murrell, J. Chem. Phys. *46*, 1759 (1967).
8. K. Morokuma and L. Pedersen, J. Chem. Phys. *48*, 3275 (1968), and literature given in this paper; K. Morokuma and J. Winick, ibid. *52*, 1301 (1970); P. Schuster, Intern. J. Quantum Chem. *3*, 851 (1969), Theoret. chim. Acta (Berl.) *19*, 212 (1970).
9. W. A. P. Luck, Naturwissenschaften *54*, 601 (1967).
10. P. A. Kollman and L. C. Allen, J. Chem. Phys. *51*, 3286 (1969); J. Am. Chem. Soc. *92*, 753 (1970).
11. See the discussion on the acetic acid dimer by G. Statz and E. Lippert, Ber. Bunsenges. *71*, 673 (1967).
12. I. N. Godnew, Berechnung thermodynamischer Funktionen aus Moleküldaten, DVW, Berlin 1963, chapter 10.
13. G. C. Pimentel and A. L. McClellan, The Hydrogen Bond, Freeman, San Francisco 1960, chapter 7.
14. E. Grunwald, R. L. Lipnick, and E. K. Ralph, J. Am. Chem. Soc. *91*, 4333 (1969) (amin-water complexes by n.m.r. relaxation); M. Eigen and L. de Maeyer in W. J. Hamer (Ed.), Structure of Electrolyte Solutions, Wiley 1959 (neutralization reaction); cf. also M. Eigen in A. Weissberger (Ed.), Technique of Organic Chemistry, Vol. VIII/2, Interscience, New York 1961.

15. N. Hill, W. E. Vaughan, A. H. Price, and M. Davies, Dielectric Properties and Molecular Behaviour, van Nostrand-Reinhold, London 1969; M. Davies, J. Chem. Education 46, 19 (1969).
16. H. Fröhlich, Trans. Faraday Soc. 44, 238 (1948).
17. C. J. F. Böttcher, Theory of Electric Polarisation, Elsevier, London 1952; for an improved version see T. G. Scholte, Thesis, University of Leiden 1950, and Physica 15, 437, 450 (1949).
18. A. Piekara and A. Chelkowski, J. Chem. Phys. 25, 794 (1956).
19. A. Piekara, J. Chem. Phys. 36, 2145 (1962); J. Malecki, J. Chem. Phys. 36, 2144 (1962); 43, 1351 (1965); J. Malecki and Z. Dopierala, Acta Phys. Polon. 34, 385, 401, 409 (1969).
20. J. R. Johnson, S. D. Christian, and H. E. Affsprung, J. Chem. Soc. (London) A 1965, 1; M. Saunders and J. B. Hyne, J. Chem. Phys. 29, 1319 (1958); E. E. Tucker, S. B. Farnham, and S. D. Christian, J. Phys. Chem. 73, 3820 (1969); but see A. N. Fletcher, to be published.
21. Electric Dipole Moments of Gases, U.S. Natl. Bur. Stand., Circular 537.
22. R. M. Meighan and R. H. Cole, J. Phys. Chem. 68, 503, 509 (1964); C. M. Lee and W. D. Kumler, J. Am. Chem. Soc. 84, 571 (1962); S. B. Brummer, J. Chem. Phys. 42, 1636 (1965); K. Quitzsch, Z. Phys. Chem. 233, 321 (1966); G. R. Leader and J. F. Gormley, J. Am. Chem. Soc. 73, 5731 (1951).
23. H. E. Affsprung, G. H. Findenegg, and F. Kohler, J. Chem. Soc. (London) A 1968, 1364.
24. Table of Dielectric Constants of Pure Liquids, U.S. Natl. Bur. Stand., Circular 514.
25. A. L. Vierk, Z. anorg. Chem. 261, 279 (1950); Gmelin, System Nr. 9, Teil B Lieferung 1, 8. Auflage (1953), 212; A. W. Francis, J. Chem. Eng. Data 5, 534 (1960).
26. R. Mecke and K. Rosswog, Z. Elektrochem. 60, 47 (1956).
27. J. Timmermans, A. M. Piette, and R. Philippe, Bull. Soc. Chim. Belg. 64, 5 (1955); R. Philippe and A. M. Piette, ibid. 600.
28. E. Constant and A. Lebrun, J. Chim. Phys. 61, 163 (1964).
29. D. W. Davidson, Can. J. Chem. 35, 458 (1957).
30. K. Heger, Dissertation, Technische Universität Karlsruhe, 1969 (ε of water and methanol up to 550°C, 5 kbar); S. Maier and E. U. Franck, Ber. Bunsenges. 70, 639 (1966); H. Köster, Dissertation, Technische Universität Karlsruhe, 1969 (ρ of water from $25-600$°C, $6-10$ kbar).
31. J. Timmermans, Physico-Chemical Constants of Pure Organic Compounds, Vol. 1 (1950), Vol. 2 (1965), Elsevier, Amsterdam.
32. Landolt-Börnstein, Zahlenwerte und Funktionen, II. Band, 8. Teil, Springer, Berlin 1962.
33. K. F. Herzfeld and T. A. Litovitz, Absorption and Dispersion of Ultrasonic Waves, Academic Press, New York 1959.
34. E. Freedman, J. Chem. Phys. 21, 1784 (1953).
35. J. Karle and L. O. Brockway, J. Am. Chem. Soc. 66, 574 (1944).
36. C. C. Costain and G. P. Srivastava, J. Chem. Phys. 41, 1620 (1964).
37. R. E. Jones and D. H. Templeton, Acta Cryst. 11, 484 (1958); F. Holtzberg, B. Post, and I. Fankuchen, ibid. 6, 127 (1953).
38. P. Waldstein and L. A. Blatz, J. Phys. Chem. 71, 2271 (1967).
39. C. Moriamez, M. Moriamez, and A. Moreaux, J. Chim. Phys. 63, 615 (1966); J. E. Piercy and L. Lamb, Trans. Faraday Soc. 52, 930 (1956); H. Posch and F. Kohler, Monatsh. Chem. 98, 1451 (1967).
40. G. Becker and F. Kohler, Monatsh. Chem. 103, (1972), in press.
41. J. Liszi, Acta Chim. Acad. Sci. Hung. 62 (3), 263 (1969) and 63 (3), 293 (1970).
42. R. W. G. Wyckoff, Crystal Structures, 2nd ed. (1966), Vol. 5, Interscience, New York.
43. W. H. Zachariasen, J. Chem. Phys. 3, 158 (1935).
44. P. Huyskens and F. Cracco, Bull. Soc. Chim. Belg. 69, 422 (1960); P. Huyskens, G. Gillerot, and T. Zeegers-Huyskens, Bull. Soc. Chim. Belg. 72, 666 (1963); P. Huyskens, R. Henry, and G. Gillerot, Bull. Soc. Chim. France 1962, 720; D. A. Ibbitson and L. F. Moore, Chem. Commun. 1965, 339; H. E. Affsprung, J. K. Morrow, R. Schano, and F. Kohler, to be published.
45. M. van Thiel, E. D. Becker, and G. C. Pimentel, J. Chem. Phys. 27, 95 (1957).
46. M. P. McDonald, Progress in Chemistry Series 63, 125 (1967), and literature given in that paper.
47. J. dos Santos, P. Pineau, and M. L. Josien, J. Chim. Phys. 1965, 628.
48. Ref. 33, table 73-1.
49. W. Maier and A. Mez, Z. Naturforsch. 7a, 300 (1952) and 10a, 167 (1955); A. B. Bhatia, Ultrasonic Absorption, Clarendon Press, Oxford 1967.
50. S. K. Garg and C. P. Smyth, J. Phys. Chem. 69, 1294 (1965).
51. P. Bordewijk, F. Gransch, and C. J. F. Böttcher, J. Phys. Chem. 73, 3255 (1969).

52. M. Magat in D. Hadzi (Ed.), Hydrogen Bonding, Pergamon Press, London 1957.
53. D. Eisenberg and W. Kauzmann, The Structure and Properties of Water, Clarendon Press, Oxford 1969; J. L. Kavanau, Water and Solute-Water Interactions, Holden-Day, San Francisco 1964; O. Samoilow, Die Struktur wässriger Elektrolytlösungen und die Hydratation von Ionen, Teubner, Leipzig 1961.
54. C. G. Malmberg and A. A. Maryott, J. Res. Natl. Bur. Stand. 56, 1 (1956).
55. A. Eucken, Nachr. Akad. Wiss. Göttingen 1946, 38.
56. L. Pauling in D. Hadzi, ref. 52; H. S. Frank and A. S. Quist, J. Chem. Phys. 34, 604 (1961).
57. M. Eigen and L. de Maeyer, ref. 14; F. Kohler, H. Arnold, and R. J. Munn, Monatsh. Chem. 92, 876 (1961).
58. M. Falk and T. A. Ford, Can. J. Chem. 44, 1699 (1966); E. U. Franck and K. Roth, Disc. Faraday Soc. 43, 108 (1967).
59. Eisenberg and Kauzmann, ref. 53, p. 132.
60. G. E. Walrafen, J. Chem. Phys. 47, 114 (1967) and 48, 244 (1968).
61. H. A. Lindner, Dissertation, Technische Universität Karlsruhe 1971.
62. T. T. Wall and D. F. Hornig, J. Chem. Phys. 43, 2079 (1965).
63. C. H. Collie, J. B. Hasted, and D. M. Riston, Proc. Phys. Soc. (London) 60, 145 (1948).
64. H. S. Frank and W. Y. Wen, Disc. Faraday Soc. 24, 133 (1957).
65. Eisenberg and Kauzmann, ref. 53, p. 208.
66. J. D. Bernal and R. H. Fowler, J. Chem. Phys. 1, 515 (1933); C. M. Davis and T. A. Litovitz, J. Chem. Phys. 42, 2563 (1965); B. Kamb in A. R. Rich and N. Davidson (Eds.), Structural Chemistry and Molecular Biology, Freeman, San Francisco 1968.
67. See the discussion in A. H. Narten and H. A. Levy, Science 165, 447 (1969).
68. W. A. P. Luck, Ber. Bunsenges. 69, 626 (1965).
69. J. A. Barker and R. O. Watts, Chem. Phys. Letters 3, 144 (1969).
70. A. Rahman and F. H. Stillinger, J. Chem. Phys. 55, 3336 (1971).
71. J. S. Rowlinson, Trans. Faraday Soc. 47, 120 (1951); A. Ben-Naim and F. H. Stillinger, in R. A. Horne (Ed.), Structure and Transport Processes in Water and Aqueous Solutions, Interscience, New York, to be published.
72. J. Del Bene and J. A. Pople, J. Chem. Phys. 52, 4858 (1970); D. Hankins, J. W. Moskowitz, and F. H. Stillinger, ibid. 53, 4544 (1970); A. P. Minton, Trans. Faraday Soc. 67, 1226 (1971).
73. Eisenberg and Kauzmann, ref. 53, tables 4.5 and 4.6.
74. E. U. Franck, Discussion Meeting of Deutsche Bunsengesellschaft at Lindau, Sept. 1971, to be published in Ber. Bunsenges.

Appendix

A. Fourier-Transformation

A function $f(x)$ may be considered as a superposition of sine- and cosine-functions and hence be written as

$$f(x) = \frac{1}{2\pi} \int_{-\infty}^{+\infty} \tilde{f}(Q) e^{iQx} dQ, \tag{A1}$$

where the spectral function $\tilde{f}(Q)$ determines the contribution of e^{iQx}. Fourier's theorem states that $\tilde{f}(Q)$ is then given by

$$\tilde{f}(Q) = \int_{-\infty}^{+\infty} f(x) e^{-iQx} dx. \tag{A2}$$

Usually $\tilde{f}(Q)$ is called the Fourier-transform of $f(x)$. This is a reciprocal property, because eq. (A1) shows that $f(x)$ is the Fourier-transform of $\frac{1}{2\pi} \tilde{f}(-Q)$. The conditions which must be imposed on an ordinary function $f(x)$ in order that Fourier's theorem holds are not important for our purpose, as in physical applications frequently an extension to generalized functions is necessary. This will be discussed below.

The connection between Fourier-transformation and folding was used sometimes in this book. The folding of two functions $g(x)$ and $h(x)$ is defined by

$$F(x) = \int_{-\infty}^{+\infty} g(x-t) h(t) dt. \tag{A3}$$

It is obvious that the above relation can also be written as

$$F(x) = \int_{-\infty}^{+\infty} h(x-t) g(t) dt. \tag{A3'}$$

We shall prove now that the Fourier-transform of a folding is given by

$$\tilde{F}(Q) = \tilde{g}(Q) \tilde{h}(Q). \tag{A4}$$

By a straightforward calculation one has

$$\tilde{F}(Q) = \int_{-\infty}^{+\infty} F(x) e^{-iQx} dx = \int_{-\infty}^{+\infty} dx\, e^{-iQx} \int_{-\infty}^{+\infty} dt\, g(x-t) h(t)$$

$$= \int_{-\infty}^{+\infty} dt\, h(t) \int_{-\infty}^{+\infty} dx\, g(x-t) e^{-iQx}$$

$$= \int_{-\infty}^{+\infty} dt\, h(t) e^{-iQt} \int_{-\infty}^{+\infty} g(x-t) e^{-iQ(x-t)} d(x-t) = \tilde{g}(Q) \tilde{h}(Q).$$

By going back these formulae the inverse theorem can be seen also: if eq. (A4) holds, eq. (A3) follows.

The Fourier-transformation can be generalized to functions of three independent variables (F.T. in space) and to four independent variables (F.T. in space and time). In the first case we have

$$f(x_1, x_2, x_3) = \frac{1}{(2\pi)^3} \int_{-\infty}^{+\infty} \int_{-\infty}^{+\infty} \int_{-\infty}^{+\infty} \tilde{f}(Q_1, Q_2, Q_3) e^{iQ_1 x_1} e^{iQ_2 x_2} e^{iQ_3 x_3} dQ_1 dQ_2 dQ_3, \tag{A5}$$

with
$$\tilde{f}(Q_1,Q_2,Q_3) = \int_{-\infty}^{+\infty}\int_{-\infty}^{+\infty}\int_{-\infty}^{+\infty} f(x_1,x_2,x_3) e^{-iQ_1x_1} e^{-iQ_2x_2} e^{-iQ_3x_3} dx_1 dx_2 dx_3. \tag{A6}$$

By using a vector notation $\vec{r} = \{x_1,x_2,x_3\}$, $\vec{Q} = \{Q_1,Q_2,Q_3\}$, $d\vec{r} = dx_1 dx_2 dx_3$ and $d\vec{Q} = dQ_1 dQ_2 dQ_3$, these equations are written more elegantly as

$$f(\vec{r}) = \frac{1}{(2\pi)^3} \int \tilde{f}(\vec{Q}) e^{i\vec{Q}\cdot\vec{r}} d\vec{Q} \tag{A5'}$$

and

$$\tilde{f}(\vec{Q}) = \int f(\vec{r}) e^{-i\vec{Q}\cdot\vec{r}} d\vec{r}. \tag{A6'}$$

The folding theorem is again valid: if

$$F(\vec{r}) = \int g(\vec{r} - \vec{s}) h(\vec{s}) d\vec{s} \tag{A7}$$

holds, then also

$$\tilde{F}(\vec{Q}) = \tilde{g}(\vec{Q}) \tilde{h}(\vec{Q}), \tag{A8}$$

and conversely.

If a function $f(\vec{r},t)$ is space and time dependent, it can be represented quite analogously by a superposition of plane waves

$$f(\vec{r},t) = \frac{1}{(2\pi)^3} \int \tilde{f}(\vec{Q},\omega) e^{i(\vec{Q}\cdot\vec{r} - \omega t)} d\vec{Q} d\omega \tag{A9}$$

with

$$\tilde{f}(\vec{Q},\omega) = \frac{1}{2\pi} \int f(\vec{r},t) e^{-i(\vec{Q}\cdot\vec{r} - \omega t)} d\vec{r} dt. \tag{A10}$$

(For the factor 2π compare the convention adopted by eq. (A1) and (A2).) It should be noted that the sign of the phase is not always taken in the same way as above. So, e.g., it has become customary in the study of non-equilibrium properties of liquids to define $S(\vec{Q},\omega)$ as in eq. (9.53). But as all functions of interest are real, this has no practical consequence.

If the Fourier-transformation is restricted to ordinary functions, it has the great disadvantage that for some simple functions no Fourier-transform exists. E.g., the Fourier-transform of the function $f(x) = 1$ would be given by

$$\tilde{f}(Q) = \int_{-\infty}^{+\infty} e^{-iQx} dx,$$

an integral, which does not converge. This difficulty is overcome by introducing generalized functions [1, 2], which will be explained now in a rather plausible way without any mathematical rigour.

Out of the ordinary functions, a certain group is called good functions for which Fourier's theorem is certainly valid. The good functions are everywhere differentiable any number of times and the functions and all their derivatives vanish with $O(|x|^{-N})$ for $|x| \to \infty$ regardless how big the value of N. An example of a good function is $f(x) = \exp\{-x^2\}$.

In order to define now the generalized functions, let us assume that for a sequence of good functions $f_n(x)$ the limit

$$\lim_{n\to\infty} \int_{-\infty}^{+\infty} f_n(x) F(x) dx$$

exists for any good function $F(x)$. Then this sequence $f_n(x)$ is called the generalized function $f(x)$, and its integral is understood as

$$\int_{-\infty}^{+\infty} f(x)F(x)dx = \lim_{n\to\infty} \int_{-\infty}^{+\infty} f_n(x)F(x)dx. \tag{A11}$$

In a very sloppy way, the generalized function itself can be regarded as the limit of the sequence of good functions:

$$f(x) = \lim_{n\to\infty} f_n(x). \tag{A12}$$

For all the good functions $f_n(x)$ the Fourier-transforms $\tilde{f}_n(Q)$ exist. If the generalized function

$$\tilde{f}(Q) = \lim_{n\to\infty} \tilde{f}_n(Q) \tag{A13}$$

also exists, it is called the Fourier transform of the generalized function $f(x)$. In this case Fourier's theorem always holds.

An important example of a generalized function is Dirac's δ-function,

$$\delta(x) = \lim_{n\to\infty} \sqrt{\frac{n}{\pi}} e^{-nx^2}, \tag{A14}$$

as one can show that

$$\lim_{n\to\infty} \int_{-\infty}^{+\infty} \sqrt{\frac{n}{\pi}} e^{-nx^2} F(x)dx = F(0), \tag{A15}$$

which is just the essential property of the δ-function

$$\int_{-\infty}^{+\infty} \delta(x)F(x)dx = F(0). \tag{A16}$$

In order to find its Fourier-transform, one has to calculate

$$\tilde{f}_n(Q) = \int_{-\infty}^{+\infty} \sqrt{\frac{n}{\pi}} e^{-nx^2} e^{-iQx} dx. \tag{A17}$$

The result obtained is

$$\tilde{f}_n(Q) = e^{-Q^2/4n}. \tag{A18}$$

Thus the Fourier-transform of $\delta(x)$ is given by

$$\lim_{n\to\infty} \tilde{f}_n(Q) = \lim_{n\to\infty} e^{-Q^2/4n} = 1, \tag{A19}$$

or

$$1 = \int_{-\infty}^{+\infty} e^{-iQx} \delta(x)dx. \tag{A20}$$

Naturally this result could have been obtained also with the δ-function property eq. (A16). But application of Fourier's theorem to generalized functions leads also to its reciprocal,

$$\delta(x) = \frac{1}{2\pi} \int_{-\infty}^{+\infty} e^{iQx} dQ, \tag{A21}$$

a result which could not have been derived otherwise.

The δ-function concept can be extended easily to three dimensions. By definition we have

$$\delta(\vec{r}) = \delta(x_1)\delta(x_2)\delta(x_3), \tag{A22}$$

and hence

$$\delta(\vec{r}) = \frac{1}{(2\pi)^3} \int e^{i\vec{Q}\cdot\vec{r}} d\vec{Q}, \tag{A23}$$

or, because in the latter equation only even functions are involved,

$$\delta(\vec{r}) = \frac{1}{(2\pi)^3} \int e^{-i\vec{Q}\cdot\vec{r}} d\vec{Q}. \tag{A23'}$$

References

1. L. Schwartz, Théorie des Distributions, Vol. I and II, Hermann et Cie., Paris 1950 and 1951.
2. M. J. Lighthill, Introduction to Fourier Analysis and Generalised Functions, Cambridge University Press 1959.

B. Liouville Operator; Formal Integration of the Liouville Equation

Let us consider a dynamical variable $A(\vec{r}^N, \vec{p}^N)$ which depends on time only implicitly over the $6N$ space and momentum coordinates in phase space. The equation of motion for A is then written

$$\frac{dA}{dt} = \sum_{k=1}^{N} (\nabla_{\vec{p}_k} H \cdot \nabla_{\vec{r}_k} A - \nabla_{\vec{r}_k} H \cdot \nabla_{\vec{p}_k} A), \tag{B1}$$

since $\partial A/\partial t = 0$. The right hand side of this equation is also called the Poisson bracket $\{A, H\}$, where H is the Hamiltonian of the system. The Liouville operator is now defined by

$$-iL = \{\ldots, H\} = \sum_k (\nabla_{\vec{p}_k} H \cdot \nabla_{\vec{r}_k} - \nabla_{\vec{r}_k} H \cdot \nabla_{\vec{p}_k}). \tag{B2}$$

With this notation the equation of motion (B1) becomes

$$\frac{dA}{dt} = -iLA. \tag{B3}$$

A formal solution in operator language is obtained if one makes the ansatz

$$A(\vec{r}^N(t), \vec{p}^N(t)) = S(\vec{r}^N(0), \vec{p}^N(0); t) A(\vec{r}^N(0), \vec{p}^N(0)), \tag{B4}$$

where the implicit time dependence of A is contained in the operator S. Insertion into eq. (B3) yields an operator equation

$$\dot{S}(t) = -iLS(t), \tag{B5}$$

which gives as solution

$$S(t) = e^{-itL_0}, \tag{B6}$$

where the notation L_0 emphasizes that there is no explicit time dependence. Thus the solution of eq. (B3) may be explicitly written

$$A(\vec{r}^N(t), \vec{p}^N(t)) = \exp\{-itL_0\} A(\vec{r}^N(0), \vec{p}^N(0)). \tag{B7}$$

This equation means that the quantity A at time t is obtained if the operator $\exp\{-itL_0\}$ acts on A evaluated at the initial time zero. With other words, the operator $\exp\{-itL_0\}$ produces a time shift of amount t in the dynamical quantities. In order to use the operator $\exp\{-itL_0\}$ for real computations, the exponential must be expanded into powers of itL_0.
In the special case that A equals $\vec{r}_k(t)$ or $\vec{p}_k(t)$ one finds

$$\vec{r}_k(t) = e^{-itL_0} \vec{r}_k(0) \tag{B8}$$

$$\vec{p}_k(t) = e^{-itL_0} \vec{p}_k(0). \tag{B9}$$

The Liouville operator also simplifies considerably the notation of the Liouville equation (8.13):

$$\frac{\partial f^{(N)}}{\partial t} = iLf^{(N)}. \tag{B10}$$

Its close similarity with the Schrödinger equation in quantum mechanics again suggests the use of operator techniques.
To apply this formulation to the perturbation treatment of section 9.2.2 we split the Liouville operator of the system, viz.

$$L = L_0 + L_1, \tag{B11}$$

where $iL_0 = \{H_0, \ldots\}$ is formed with the unperturbed Hamiltonian H_0 and thus describes the natural evolution in time of the unperturbed system. On the other hand, $iL_1 = \{H', \ldots\}$ gives the changes of the natural motion due to the external perturbation H'. With this notation the inhomogeneous differential equation (9.20) is written as

$$\frac{\partial f_1^{(N)}(t)}{\partial t} = iL_0 f_1^{(N)}(t) + \frac{v f_0^{(N)}}{kT} F(t) \dot{A}_0 . \tag{B12}$$

An exact differential can be formed by multiplying eq. (B12) by an integrating factor e^{-itL_0}, which leads to

$$du(f_1^{(N)}, t) = P df_1^{(N)} + Q dt = 0, \tag{B13}$$

where $$P = \frac{\partial u}{\partial f_1^{(N)}} = -e^{-itL_0},$$

$$Q = \frac{\partial u}{\partial t} = iL_0 e^{-itL_0} f_1^{(N)}(t) + \frac{v f_0^{(N)}}{kT} F(t) e^{-itL_0} \dot{A}_0 , \tag{B14}$$

and for which

$$\partial P/\partial t = \partial Q/\partial f_1^{(N)} . \tag{B15}$$

Since the total differential du equals zero, we have

$$u = \int_0^{f_1^{(N)}} P df_1^{(N)} + p(t) = -e^{-itL_0} f_1^{(N)} + p(t) = 0 , \tag{B16}$$

from which

$$\frac{\partial u}{\partial t} = iL_0 e^{-itL_0} f_1^{(N)} + \frac{\partial p(t)}{\partial t} = 0 . \tag{B17}$$

Comparison of eq. (B17) with Q of eq. (B14) shows that

$$\frac{\partial p(t)}{\partial t} = \frac{v}{kt} f_0^{(N)} F(t) e^{-itL_0} \dot{A}_0 , \tag{B18}$$

and after integration

$$p(t) = \frac{v f_0^{(N)}}{kT} \int_{-\infty}^{t} F(t') e^{-it'L_0} \dot{A}_0 dt' \tag{B19}$$

or, using eq. (B16),

$$f_1^{(N)}(t) = \frac{v f_0^{(N)}}{kT} \int_{-\infty}^{t} F(t') e^{-i(t'-t)L_0} \dot{A}_0 dt' . \tag{B20}$$

In eq. (B19) and (B20) the integration limits are set such that $f_1^{(N)}(-\infty) = 0$. It follows from eq. (B7) that the operator $e^{-i(t'-t)L_0}$ produces a time shift of amount $t' - t$ in the dynamical quantity $\dot{A}_0$. Eq. (9.21) is thus recovered.

List of General References

1. J. A. Barker, Lattice Theories of the Liquid State, Pergamon Press, Oxford 1963.
2. G. H. A. Cole, An Introduction to the Statistical Theory of Classical Simple Dense Fluids, Pergamon Press, Oxford 1967.
3. P. A. Egelstaff, An Introduction to the Liquid State, Academic Press, London 1967.
4. H. L. Frisch and Z. W. Salsburg (Eds.), Simple Dense Fluids, Academic Press, New York 1968.
5. T. L. Hill, Statistical Mechanics, McGraw-Hill, New York 1956.
6. J. O. Hirschfelder, C. F. Curtiss, and R. B. Bird, Molecular Theory of Gases and Liquids, John Wiley, New York 1954.
7. T. J. Hughel (Ed.), Liquids; Structure, Properties and Solid Interactions, Elsevier, Amsterdam 1965.
8. N. H. March, Liquid Metals, Pergamon Press, Oxford 1968.
9. I. Prigogine, The Molecular Theory of Solutions, North Holland, Amsterdam 1957.
10. S. A. Rice and P. Gray, Statistical Mechanics of Simple Liquids, Interscience, New York 1965.
11. J. S. Rowlinson, Liquids and Liquid Mixtures, 2nd ed., Butterworths, London 1969.
12. H. N. V. Temperley, J. S. Rowlinson, and G. S. Rushbrooke (Eds.), Physics of Simple Liquids, North-Holland, Amsterdam 1968.
13. The Structure and Properties of Liquids, Discussions of the Faraday Society, No. *43*, London 1967.
14. Theory of Condensed Matter, Lectures, International Atomic Energy Agency, Vienna 1968.

List of Symbols

Only symbols occuring repeatedly are listed. When a special meaning is given to a symbol by using a subscript or superscript, this is usually not listed separately. Extensive thermodynamic quantities are denoted by small letters (p. 6). The corresponding molar quantities, denoted by capital letters, are not included in this list.
The page numbers given below refer to the first occurrence of the symbol. Page numbers given in parenthesis refer to a defining equation.

A	amplitude of scattered waves 48	f_{ij}	Mayer's f-factor 117
A	property density, to which a perturbing force is coupling 158	f_{aj}	oszillator strength (for transition $a \to j$) 61
A	magnitude of ultrasonic relaxation 216	$f^{(N)}$	N-particle distribution function 135
a	distance of nearest neighbours 28		
a	thermodynamic activity 222	g	Gibbs free energy or free enthalpy 6
a	van der Waals' constant 22	$g(R)$	pair distribution function 2 (42)
		$g(x)$	increased number of arrangements (due to defects) 26
B	number of different states 31		
B	response of a dynamical quantity to a perturbing force 158	$g(\vec{r}_1, ..., \vec{r}_s)$	s-tuplet distribution function 42
		g	dipole correlation parameter 224
B_2	second virial coefficient 72	$G(\vec{r}, t)$	van Hove correlation function 56
b	scattering length 54	$G(\Omega_0, \Omega, t)$	orientational part of van Hove correlation function 205
c	concentration 133		
c_L	velocity of light 53	h	enthalpy 3 (6)
c	sound velocity 8	H	Hamiltonian 10
c_p	heat capacity at constant pressure 7	h	Planck's constant 10 ($\hbar = h/2\pi$ 48)
c_v	heat capacity at constant volume 7	$h(R)$	total correlation function 45
$c(R)$	direct correlation function 46	$\tilde{h}(Q)$	Fourier transform of $h(R)$ 47
$\tilde{c}(Q)$	Fourier transform of $c(R)$ 47	$\vec{H}$	magnetic field strength 212
D	self-diffusion coefficient 1 (133)	$i(Q)$	scattering function 1 (50)
D_T	thermal diffusivity 100 (180)	$I(Q)$	scattered intensity 49
D_r	rotational diffusion coefficient 207	I	moment of inertia 183
$|d|^2$	differential cross-section per atom 53		
$\vec{D}$	dielectric displacement 208	$\vec{j}$	electron flux 163
		$\vec{j}_m$	mass flux 133
E	potential or kinetic energy as specified in the text 10	J	rotational quantum number 203
e	electronic charge 53	$J(\omega)$	spectral density (in n.m.r.) 213
$\vec{E}$	electric field strength 163		
		k	Boltzmann's constant 10
f	(Helmholtz) free energy 6	$\vec{k}$	wave vector 47
F	perturbing generalized force 157	k_B	corresponding states parameter 193
$F_l(t)$	orientational time correlation functions 205	k	reaction rate constant 227
		$\vec{K}$	external force 139
$F(Q, t)$	intermediate scattering function 165 (166)	K	reaction equilibrium constant 222
$\vec{F}$	force 135	l	distance between atomic centers in diatomic model 187
$\vec{F}$	electric field strength 62		
F	functional 122	L	Liouville operator 161 (241)
f	reaction field factor 225	m	mass 8
$f(\vec{Q})$	atomic scattering factor 48		
$\vec{f}$	volume force density 134	N	number of particles 7

List of Symbols

Symbol	Description
N_A	Avogadro's number 72
n	number of moles 6
n_D	refractive index of Na-D-line 225
$n(\vec{r}_1,...,\vec{r}_s)$	s-particle distribution function 41
$\tilde{n}(Q)$	Fourier transform of the one particle distribution function 166
P	pressure 6
P_l	Legendre polynomial 187
$\vec{P}$	polarization 208
p_{ij}, P_{mn}	transition probability 32, 212
$\vec{p}$	momentum 10
$p(\vec{r}_1,...,\vec{r}_N)$	probability density of a certain state 111
$\vec{Q}$	change of wave vector at scattering 1 (48)
Q	configurational partition function 11
q	degree of positional order 195
q	generalized coordinate 10
q	electric charge 67
$\vec{q}$	energy flux 133
R	pair distance 2
R	gas constant 3
$S(Q,\omega)$	scattering law 56
$S(Q)$	structure factor 53
$S(\omega)$	spectral density of a process 163
s	degree of orientational order 195
sym $\vec{\tau}$	symmetrical part of a tensor $\vec{\tau}$ 134
T	absolute temperature 2
t	time or time interval 35
T_0	macroscopic decay time (dielectric relaxation) 209
T_1, T_2	nuclear magnetic relaxation times 212
u	internal energy 3 (6)
u_i	probability of state i 32
u	external potential 122
$\tilde{u}$	mean velocity of an assembly 134
v	volume 3 (6)
v_f	free volume 11
$\tilde{v}$	particle velocity 35
V	scattering potential 52
W	virial of intermolecular forces 13
W	cell volume (in cell theory) 27
w	repulsive pair energy parameter 195
x	mole fraction 26
$X^{ll'm}$	coefficients in the spherical harmonics expansion 67
Y_l^m	spherical harmonics 66
Y	corresponding states parameter 193
Z	partition function 10
Z	compressibility factor 192
z	coordination number 25
$z(t)$	velocity correlation function 165
$\tilde{z}(\omega)$	frequency spectrum 168
$Z(\tau)$	n.m.r. correlation function 212
α	(thermal) expansion coefficient 7
α	polarizability 61
α	critical index for c_v 100
α	isotopic separation factor 150
α	ultrasonic absorption coefficient 214
β_s	compressibility (adiabatic) 7
β_T	compressibility (isothermal) 3 (7)
β	critical index for coexistence curve 100
β_k	cluster integral 111
γ_v	thermal pressure coefficient 7
γ	critical index for β_T 99
γ_k	1,2-cluster integral 119
$\gamma(t)$	dipole autocorrelation function 210
δ	Dirac's δ-function 43
δ	critical index for P-ρ-isotherm 99
$\delta(T)$	parameter characterizing fluids of non-spherical molecules 191
ε	pair energy 12
ε^*	depth of minimum of pair potential 18
ε	(relative) dielectric permittivity 209 (ε'' loss factor)
ζ	friction constant 147
η	critical index for $h(R)$ 101
η	shear viscosity 134
ϑ	orientation coordinate 65 (67)
θ	scattering angle 1
λ	wavelength 47
λ	heat conductivity 133
μ	chemical potential 6
$\vec{\mu}$	dipole moment 61
μ	ultrasonic absorption per wavelength 216
ν	frequency 61
ν	critical index for ξ 101
ν	parameter in melting model 195
Ξ	grand partition function 14

List of Symbols

ξ	characteristic correlation length 101 (103)	$\varphi(\vec{r}_i)$	probability density 114
ζ	degree of advancement of reaction 227	ϕ	bulk viscosity 134
		ϕ_{BA}	after effect function 158
		φ	orientation coordinate 65 (67)
ρ	number density N/v 2		
ρ_m	mass density 134	χ_{BA}	(generalized) susceptibility 159
		$\chi(T)$	parameter characterizing polar liquids 190
σ, σ_H	hard sphere diameter or zero of pair potential 18		
σ	disorder factor in cell model 27	ψ	eigenfunction 52
σ	cross-section in scattering 53	ψ	potential energy of a molecule 112
σ	electrical conductance 163		
$\vec{\sigma}$	stress tensor 134	ω	circular frequency 11
		Ω	solid angle 52
τ	time interval 145	ω	orientation coordinate 13
	molecular relaxation time 210	ω	acentric factor 185

Subject Index

absorption coefficient, ultrasonic 100, 214, 216, 227, 231, 232
– per wavelength, ultrasonic 216, 227, 228
acentric factor 185, 192, 194
activation enthalpy 223, 226, 232
activity 222, 224, 227–230
– coefficient 227, 229
adiabatic compressibility, see compressibility
additivity, pairwise, of interaction energy 12, 16, 34, 46, 61, 71, 75–77, 111, 234
after effect function 158–164
alkyl bromides 211, 232
alkali halides 92
alcohols 225, 226, 230–232
alloy 55
angle correlation functions 203–215
angular momentum 36, 205
anisotropy of molecules 17, 74, 183–234
articulation point 118
associated liquids 17, 217–234
association equilibrium constant 222, 224, 227–230
– mechanism 219, 224, 228–234
– kinetics 224, 228
atomic scattering factor 48
attractive pair interaction 16–19, 22, 61–74, 78, 80, 102, 141, 210, 219
averaging over orientations 65, 188–191

background potential, see van der Waals model
back scattering 143, 145
barrier, to intramolecular rotation 198–200, 217
– to molecular rotation in external field 203, 204
base point 118
Bernal's model 23–25, 29, 93
Bessel function 205
binomial distribution 29
boiling point 4, 5, 83, 185
Boltzmann equation 133, 138–142
– factor 10, 31, 65, 111, 114
Born approximation 52, 53, 74
Boyle temperature 194
bracket notation 61–63, 66, 68, 135
Bravais lattice 44
bridge clusters 118–121
Brillouin scattering 15, 98–101, 174, 175
Brownian motion 133, 145–153
Buckingham potential 70, 73, 74, 76
bulk viscosity 134, 142, 148, 152–155, 180, 214

canonical ensemble 7, 11–13, 31, 41, 42, 46, 111
carboxylic acids 223, 225, 228–230
cell cluster theory 28

– theory 27–30, 38, 88, 94, 111–116, 171, 195
chain molecules, see flexible molecules
chaos assumptions 133, 137, 139, 142
charge transfer 219, 221
charged particles (see also ionic melts) 66
chemical potential, see potential
Clausius-Clapeyron equation 9, 86, 89
cluster integrals 75, 111, 117–122, 125
coarse graining 141, 147
coexistence curve (gas/liquid) 85–89, 97, 100, 131
coexistent phases 9, 10, 13
cohesion energy 4, 17, 64
Cole-Cole plot 209, 210
collective particle motion 174–180
collision (see also hard core – and soft –) 35, 36, 73, 138–145, 168
– complex 204, 208, 210, 212
– -limited liquids 214
collisional transfer 142
coloured substances 62
communal entropy 12, 21
– free energy 28
Compton scattering 47
compressibility, adiabatic 7, 8, 15, 88, 89, 98, 214
–, isothermal 3, 4, 7, 8, 15–18, 42–47, 50, 77, 88–90, 97–100, 214, 217
–, local 129
– equation, see equation of state
– factor 192–194
computer experiments (see also Monte Carlo and molecular dynamics) 21–24, 31–39, 124, 125, 130, 131, 234
configurational partition function 10–13, 21, 25–28, 31, 37, 42, 45, 83, 111, 127, 183, 188, 189, 195, 196
conformations, see flexible molecules
constants of motion 36
continuity, between liquid and gaseous state 9, 97
coordination defects (see also disorder) 28–30
– number 3, 24, 28, 29, 43, 44, 57–59, 195, 196, 219, 224, 225, 232, 233
correlation functions, see direct –, total –, and time –
correlations of electronic motions 61–64
– of fluctuations 22, 46, 101–104, 128
– of molecular motions 30, 112–116, 139, 142–146, 151
corresponding states, see theorem of –
Coulomb forces 16, 17, 66, 78
coupling of rotation and translation 204
critical coefficient 83, 84, 90, 99, 185
– frequency, see dielectric relaxation time
– indices 99–102
– opalescence 102

critical parameters   5, 84, 185, 197
– point (gas/liquid) (see also reduced properties)   4, 16, 17, 22, 25, 42, 46, 47, 59, 83 – 89, 96 – 107, 126
– – (liquid/solid), possibility of   5, 96
cross-section   53, 56, 73, 74, 173
crystal defects, see defective crystal model
crystallites (see also polycrystalline solids)   26
crystallization   25, 95
crystal structure of simple solids   75, 76
current-current correlation function   177 – 179

damping of molecular motions   149, 168 – 171
Debye theory of dielectrics   208 – 210
Debye-Hückel theory   79
Debye solid   75, 76, 94, 169
defective crystal model   21, 25 – 28
dense packing, see random – and regular –
density-density correlation function   166, 167
density expansion of pair distribution function   43, 118, 119
depolarized scattering, of light   211, 212
– –, of neutrons   172
diamagnetic liquids   212
– susceptibility   63
diatomic model   187, 191
dielectric displacement   208, 209
– permittivity   208 – 210, 224 – 226, 229 – 233
– relaxation (absorption)   165, 203, 208 – 219, 224, 226, 230 – 233
– relaxation time, macroscopic   209 – 211, 230
– – –, molecular   210, 215, 232 – 234
– saturation effect   225 – 226, 231
differential cross-section, see cross-section
diffusion, see self-diffusion
dimerization (see also association)   219 – 224, 227 – 231
dipole-dipole energy   63 – 65, 68, 84, 190, 210, 220, 221
– interaction, nuclear   213
dipole moment   17, 65 – 67, 183, 184, 188 – 192, 205, 210, 220, 224 – 226, 228 – 231
Dirac's δ-function   43, 125, 136, 158, 168, 169, 239 – 240
direct correlation function   46, 47, 50, 59, 75, 102 – 104, 119 – 121, 124
Dirichlet polygons   23, 25
disorder   21, 27 – 30, 88, 92 – 94, 111, 169, 171, 195, 196
dispersion energy   16, 17, 61 – 64, 68, 75, 84, 221
– relation   159, 179
dissipation   133, 158 – 164
distribution functions   41, 116, 122, 123, 128, 135 – 142, 147, 152, 189, 191
Doppler effect   15, 55
dynamics of the liquid state   157 – 180

effective pair potential   75 – 77, 234
eigenfunction   52, 61, 62
Einstein formula for diffusion coefficient   146, 167
– solid   11, 76, 77, 94, 112, 169
elastic constants of the solid   75
electric conduction   163
– field gradient   213
– field strength   62, 65, 163, 208 – 210, 225 – 226
electrolyte solutions   79
electron scattering   1, 53, 200
electrostatic energy   64, 65, 79, 84, 220, 221, 234
end-over-end rotation   211, 213, 215
energy, internal   6, 7, 34 – 37, 45, 89
– conduction   133 – 137, 140 – 142
– exchange, at scattering   48, 53 – 56, 174 – 175
– of vaporization   3
–, storage of   158 – 160
ensemble average   10, 31 – 36, 44, 45
Enskog equation   133, 141 – 145, 151, 152
enthalpy (see also heat)   6, 7
– of formation   200, 223, 227
entropy   6 – 8, 25, 26, 31, 46, 93, 114, 115, 222
– change of melting   16 – 18, 84, 92, 96, 185, 194 – 199
– – of phase transformation   9, 10
– – of rotational isomerization   199, 200
– – of transition   197 – 199
– – of vaporization   84
equation of continuity   134, 137
– of energy transport   134 – 137, 140
– of motion (see also hydrodynamic – und Liouville –)   31, 35, 36
– of state   9, 11, 22, 23, 34 – 39, 44 – 47, 117, 124 – 127, 130, 192
equilibrium   6, 31, 37, 139
equivalent orbital representation   221
ergodicity   32, 33, 39
exchange integrals   68, 221
expansion (thermal)   58
– coefficient   7, 8, 88 – 90, 97, 100
external potential   122, 123, 136

Fermi energy   78, 80
Fick's law   133
field points   118, 119
flexible molecules   17, 198 – 200, 209, 211, 217
flickering clusters   233
fluctuations   14, 15, 24, 42, 46, 101, 128, 133, 149, 157 – 180, 210 – 214, 234
fluctuation-dissipation theorem   161 – 164, 167
fluorocarbons   62, 84
Fokker-Planck equation   145 – 149, 152
folding theorem   47, 237
force correlation function   149, 150
forward scattering   49 – 51, 166
Fourier's law   133

Subject Index

Fourier transform 47, 50, 51, 56, 57, 158, 159, 162, 166–168, 172, 178, 203–207, 210–213, 237–240
free electron gas 78, 79
– energy (Helmholtz) 6, 7, 9–11, 26, 38, 96, 113–115, 188–192
– – analycity around the critical point 97–104
– enthalpy (or Gibbs free energy) 6–9, 200, 222
– molecular rotation 203–208, 214
– volume 11, 25, 26, 112, 115, 223, 224
frequency spectrum of a liquid 167–174
– – of a solid 76, 94
friction constant 147–152
Fröhlich equation 224
functional 46, 122, 123

gas phase, difference to the liquid state 4, 21, 22, 43
– –, dilute gas properties 71–74, 138–141
– –, ideal gas 11, 168
– –, molecular properties 17, 18, 78, 200, 222–224, 228, 230
– –, real gas 111, 168
geometric neighbours 23, 24
Gibbs free energy, see free enthalpy
Gibbs-Helmholtz equation 7
glasses 16
grandcanonical ensemble 7, 13–16, 42, 45, 46, 111, 116, 128
grand partition function 14
gravitational effects 98, 99
group relaxation 210, 217

half width 101, 177, 179, 207, 231
Hamiltonian 10, 13, 14, 66, 158–161, 241, 242
hard core collisions 141–145, 149, 151–154
– core diameter 143, 144
– spheres 21–24, 28, 33, 35–39, 94, 96, 111, 123–126, 141–143
– sphere model, see van der Waals model
– sphere potential 18, 34, 35, 115, 124, 127–131
Hartree-Fock wave functions 63
heat capacity at constant pressure 7, 8, 15, 16, 88, 89, 97, 100, 180, 214, 216
– – – volume 7, 8, 15, 16, 89–91, 97, 100, 102, 180, 192, 193, 214, 216, 233
– –, difference between gas and liquid 86, 87
– – of internal rotation 199
– – of reaction 214, 216, 227
heat conductivity, see thermal conductivity
– of melting 3, 95
– of vaporization 4, 86–89
Heaviside step function 125
high polymers 16
holes 25–28
hydrodynamic equations 133, 136, 140
– model 172, 175, 179, 180

hydrogen bond 17, 219–221, 225–234
hypernetted chain theory 104, 121–127
hypersonic waves (see also Brillouin scattering) 15

ideal gas, see gas phase
importance sampling 31
important configurations 33
incoherent scattering 47, 50, 51, 54, 56, 165, 172
induction energy 64, 65, 78, 190, 221, 229
inelastic scattering 55–57, 165, 172, 175–177
infrared absorption 165, 203–205, 213, 215, 219, 229–233
intensity, of scattered radiation 47–52, 54–56, 172
intermediate scattering function 165, 166, 172, 178
intermolecular forces, see pair energy or potential energy
intramolecular vibrations 200, 217, 219, 222–223, 231, 233
internal rotation 198–200, 209, 217, 223
inversion point, of P-ρ-isotherm 106, 107
ionic melts 17, 78, 92, 234
irreducible clusters 118, 119
irreversibility 133, 137, 138, 141, 147
Ising model 22, 101–105
isotopes 54, 55
isotopic separation factor 150, 151

jumps, in diffusion 171
– in reorientation 214, 234

Kihara potential 64, 70, 71, 73–76
kinetic equations 133, 138–150, 157
Kramers-Kronig relations 159

Landau theory 99–101, 104
Landau-Placzek ratio 16, 100
Langevin equation 148–150
lasers 100, 174
lattice energy 11, 75
– gas 21, 22, 101, 104
Legendre polynomial 187, 205
– transformation 6
Lennard-Jones-Devonshire cell model 112–113
– melting model 194–196
Lennard-Jones potential 18, 35, 36, 69–77, 87, 88, 96, 112–115, 125–131, 153, 187–192, 234
life time, of associated complexes 219, 228, 230
– –, of fluctuations 178–179
light scattering (see also Brillouin –, and Rayleigh –) 15, 102–104
linear response theory 133, 157–167, 225
liquid crystals 16, 196
– range 4, 5, 16–18
Liouville equation 133–141, 147, 160–161, 241, 242

local additivity, of dispersion energies 64
London theory 61–63, 221
loss factor, see dielectric relaxation

magnetic field 212, 213
magnetization 165
many body interaction 76–77
Markov processes 145, 147, 152
Maxwell distribution 36, 74, 140
– equations 7, 90
mean free path 142, 168–171
– square displacement 36, 167–172
melting curve 92, 95, 96
melting (point) 2–5, 17, 23, 26, 33, 37–39, 91–96, 117, 125, 185, 194–200
metallic liquids 18, 59, 77–80, 84, 87, 88, 92
microscopic reversibility 32
microwave spectrum 228
model potentials 16, 18, 70–71, 74, 129, 151, 187, 188, 220, 221, 234
Moessbauer effect 165
molecular beam scattering 69–71, 74–75, 77
– dynamics 31, 35–39, 57, 94, 143, 169–173, 177–179, 210, 234
– liquids 17, 55, 183–234
– orbital calculations 221
– shape 184, 194, 210
moment of inertia 183, 204, 207, 222, 223
momentum (see also angular –) 10, 11, 36, 47
– transfer at scattering 48, 53, 55, 56, 174–175
Monte Carlo calculations 24, 29, 31–35, 37–39, 77, 94, 96, 112, 113, 126, 127, 130, 234
multiple scattering 49, 54
multipoles 64, 66–68

nearest neighbour distance 58, 219, 232
neutron scattering 1, 41, 53–59, 77, 101, 165, 172, 175–177, 203–208
non-additivity, see additivity
noble gases 16, 18, 59, 62, 75, 76, 78, 83–88, 92, 144
node clusters 118–121
non-spherical particles, see orientation of molecules
nuclear magnetic moment 213, 215
– magnetic relaxation 100, 165, 203, 212–215, 234
number of nearest neighbours, see coordination number
numerical pair potential 71, 74, 76
– statistical mechanics, see Monte Carlo
Nyquist theorem 164

Ohm's law 163
optical birefringence 194
order-disorder transition, see solid-solid –
orientation energy, see dipole-dipole or electrostatic energy

orientation of molecules 13, 16, 17, 65, 83, 183–234
Ornstein-Zernike equation 47, 102, 103, 120–123
oscillator strength 61–63
overlap integrals 68

Padé approximant 124, 126
pair distribution function 2, 3, 24, 34, 41–59, 75–77, 111–131, 142, 148, 153, 190, 200, 219, 232
– energy 12, 16–19, 21, 34, 41–45, 61–80, 83, 88, 111, 124, 150, 151, 173–174, 183–194, 219–221, 234
paracrystals 26
parallel clusters 118–121
partition function 10, 13, 14–17, 31, 151, 183, 195, 198–200, 222–224
Pauli principle 68
Percus-Yevick theory 24, 77, 121–127
periodic boundary conditions 31, 34, 35, 37
perturbation methods, for kinetic equations 140, 241, 242
– –, for monoatomic liquids 21, 37, 111, 127–131
– –, for polyatomic liquids 188–191
– –, in quantum mechanics 66–70, 77, 204, 221
phase diagram 4, 5
– space 10, 135
– transformation 8, 9, 91
phonon 16, 177
plastic crystals 194, 196
plasticine 23
Poisson bracket 241
polarizability 61–65, 78, 84, 92, 93, 183, 184, 206, 209, 210, 225, 229
polarization 208–210, 225, 226
polyatomic molecules 13, 87, 88, 183–234
polycrystalline solids 180
potential, chemical 6, 14, 38, 94, 104, 105, 164
–, thermodynamic 6
–, of mean force 117
– $(-Pv)$ 6, 14
– energy 10–13, 16–18, 22, 26, 34, 41, 44, 45, 61, 75, 114, 150, 188
pressure equation, see equation of state
probability of a certain state 10, 32, 33, 41, 111, 114, 115, 145
– theory 32
pseudo-pair potential 18, 77–80, 124

quadrupole moment, molecular 17, 66, 68, 84, 183, 184, 203, 204
– – nuclear 213, 215
quantum effects 10, 16, 43, 151, 198
– mechanical calculations 61–63, 66–70, 74, 77–80, 161, 162, 166, 185, 186, 204, 221
quasi-lattice vibrations, see collision complex and frequency spectrum

Raman scattering 165, 200, 203–208, 213, 215, 219, 229, 233
random dense packing 23, 24
– numbers 33
– walk 32, 33, 207
randomization, orientational 194, 196
rare gases, see noble gases
Rayleigh scattering 15, 100, 101, 165, 203, 208–215
reaction field factor 225
– rate constants 227–230
rectilinear diameter 25, 85, 106, 107
reduced properties 83–91, 155, 185, 192–194
refractive index 62, 225, 230
regular dense packing 24, 28, 34, 37, 75, 76
relative coordinates 35, 36, 74
relaxation, see dielectric –, n.m.r. –, and ultrasonic –
re-orientation, molecular 203–214, 219, 226, 231, 232
repulsive pair interaction 16–19, 22, 23, 68–75, 78, 80, 95, 129, 151, 194–196, 210, 219
resonance frequency 212–214
response, see linear response theory
retardation effect 64
Rice-Allnatt theory 133, 141, 151–154
rigid molecules 17, 194–198
rotation, see free molecular –, and internal –
rotational diffusion coefficient 207, 208, 211–215
– – model 207, 213, 214, 234

Salem formula 63
sampling procedures 31
saturation curve, see coexistence curve
scaled particle theory 125
scaling laws 104–107
scattering function 50–54, 57–59, 103
– law 56, 165–167, 174–180, 205
– length 45, 55, 172
– potential 52–55, 166
second moment, of scattering law 167
second-rank tensor 164
second virial coefficient 72–73, 124
self consistent cell model 113–116
self-correlation function, see van Hove –
self-diffusion 164, 165
self-diffusion coefficient 1, 2, 73, 133, 134, 143, 146, 148, 150–155, 167–171, 177
self-scattering law 56, 57, 165–167, 172–173, 176
shear viscosity 73, 134, 142, 144, 147, 148, 152–155, 180, 185, 214
short-range order 3, 24
short-time behaviour 36, 157–180, 203–217, 226–234
simplical graph 33
single particle motion 167–174
size ratio cations/anions 92, 93
small angle scattering 26, 51, 101, 233

smearing approximation 112
Smoluchowsky equation 145, 149
soft collisions 143, 145, 151–153
– spheres 96
solid state, difference to the liquid state 1–4, 16, 21, 22, 28, 38, 39, 43, 88, 94, 194–196, 199, 200, 210, 211, 230, 232; see also Bravais lattice and crystal structure
solid-solid transitions 17, 194–199
solutions 224, 225, 229–233
sound velocity, see velocity
spectral density of a process 160, 163, 203–214, 231, 233
spherical harmonics 66, 67, 187, 205
spherical shell potential 187–189
spin 54, 172
spin-lattice relaxation time 212
spin-rotational interaction 213
spin-spin interaction 212
static approximation 52–55, 57
stationarity 149, 167, 172
steady state 32
steel balls 23
sterically hindered structure-limited liquids 214
Stirling's approximation 11, 28
Stokes-Kirchhoff equation 214
stress tensor 134, 137, 140–142
structure factor 53–55, 103, 177–179
– of associated complex 219, 224, 228–232
superposition approximation 116, 117, 153
susceptibility (generalized) 158–164, 167
– theory 64, 77
submillimeter absorption, see dielectric absorption

Tait equation 90, 91
theorem of corresponding states 83–88, 155, 190–194
theories of equilibrium properties 24, 31, 37, 111–131, 188–191
thermal conductivity 73, 133, 134, 142, 144, 148, 152–155, 180, 214
– diffusivity 100, 180
– expansion, see expansion
– pressure coefficient 7, 89–91, 97
thermodynamic potential, see potential
thermodynamics of irreversible processes 158
third virial coefficient 75, 124
three body interaction 75, 76, 234
– particle distribution function, see distribution functions
time average structure 224, 231–234
– correlation functions 147, 155, 157–167, 203–217, 226–234, 238; see also angle correlation function, current-current –, density-density –, force –, velocity –, and van Hove –
torque 207

total correlation function   45–47, 50, 51, 55, 59, 75, 102–104, 119–121
trajectories   35–38, 151
transition moment   61, 62, 205
– probability   32, 33, 145, 212, 213
translational motion, effects in spectra   203–204, 212, 213
transport properties   73–74, 77, 133–155, 162–165, 175, 180, 185
trimers   226, 229, 231
triple point (see also melting)   4, 5, 22–24, 57, 83, 87, 92, 95, 185, 197
true pair potential   75
truncation of Fourier integrals   51, 52
tunnel theory   28, 29

ultrasonic relaxation (absorption)   99, 148, 200, 203, 214–217, 219, 227–232
undercooling   25, 91, 93, 94
united atom   68, 69

van der Waals equation   9, 22, 85, 97, 99
– gas   101
– model   21, 22, 89, 94–96, 99, 100, 102, 143
van Hove correlation function   56, 166, 172
– self-correlation function   56, 146, 166, 172, 204–208
vapour pressure   83, 84, 86–88, 97, 193, 229, 230

variational calculations in quantum mechanics   63, 68–70, 221
velocity, of particles   31, 35–37, 74, 149, 177
– of sound   8, 77, 88, 97–99, 174, 175, 179, 180, 214, 216, 227
– correlation function   36, 149, 150, 165, 167–173
Verdet constant   62
virial   13, 34, 36, 37, 45
– coefficient (see also second – or third –)   124, 125, 188, 194
– theorem   13
virtual transitions   62, 68
viscosity, see bulk–, and shear –
volume change of phase transformation   9
– – of melting   3, 16–18, 39, 84, 92, 93, 96, 194–198
– – of reaction   200, 217
Voronoi polyhedra   23–25
vortex flow of molecules   171–172

wavelength   47, 53, 55, 64, 174, 177
wave vector   47, 55, 214
Wiener-Khintchine theorem   163, 168

X-ray scattering   1, 41, 47–54, 57–59, 71, 75, 77, 101, 165, 231, 232

Yvon-Born-Green theory   116–123, 125

Index of Substances

The element symbols in a molecular formula are ordered as follows:
a) for compounds not containing carbon: strictly alphabetically,
b) for carbon containing compounds: C first, followed immediately by H, then the remaining symbols alphabetically,
c) substances containing one atom of the first kind are ordered before those containing two, etc.

Al (aluminum) 4
Ar (argon) 2−5, 16, 17, 24, 57−59, 63, 64, 70−80, 84−97, 103, 130, 131, 144, 148, 152−154, 169−179, 193, 204, 219

BrCs (cesium bromide) 93
BrH (hydrogen bromide) 184, 197, 198
BrK (potassium bromide) 92, 93
BrLi (lithium bromide) 92, 93
BrNa (sodium bromide) 92, 93
BrRb (rubidium bromide) 92, 93
Br_2 (bromine) 84, 197

$CBrCl_3$ (trichlorobromomethane) 215
CBr_4 (carbon tetrabromide) 197, 198
$CClF_3$ (trifluorochloromethane) 106
CCl_4 (carbon tetrachloride) 91, 174, 188, 197, 198, 210, 215
CD_4 (methane-d_4) 197
CF_4 (carbon tetrafluoride) 84, 188, 193, 197, 198
$CHBr_3$ (tribromomethane) 215
$CHCl_3$ (chloroform) 91, 215
CHN (hydrogen cyanide) 184
CH_2Cl_2 (dichloromethane) 207, 215
CH_2O_2 (formic acid) 226, 228−230
CH_3J (iodomethane) 207, 208, 215
CH_3NO_2 (nitromethane) 226
CH_4 (methane) 2, 3, 5, 17, 64, 84, 85, 92, 154, 197, 198, 204, 207, 208, 215
CH_4O (methanol) 84, 226, 230−232
CO (carbon monoxide) 84, 85, 154, 175, 184, 193, 197, 206, 207
COS (carbonyl sulfide) 197
CO_2 (carbon dioxide) 3, 5, 17, 84, 97−99, 106, 184, 193, 197
CS_2 (carbon disulfide) 184, 197, 210, 215
C_2Cl_4 (tetrachloroethylene) 91
C_2HCl_3 (trichloroethylene) 91
C_2H_2 (acetylene) 3, 5, 17, 84, 184, 193, 197
$C_2H_2Cl_2$ (1,2-dichloroethylene, trans) 91
$C_2H_2Cl_4$ (1,1,2,2-tetrachloroethane) 91
C_2H_3N (or C_2D_3N) (acetonitrile) 213, 215
C_2H_4 (ethylene) 84
$C_2H_4Br_2$ (1,2-dibromoethane) 200, 217
$C_2H_4Cl_2$ (1,2-dichloroethane) 91, 200, 217
$C_2H_4O_2$ (acetic acid) 216, 219, 226, 228−230

C_2H_5Br (bromoethane) 84, 215
C_2H_5Cl (chloroethane) 84
C_2H_6 (ethane) 84
C_2H_6O (ethyl alcohol) 230−232
$C_2H_6O_2$ (1,2-ethanediol) 91
C_3F_8 (perfluoro-n-propane) 84
C_3H_6O (acetone) 84, 91, 226
$C_3H_6O_2$ (propionic acid) 226, 229
C_3H_7NO (N,N-dimethylformamide) 226
C_3H_8 (propane) 84
C_3H_8O (propyl alcohol) 154
C_4H_9Br (1-bromobutane) 214, 215
C_4H_{10} (n-butane) 198, 199, 200, 217
$C_4H_{10}O$ (ethyl ether) 84, 225, 226
C_5H_{10} (cyclo-pentane) 154, 184, 185, 194
C_5H_{12} (n-pentane) 154, 185, 194, 199, 200, 217
C_5H_{12} (iso-pentane) 185, 194
C_5H_{12} (neo-pentane) 154, 184, 185, 194, 197, 198
C_6F_6 (hexafluorobenzene) 215
C_6F_{14} (perfluoro-n-hexane) 84
C_6H_5Br (bromobenzene) 91, 226
C_6H_5Cl (chlorobenzene) 4, 91, 210, 211, 214, 215, 226
$C_6H_5NO_2$ (nitrobenzene) 91, 225
C_6H_6 (benzene) 3−5, 17, 84, 87, 90, 91, 210, 215
C_6H_6O (phenol) 226, 231
C_6H_7N (aniline) 91
C_6H_{12} (cyclo-hexane) 2, 84, 154, 208, 210, 215
C_6H_{14} (n-hexane) 3, 5, 84, 91, 199, 200
$C_6H_{15}N$ (triethylamine) 226
C_7H_{16} (n-heptane) 91, 199, 200
C_8H_{18} (n-octane) 3, 5, 91, 199
C_9H_{20} (n-nonane) 199
$C_{10}H_{22}$ (n-decane) 199
$C_{14}H_{10}$ (anthracene) 4
$C_{15}H_{32}$ (n-pentadecane) 91
ClCs (cesium chloride) 92, 93
ClH (or ClD) (hydrogen chloride) 3, 5, 84, 184, 197, 198
ClK (potassium chloride) 3−5, 84−87, 90−95, 154
ClLi (lithium chloride) 92, 93
ClNa (sodium chloride) 2, 92, 93
ClRb (rubidium chloride) 92, 93
Cl_2 (chlorine) 84, 184, 197
Cl_4Si (silicon tetrachloride) 197, 198

Index of Substances

Cl_4Sn (tin tetrachloride)   197, 198
Cl_4Ti (titanium tetrachloride)   197, 198
Cs (cesium)   5, 84, 87
CsJ (cesium iodide)   93

DH (hydrogen-d)   204
D_2 (deuterium)   203, 204

FK (potassium fluoride)   92, 93
FLi (lithium fluoride)   92, 93
FNa (sodium fluoride)   92, 93
F_2 (fluorine)   184, 197
F_4Si (silicon tetrafluoride)   197
F_6S (sulfur hexafluoride)   84, 92, 95–97, 106, 197

GeH_4 (germanium hydride)   197, 198

HJ (or DJ) (hydrogen iodide)   184, 197, 198
H_2 (hydrogen)   10, 17, 62, 64, 184–187, 203, 204
H_2O (water)   3, 84, 219, 221, 226, 232–234
H_4Si (silicon hydride)   197, 198
He (helium) 10, 16, 62, 68–70, 75, 99
Hg (mercury)   3, 5, 74, 84, 86, 87, 90, 92, 95, 154

JK (potassium iodide)   92, 93
JLi (lithium iodide)   93
JNa (sodium iodide)   92, 93
J_2 (iodine)   197

K (potassium)   74
Kr (krypton)   57, 58, 74, 84, 85, 144, 154

Li (lithium)   74

N_2 (nitrogen)   3, 4, 17, 85, 92, 154, 184, 193, 197
N_2O (dinitrogenoxide)   106, 184, 197, 207
Na (sodium)   2–5, 79, 92, 172–174
Ne (neon)   10, 16, 144

O_2 (oxygen)   85, 154, 184, 197
O_2S (sulfur dioxide)   3, 5, 17, 84, 226

Pb (lead)   59

Sn (tin)   2

Xe (xenon)   2, 74, 84, 98–101, 106, 107, 144, 154

QD
541
K64

MAY 3 1974